Fundamentals of Ground Improvement Engineering

Fundamentals of Ground Improvement Engineering

Jeffrey Evans
Daniel Ruffing
David Elton

CRC Press
Taylor & Francis Group
Boca Raton London New York

CRC Press is an imprint of the
Taylor & Francis Group, an **Informa** business

First edition published 2022 by
CRC Press
2 Park Square, Milton Park, Abingdon, Oxon, OX14 4RN

and by
CRC Press
6000 Broken Sound Parkway NW, Suite 300, Boca Raton, FL 33487-2742

British Library Cataloguing-in-Publication Data

A catalogue record for this book is available from the British Library

---

Library of Congress Cataloging-in-Publication Data

Names: Evans, Jeffrey C., author. | Elton, David J., author. | Ruffing, Daniel, author.
Title: Fundamentals of ground improvement engineering / Jeffrey Evans, David Elton, Daniel Ruffing.
Description: First edition. | Boca Raton : CRC Press, 2021. | Includes bibliographical references and index.
Identifiers: LCCN 2021002848 (print) | LCCN 2021002849 (ebook) | ISBN 9780367419608 (hbk) | ISBN 9780415695152 (pbk) | ISBN 9780367816995 (ebk)
Subjects: LCSH: Soil stabilization.
Classification: LCC TA749 .E94 2021 (print) | LCC TA749 (ebook) | DDC 624.1/51363--dc23
LC record available at https://lccn.loc.gov/2021002848
LC ebook record available at https://lccn.loc.gov/2021002849

ISBN: 978-0-367-41960-8 (hbk)
ISBN: 978-0-415-69515-2 (pbk)
ISBN: 978-0-367-81699-5 (ebk)

Typeset in Sabon
by Deanta Global Publishing Services, Chennai, India

Contents

Preface and Acknowledgments: Fundamentals of Ground Improvement Engineering

OVERVIEW

Engineers have long known that the properties of soil and rock can be improved. The modern field of ground improvement began to coalesce in the 1960s and has since grown enormously. This textbook synthesizes ground improvement literature and practice in a way that allows students to begin their studies of ground improvement engineering and helps professionals dig deeper into specific topics of relevance to their work.

Fundamentals of Ground Improvement Engineering is intended to explain key topics and fundamentals of ground improvement engineering and construction for students and professionals. This book is structured to broadly introduce each topic and then delve into the details. The authors approach the topic from the balanced viewpoints of both academics and professional practice.

Overall, this book provides a comprehensive introduction to the field of ground improvement to provide readers with sufficient background to understand and apply the techniques presented. It is the intention of the authors to provide the users of this book with both the current practices in ground improvement as well as the fundamental understanding of the principles to allow users to adapt to inevitable new developments in the field.

Readers are expected to already have an understanding of basic geology, the fundamentals of soil mechanics, and the mathematical and natural science training that accompanies the first few years of undergraduate education in civil engineering. In order to accomplish the objectives, this book contains the following elements:

- Balanced presentation of academic and practical aspects of ground improvement engineering.
- Example problems with solutions and practice problems so readers can see the application of theory.
- Information to meet needs in both university and professional markets.

From the perspective of the student, the book provides:

- A new, up-to-date, comprehensive text which blends the study of current ground improvement technologies with theoretical principles and applicable design and construction information.
- Example problems with solutions, and practice problems for additional learning opportunities.
- Improved ground improvement courses and offerings as faculty adopt a well-prepared textbook with instructor resources.

From the perspective of practicing professionals, the book provides:

- A resource allowing practicing professionals to understand and select ground improvement techniques with confidence.
- Up-to-date and thorough reference lists, enabling practicing engineers to access original materials used to evaluate alternatives and prepare designs.
- Photos to enable practitioners to use this material in presentations to clients allowing improved communications about ground improvement in the engineering and industrial/commercial environments.

PEDAGOGY

This new book, *Fundamentals of Ground Improvement Engineering*, has been written for advanced undergraduate and graduate students and practicing professionals. Most topics are organized on the basis of construction methods rather than a theoretical or analytical organization. In this manner, the goals and means of construction are first presented followed by the underlying geotechnical engineering principles and design considerations. This method of presentation is adopted under the ideology that most people learn best when the material is presented from the general progressing to the specific. This book also includes thorough and up-to-date literature citations as well as an abundance of graphics including photographs, schematics, charts, and graphs.

LIMITATIONS

Each and every topic in this text is the subject of hundreds of technical papers published in journals, conferences, or even other textbooks. As a result, each topic could easily be the subject of a complete text. The authors encourage readers interested in a given topic to delve more deeply into the literature and citations provided in this text.

ACKNOWLEDGMENTS

The authors thank their supportive wives and families. Without encouragement and support on the home front, an undertaking such as this simply could not have happened. Thank you, Laurel Evans, Megan Ruffing, and Linda Elton.

Bucknell University, Geo-Solutions, Inc. and retirement from full-time teaching all provide an atmosphere where the scholar can flourish. For this, the authors are grateful.

The authors have enjoyed working with, and appreciate the assistance of, numerous Bucknell University students that have contributed to this work. Students who reviewed and edited various chapters include Jeff Ayers, E. J. Barben, Landon Barlow, Tim Becker, Mark Beltamello, Bradley Bentzen, Dan Bernard, Paul Bortner, Conner Briggs, Jeremy Byler, Minwoo Cho, John Conte, Michael Cortina, Kate Courtein, Loujin Daher, Akmal Daniyarov, Louis DeLuca, Ben Downing, Jonathan Eberle, Sarah Ebright, Johnna Emanuel, Jack Foley, Jake Hodges, Orman Kimbrough IV, Roger Knittle, Chris Kulish, Rich LaFredo, Muyambi Muyambi, Rachel Schaffer, Chandra Singoyi, Matthew Geiger, Jason McClain, Matthew McKeehan, Kelsey Meybin, Ryan Orbison, Brendan O'Neal, Nolan O'Shea, Michael Pontisakos, Max Pucciarello, Melissa Replogle, Kyle Rindone, Shelby Roberts, Joe Sangimino, Joseph Scalia, Brian Schultz, John Skovira, Ben Stodart, Michael Stromberg,

Benjamin Summers, Brendan Swift, Dan Tischinel, Curtis Thormann, Kirsten Vaughan, Brian Ward, Nathaniel Witter, Nikki Woodward, Seungcheol Yeom, Gregory Zarski, and Tyler Zbytek. Special thanks go to Zach Schaeffer and Jeremy Derricks for their contributions. We offer apologies for students overlooked in this listing.

The authors also appreciate the review and assistance of Geo-Solutions employees Ken Andromalos, Nathan Coughenour, Wendy Critchfield, and Mark Kitko for their contributions to this effort.

The authors also appreciate the assistance of James Pease of McCrossin Engineering, Inc., Paul Marsden and Richard Holmes of Keller UK, Greg Stokkermans of GFL Environmental Inc., and Paul Schmall of Keller NA. Special thanks go to Jennifer A. E. Shields of Cal Poly San Luis Obispo for her work on the cover collage.

Many of the figures in this text are original art created by the authors. Some were prepared with the assistance of those contributors listed above. Some photographs were provided by industry professionals as credited in the text. The authors appreciate their willingness to contribute to our efforts. Photographs and artwork not attributed to others are products of the authors and their student assistants.

Lastly, the authors are appreciative of the undying patience and guidance of the publishers/editors: Tony Moore, Siobhan Poole, Scott Oakley, Gabriella Williams, and Frazer Merritt of Taylor and Francis.

Jeffrey C. Evans, P. E., Professor Emeritus,
Bucknell University, Lewisburg, Pennsylvania, USA

Daniel G. Ruffing, P. E., Vice-President,
Geo-Solutions, New Kensington, Pennsylvania, USA

David J. Elton, P. E. Professor Emeritus,
Auburn University, Auburn, Alabama, USA

Chapter 1

Introduction to ground improvement engineering

1.1 INTRODUCTION

Ground modification in the constructed environment is not a new idea. For instance, the method of wattle and daub has been used for thousands of years to provide tensile reinforcement to clayey materials in buildings. The process of adding straw to clay and baking it in the sun improved the strength properties of the clay creating a building material that has been used for thousands of years. In another ancient application, the Romans used timber as a base layer for roads. In modern times, inclusions (such as geogrids and geotextiles) are commonly employed for ground improvement. Similarly, the addition of lime to clay (a chemical admixture in modern terminology) has long been used to create a weak binder in stone foundations. The Roman road, Via Appia, now in modern-day Italy, is the earliest known example of the use of lime in ground improvement engineering (Berechman 2003).

The terms *ground improvement, ground modification,* and similar terms are lexicon of the late 20th century. The first conference on the subject was "Placement and Improvement of Soil to Support Structures" and was held in Cambridge, Massachusetts, in 1968, sponsored by the Division of Soil Mechanics and Foundation Engineering of the American Society of Civil Engineers (ASCE 1968). The first comprehensive textbook on the subject was by Hausmann (1990). University courses on the subject began at about the same time. In many ways, ground improvement engineering is a relatively new field within geotechnical engineering. New developments are occurring at a rapid pace and no doubt will have occurred throughout the life of this book. Thus, this book focuses on fundamentals, enabling the user to understand and adapt to the latest ground improvement developments.

How might ground modification/improvement be defined? In the proceedings on the Conference on Soil Improvement (ASCE 1978), the introduction succinctly states that one of the alternatives available when poor soil conditions are encountered is to "treat the soil to improve its properties." Moseley and Kirsch (2004) in the second edition of their book, *Ground Improvement*, note that

> All ground improvement techniques see to improve those soil characteristics that match the desired results of a project, such as an increase in density and shear strength to aid problems of stability, the reduction of soil compressibility, influencing permeability to reduce and control groundwater flow or to increase the rate of consolidation, or to improve soil homogeneity.

Schaefer et al. (2017) define ground modification as "the alteration of site foundation conditions or project earth structures to provide better performance under design and/or operational loading conditions." For the purposes of this book, ground improvement is defined as the application of construction means and methods to improve the properties of soil.

Note that some improvements are of the first order. For example, compaction will increase the density of soil. However, density increases can lead to second order effects such as increased strength and reduced compressibility. Finally, these second order improvements can result in third order effects such as increased bearing capacity and reduced settlement and/or improved liquefaction resistance. By beginning with the fundamentals of ground improvement engineering, the text is designed to provide an understanding of both the fundamental first-order effects as well as those second- and third-order effects that are often the actual desired outcome of the application of ground improvement. As there are many definitions of ground improvement and further much gray area within each definition, the authors used this definition as a guide to define the scope of this book.

Finally, for the purposes of the selection of the content in this book, the authors use the term ground improvement rather than ground modification. Ground modification is a neutral term meaning the modification could either improve or worsen the ground whereas ground improvement is unambiguous.

Prior to in-depth study of ground improvement, what are the alternatives to ground improvement? Imagine a site where the subsurface conditions are not suitable for the anticipated project. While ground improvement is the option to be considered in detail in this book, what are the alternatives? Some common alternatives to the application of ground improvement include:

1. **Avoid the site or area:** There are many circumstances where the owner/developer has options regarding the location of the proposed facility and finding an alternative site or a different area of the same site is a viable option.
2. **Remove and replace:** If the unsuitable materials are limited in aerial and/or vertical extent, the best (and most economical) option may be to simply excavate the unsuitable soils and replace them with more suitable materials having more predictable properties, such as crushed stone. This is a commonly chosen alternative when a localized fill is encountered.
3. **Transfer load to deeper strata:** The use of deep foundations, such as piles or drilled shafts, has long been the option of choice in locations where unsuitable bearing materials are present near the ground surface. Deep foundations affect load transfer through the use of stiff structural members placed between the structure and competent bearing materials found at deeper depths. Although significantly more sophisticated today, this technique has existed for centuries with ample evidence including ancient Roman bridges supported on timber piles.
4. **Design structure accordingly:** Some sites and structures, in combination, may lend themselves to structural redesign to accommodate the site conditions. For instance, it may be possible to stiffen the structure to redistribute stresses within the structure and minimize differential movement. In a specific application, sinkhole prone areas such as solution-prone geologic settings, grade beams can be used to connect spread footings in order to redistribute loads in case of loss of support beneath any single footing. Likewise, structures can incorporate construction joints, allowing some differential settlement without causing distress.

1.2 IMPROVEMENTS IN SOIL BEHAVIOR

Ground improvement may be viewed from the perspective of system performance. For example, it may be necessary to improve the ground to increase the allowable bearing value of a footing supported on the soils beneath a structure. From the system perspective, ground improvement alternatives would be evaluated for their ability to increase bearing capacity and decrease settlement, i.e. increase the allowable bearing value. More precisely, the allowable bearing value can be increased by:

1. increasing the stiffness of the soil (decreases settlement),
2. increasing the shear strength of the soil (increases bearing capacity), and/or
3. decreasing soil property variability (decreases differential settlement).

Densifying granular materials or consolidating cohesive materials can increase soil strength and stiffness.

Using these definitions, there are many ways ground improvement can be viewed. For the purposes of understanding ground improvement, this text will focus on a fundamental understanding of the interactions between ground improvement techniques and the resulting changes in soil and/or soil system behavior. This text also provides insight into the means and methods used by contractors to implement ground improvement techniques with most of the chapters and information segmented by construction techniques.

In this chapter, it is useful to articulate the improvements in soil behavior that may result from the ground improvement methods employed. These fundamental soil behavior characteristics include shear strength, compressibility, hydraulic conductivity, liquefaction potential, shrink and swell behavior, and reduction in variability in any of the aforementioned behavioral characteristics. Details of soil behavior principles related to ground improvement are provided in Chapter 2.

1.2.1 Shear strength

Shear strength is a fundamental engineering property of soils that can be increased through the application of numerous ground improvement techniques. Shear strength is a measure of the soil's ability to resist failure under the application of a load that induces shear stresses in the soil. Shear strength can be increased through ground improvement techniques that decrease the void ratio (Chapters 4, 5, and 11), and/or adding a cohesive (cementing) component (Chapter 6 and 7). There are many applications that benefit from improved shear strength including increased bearing capacity, improved slope stability, and reduced liquefaction potential.

The shear strength of soils is a sophisticated concept. There are entire texts devoted solely to this topic. Unconfined compression tests (see Figure 1.1) are a common means to quantitatively judge the benefit of ground improvement efforts. For some projects, more sophisticated testing may be needed. Principles of shear strength, both drained and undrained, are reviewed in Chapter 2.

1.2.2 Compressibility

Soil stiffness is a measure of the deformation of soils associated with the application of a load. Compressibility is not a unique value, since it depends on the nature of the load application and the initial stress state of the soil. The soil stiffness can be increased, i.e. decreased compressibility, through ground improvement techniques that reduce void ratio or add a cohesive or cementing component. Cohesive soil stiffness can be increased by compaction (Chapter 4) and consolidation (Chapter 5). Granular soil stiffness is generally increased by densification (Chapter 4). Cohesive and granular material compressibility can also be reduced via increasing cohesiveness through soil mixing (Chapter 6) or grouting (Chapter 7).

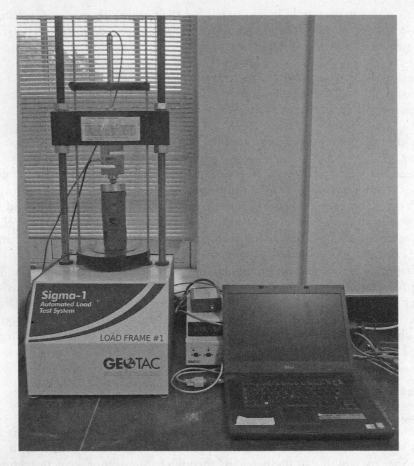

Figure 1.1 Unconfined compressive shear strength apparatus.

One of the most well-known cases of excessive deformation (aka settlement) is the campanile (bell tower) in Pisa (see Figure 1.2), aka the "Leaning Tower of Pisa." Differential movement of the ground below the tower has been the subject of numerous studies and there have been multiple attempts to stabilize the tower. The differential movement results from non-uniform subsurface conditions and is exacerbated by the uneven load application once tilting began. In Figure 1.2, notice the cables extending outward from the left side of the tower. This photograph was taken in 1999 at which time a pulley and counterweight system were in place coupled with lead weights placed directly on the foundation acting as a counterweight employed as an emergency measure to stabilize the tower. Subsequently, ground extraction beneath the high side of the tower proved successful in arresting the movements (Burland et al. 2009). This famous landmark remains a reminder that controlling deformation and preventing strength failures are two key performance criteria for geotechnical engineering projects.

1.2.3 Hydraulic conductivity

In most cases, improved ground is ground that is modified to produce a zone of reduced permeability in order to control the detrimental effects of groundwater. For example, flow beneath a dam can lead to soil particle movement (piping) and/or instability. Construction

Figure 1.2 The Leaning Tower of Pisa.

projects also frequently require construction below grade and often below the water table. In these cases, construction dewatering is needed. Ground improvement in such cases might include dewatering, installation of a low permeability vertical barrier (Chapter 8), or reduction in permeability by grouting (Chapter 7). As is often the case in practice, hydraulic conductivity and permeability are used interchangeably in this book.

1.2.4 Liquefaction potential

Loose granular materials below the groundwater level can be subject to liquefaction (see Figure 1.3) upon the application of a dynamic load, such as during an earthquake. During shaking, loose granular soil deposits generally decrease in volume (i.e. loose soils densify). If these loose soils are located below the water table, drainage would be needed for the soils to actually densify. This drainage requires sufficient time, which for granular materials, is normally not a problem during static loading. However, during earthquake loading, there is insufficient time for drainage which results in an increase in porewater pressure and a reduction in the effective shear strength of the granular soil. These principles of shear strength and liquefaction potential are presented in more detail in Chapter 2. The most common mitigation of this risk is to densify the soils, which reduces their liquefaction potential. Common tools for densifying granular materials are described in Chapter 4. Other ground improvement techniques to reduce liquefaction potential include groundwater control (Chapters 7 and 8) and in situ mixing (Chapter 6).

Figure 1.3 Liquefaction of road foundation in New Zealand (photo courtesy National Environmental Satellite, Data, and Information Service).

1.2.5 Shrink/swell behavior

Soils containing smectitic clays are subject to substantial volume changes in response to cycles of wetting and drying. The shrink/swell behavior of these expansive soils can have detrimental effects and can progressively damage a building or cause a retaining wall to fail. Figure 1.4 illustrates road damage due to expansive soils. Understanding clay mineralogy and the resulting expansive behavior (Chapter 2) prior to selecting and designing ground improvement methodologies is important. Ground improvement, through the use of admixtures and in situ mixing (Chapter 6), can minimize the propensity for these materials to change volume with wetting and drying.

1.2.6 Variability

Physical and engineering properties of soils are naturally variable. At times, this variability can affect the performance of a planned structure. For example, if the compressibility varies enough from location to location, an excessive differential settlement could be expected. Ground improvement can modify the properties of subsurface materials to provide a more uniform performance. For example, consider the settlement sensitive structure shown in Figure 1.5. Here, the depth to bedrock increased in the downslope direction along the axis of the building. Overlying the bedrock were unconsolidated materials of increasing thickness

Figure 1.4 Structural damage due to expansive soils (photo courtesy of Anand Pupala).

Figure 1.5 Settlement sensitive brick structure with variable subsurface conditions.

from one end of the building to the other. Unsurprisingly, concerns with differential settlement arose and a deep foundation system was chosen for the structure (drilled shafts into pinnacled limestone). However, the chosen foundation system was very costly. This short case study serves to illustrate that, under variable site conditions, ground improvement could reduce site variability, permitting an inexpensive shallow foundation system rather than requiring an expensive deep foundation system. For this site, vibro methods (Chapter 5) could have both densified the soils and reduced variability in compressibility across the site. In cases such as this, ground improvement can prove to be significantly less costly and provide performance equivalent to a deep foundation system.

1.3 OVERVIEW OF GROUND IMPROVEMENT TECHNIQUES

Ground improvement principles have certain fundamental mechanistic characteristics that are used to develop a classification system for ground improvement. Accordingly, this book uses four defining principles, in order of increasing complexity:

1. control of water – removal or control of groundwater,
2. mechanical modification – rearrangement of soil or water particles,
3. modification by additives – addition of chemicals and,
4. modification by inclusions or confinement – system behavior modification through rigid or flexible element inclusion or soil confinement.

Assigning a particular ground improvement technique to a particular category is imperfect since some techniques possess multiple behavioral characteristics or provide improvement via multiple principles. This results in some techniques having characteristics from more than one category. Nonetheless, such classification system is useful in understanding how particular techniques work on a fundamental level. Based upon how a given ground modification technique improves the soil, this book is structured according to Figure 1.6.

Some of the important principles, engineering considerations, and construction methods that are the focus of this book are discussed further in the subsections below.

1.3.1 Compaction: shallow methods

Compaction is the densification of soils at constant water content. Consolidation, in contrast, is differentiated from compaction by the decrease in water content due to the application of load to a saturated soil. Compaction (densification) is achieved through the application of mechanical energy to soil such that the air void volume is decreased, increasing soil density. Surface compaction with equipment, such as the pad foot self-propelled roller pictured in Figure 1.7, has long been used to increase strength, reduce compressibility, and reduce the permeability of soils. Examples of ground improvement techniques that use mechanical energy as the principal means to improve soil behavior include surface compaction, deep dynamic compaction, and rapid impact compaction. These are all surface applications of mechanical energy that dissipate with depth. In doing so, the mechanical energy causes a rearrangement of the soil structure into a denser configuration. Shallow (surface) methods of compaction for ground improvement are presented in Chapter 4.

1.3.2 Compaction: deep methods

Occasionally, the effective depth of surface compaction is insufficient compared to the depth of material targeted for compaction. Here, deep compaction methods, which apply mechanical energy below the surface, are needed. In most cases, deep compaction methods also employ vibration and often involve the addition of stone, grout, or concrete during the process to fill the space created by the densification. Depending upon the details of the process and the contractor completing the work, various names are given to these deep methods. Such names include, but are not limited to, vibroflotation, vibrocompaction, vibroreplacement, Geopiers®, and rammed aggregate piers® (RAP). For example, Figure 1.8 shows a vibrator used for deep vibrocompaction or vibroreplacement and Figure 1.9 shows the hopper being filled with sand during a vibrocompaction project.

These techniques evolved from work done over 70 years ago by the Keller Company (Kirsch and Kirsch 2016). Deep compaction techniques began to flourish in the 1950s and

Ground improvement principle		Engineering principles		Construction methods	
Removal or control of water	1	Consolidation	2	Ground freezing	11
		Dewatering	2	Preloading	5
		Drainage	2	Slurry trenching	8
		Filtration	2	Vacuum consolidation	5
		Preloading	2	Vertical drains	2
		Seepage control	2	Vertical wells	2
				Geosynthetics	9
Particle rearrangmenet	1	Compaction	2	Compaction grouting	11
		Densification	2	Deep dynamic compaction	4
		Thermal treatment	2	Rammed aggregate piers	4
				Rapid impact compaction	4
				Soil fracturing	7
				Stone columns	4
				Vibrocompaction	4
Additives	1	Bio-Treatment	12	Solution grouting	7
		Chemical Treatment	2	Direct injection	7
		Hydrophobicity	0	Jet grouting	11
		Chemical Oxidation	2	Suspension grouting	7
		Chemical Reduction	2	Permeation grouting	7
		Stabilization / Solidification	2	Soil fracturing	7
				Soil mixing	6
System behavior	1	Confinement	2	Secant piles	11
		Reinforcement	2	Compaction grouting	11
		Retaining Walls	2	Fracture grouting	7
		Rigid Inclusions	2	Geosynthetics	9
		Shear Walls	2	Soil mixed columns	6
		Underpinning	2	Mechanically stabilized earth	10
				Rammed aggregate piers	4
				Sheetpiles	8
				Slurry walls	8
				Soil nails	10
				Vibroreplacement	4

#	Chapter Title
	Preface
1	Introduction to ground improvement engineering
2	Geotechnical fundamentals
3	Fundamentals of geosynthetics in ground improvement
4	Compaction
5	Consolidation
6	Soil mixing
7	Grouting
8	Slurry trench cutoff walls
9	Ground improvement using geosynthetics
10	Reinforcement in walls, embankmens on stiff ground, and soil nailing
11	Additional techniques in ground improvement
12	The future of groundi mprovement engineering

Figure 1.6 Organization of the book.

Figure 1.7 Soil compaction with a pad foot compactor.

Figure 1.8 Vibrator used for deep vibratory compaction (courtesy of Keller North America).

Figure 1.9 Adding sand to the hopper for vibrator during deep vibratory compaction.

1960s. Early projects used large, torpedo-like vibrators operating between 1,500 rpm and 3,000 rpm and that developed horizontal forces in the range of 100 kN to 150 kN to effectively compact loose sands. Initially, sand was added at the surface to compensate for the volume change resulting from the densification of the in situ sand. As time passed, bottom-feed vibrators were developed for the addition of sand or stone, enabling the construction of stone columns. For a more detailed history, particularly European history, of the development of deep vibratory technics, see Kirsch and Kirsch (2016). Deep compaction methods are addressed in Chapter 4.

1.3.3 Soil mixing and injection methods

Soil mixing methods, such as those described in Chapter 6, are methods of ground improvement wherein the soil properties are improved in situ via the addition of one or more reagents. Injecting or mixing in reagents such as lime, portland cement, slag cement, or combinations of reagents, can result in increased shear strength, reduced compressibility, and reduced hydraulic conductivity. In addition to understanding the means and methods of soil mixing and injection, an understanding of the mechanisms by which the additives work is critical to the successful choice and use of any particular soil mixing and injection method. For example, reagents can be added in a dry mix method (Figure 1.10) or a wet mix method (Figure 1.11). The process of selecting the best method, mix designs, and field configurations depends on the knowledge of soil conditions and the desired outcomes. To

Figure 1.10 Soil mixing using dry mix method.

Figure 1.11 Soil mixing using wet mix method.

this end, common materials and the mechanisms of addition along with construction and testing methods to verify performance are presented in Chapter 6.

1.3.4 Stabilization and solidification

The improvement of the ground at contaminated land and hazardous waste sites involves additional considerations, materials, and methods beyond those that might be used for ground improvement for geotechnical purposes. Much of the equipment and many of the construction methods are the same as, or similar to, those discussed in Chapter 6. For

Figure 1.12 Stabilization/solidification of a contaminated site.

example, Figure 1.12 shows stabilization and solidification of a contaminated site. Like many contaminated sites, multiple methods of site remediation were employed as a system to contain the contaminants and mitigate the risk to public health and the environment. At this site, stabilization and solidification were used for the upper portion of the area of the disposal pits along with a vertical cutoff (Chapter 8) to control and contain the remaining contaminated soil, sludge, and groundwater.

The special nature of contaminated ground as well as the protection of public health and the environment requires additional reflection. For these applications, topics such as contaminant transport and bonding mechanisms need to be coupled with traditional considerations such as strength, permeability, and compressibility. These topics are presented in Chapter 2 and discussed in Chapters 6 and 8.

1.3.5 Grouting

Grouting, as a means of ground improvement, consists of injecting, usually under pressure, a fluidized material (grout) into the subsurface. The grout then either fills pore space or displaces soil, producing stronger a soil formation. Grouting techniques include permeation grouting, fracture grouting, compaction grouting, and jet grouting (a form of soil mixing). Mechanistically, each technique is different, using different materials, methods, and design methodologies.

Grout materials often "set" or harden after injection. Chemical grouts, such as silicate grouts, can have low viscosities and penetrate small void spaces. Most cement grouts, particularly those made with ordinary portland cement, cannot penetrate small voids but work well in rock containing open fractures and voids. Successful grouting programs are developed with an in-depth understanding of the rheological properties of the grout (viscosity, set time, and stability) to predict the movement of the grout in the subsurface.

Compensation grouting is of special importance in urban areas. For example, the construction of the CrossRail project in London included the construction of new railway tunnels and stations in an already crowded subsurface environment. Given the above-ground environment that includes many historic and aesthetic structures along the route, techniques to avoid damage to existing structures were required. Excavations for tunnels and stations below grade would inevitably cause surface movements if not for the ability to "compensate" for the subsurface movements via grouting. Thus, surface movements are regularly anticipated, monitored, and corrected by subsurface compensation grouting. Figure 1.13 schematically illustrates the benefits of compensation grouting to the minimization of the settlement of buildings along a tunnel alignment. Analysis of monitoring data to detect movements can lead to the decision to inject grout under pressure at specified locations to compensate for the detected movements. Compensation, and other types of grouting, are discussed in Chapters 7 and 11.

1.3.6 Dewatering

There are times that ground is unstable only because groundwater is present or flowing in such a way as to destabilize the soil. While grouting (Chapter 7) and cutoff walls (Chapter 8) are two ground improvement methods that can reduce hydraulic conductivity and improve stability, there are numerous occasions when dewatering may be a better choice. Without proper groundwater control, flowing groundwater can result in bottom heave, unstable slopes, and difficult or impossible working conditions. Figure 1.14 shows an excavation below the water table in a stratigraphy of sand overlying silt of lower permeability. Even with deep dewatering wells, three meters on center, seepage between the wells at the interface between the sand and the silt resulted in localized and progressive slope instability.

Ground improvement by dewatering is a widely used, but often difficult, technique that requires detailed knowledge of subsurface conditions, theoretical understanding of groundwater flow, and experience. Dewatering is well covered in many texts, including Powers et al. (2007).

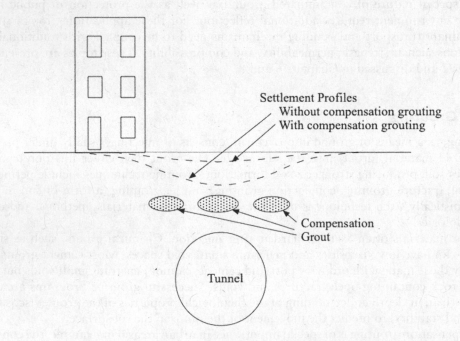

Figure 1.13 Compensation grouting to minimize settlement during tunneling.

Figure 1.14 Improperly dewatered excavation.

1.3.7 Consolidation

While compaction (Chapter 4) is densification at constant water content, consolidation is densification at decreasing water content (Chapter 5). As a result, consolidation is a time-dependent process, as it takes significant time for water to leave clay. During consolidation, soils gain strength and their compressibility is reduced. Soft, compressible, fine-grained soils are prime candidates for ground improvement by consolidation. Soft cohesive soils generally have low hydraulic conductivities and, since the time-rate of consolidation depends upon soil permeability, the time required to consolidate soft cohesive soil may exceed the time available in the construction schedule. In these cases, consolidation can be enhanced by inserting vertical drains. Traditionally sand drains were installed to shorten the drainage path and speed up the consolidation process. Prefabricated vertical drains are now more commonly used. Figure 1.15 shows schematically a typical use of vertical drains to speed consolidation of soft ground beneath an embankment. Consolidation, to improve the properties of ground using techniques such as vertical drains, preloading, and vacuum consolidation, is discussed in Chapter 5.

1.3.8 Mechanically stabilized earth

For thousands of years, masonry structures were built in such a way as to impart compressive stresses on the stone building materials. Arches were commonly used to span openings as this configuration assured the masonry materials were in compression. This building approach

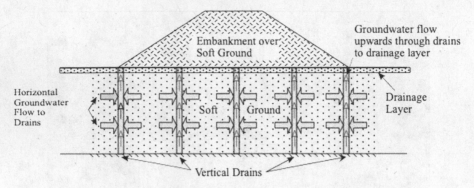

Figure 1.15 Vertical drains to speed consolidation of soft ground.

Figure 1.16 Mechanically stabilized earth (courtesy of Robert Barrett, GeoStabilization International).

was used because stone has little tensile strength but large compressive strength. Similarly, soils have negligible tensile strength but large compressive strength. Soils work well to support structures and serve as earthen structures when in compression. The introduction of tensile reinforcement, first popularized as Reinforced Earth™, in the 1960s, opened the door to a wide range of applications including the now widely used mechanically stabilized earth (MSE) retaining walls. The enormous benefit of reinforcement is illustrated in Figure 1.16. Not only can a vertical face of fill be achieved but a reverse batter as well.

The benefit of reinforcing is further illustrated to students via the ASCE GeoChallenge, a student competition. Shown on the left side of Figure 1.17 is a sheet of construction paper (the retaining wall face) with strips of brown wrapping paper attached (the reinforcement). Shown on the right side of Figure 1.17 is the completed retaining wall 0.5 m high supporting a sandy backfill and a 22 kg surcharge. This laboratory experiment demonstrates the important improvement in granular soil strength by the addition of even modest tensile reinforcement. Chapter 10 discusses the forms and uses of geosynthetic reinforced soil.

Figure 1.17 Laboratory-scale mechanically stabilized earth wall during load testing.

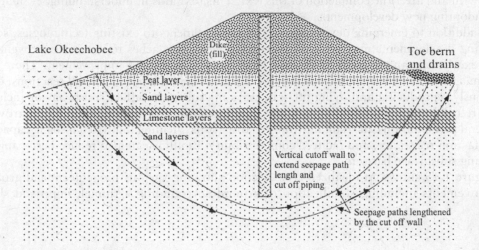

Figure 1.18 Herbert Hoover Dike cutoff wall.

1.3.9 In situ barriers

In situ vertical barriers (cutoff walls) have been used for over 40 years to control the horizontal flow of groundwater in the subsurface. Improving the ground conditions, by reducing the flow in the horizontal direction, has been commonly used for dewatering to improve slope stability and reduce water flow into excavations. In the 1980s, these same barriers came into widespread use for environmental applications to control contaminant transport in the subsurface. Engineers also know and acknowledge that many of the dams and levees constructed over the last 100 or more years need improvement. Issues with seepage and stability jeopardize their performance, particularly during flood events. Thus, in situ barriers (cutoff walls) have found widespread use to improve the properties of the underlying materials and improve the properties of the dam or levee.

The Herbert Hoover Dike in Florida, USA, is a prime example of the use of a barrier wall for levee rehabilitation in response to seepage and piping problems. In order to cut off seepage through and beneath the dam, a cutoff wall was installed (Figure 1.18). The wall, 0.7 m wide and averaging 22 m deep, penetrated the dike and the underlying layers of peat, sand, and limestone. As a result, existing piping paths were cut off, the seepage path was lengthened, and exit gradients were reduced.

There are myriad materials that may be used in cutoff walls and numerous ways to construct them. The desired final product is usually a cutoff wall that is homogeneous and has a low permeability (hydraulic conductivity). Often there is a moderate strength requirement as well. Special considerations of compatibility between the barrier and the contaminants are needed when these barriers are used to control contaminant transport around waste or contaminated land sites. Materials, methods, designs, and analyses of cutoff walls are discussed in detail in Chapters 6, 7 and 8.

1.3.10 Future developments in ground improvement

While many ground improvement techniques are tried and proven, there are continuous developments in design, equipment, and construction techniques for these established methods. No doubt there will be new publications reporting on these developments coincident with and after the completion of this text. This text aids in understanding, evaluating, and adopting new developments.

In addition to emerging developments and improvements to existing technologies, some pending developments may prove to be entirely new approaches to ground improvement. One excellent example might be termed biogeotechnical ground improvement. There is a rich microbial environment in soils. Microbes participate in biogeochemical reactions, continuously reproducing and dying off. The ways microbes can affect soil behavior include, but are not limited to, mineral precipitation, mineral transformation, and biofilm growth. Mineral precipitation can result in stronger, stiffer soils, yielding improved bearing capacity, liquefaction resistance, and reduced compressibility. Biofilms can also reduce permeability, forming subsurface barriers.

Figure 1.19 shows an idealized cross-section showing various biogeotechnical ground improvements including stabilization of ground surrounding a tunnel, improved slope

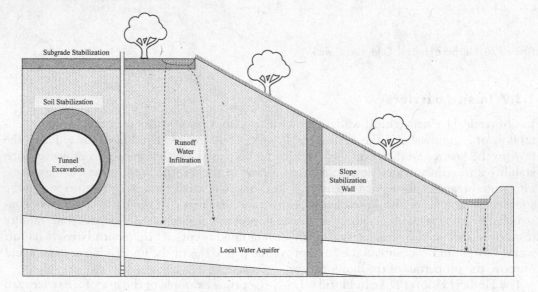

Figure 1.19 Biogeotechnical ground improvement.

stability, low permeability barrier to control subgrade water, and improved erosion control.

In addition to developments like biogeotechnical ground improvement, the future is likely to reveal the development and use of existing materials and methods in ways that are not currently used. While not in widespread use, mixing plastic fibers to increase the strength of sand (Park and Tan 2005; Gray and Ohashi 1983) and the use of geofoam to reduce earth pressures on retaining walls (Horvath 2010; Dasaka et al. 2014) are gaining use. The benefits of reusing a variety of waste materials, such as recycled gypsum (Ahmed and Issa 2014) and electrokinetics for the stabilization of soft clay (Lamont-Black et al. 2012; Malekzadeh and Sivakugan 2017), are being studied.

It is likely that the future of ground improvement will provide for explicit considerations of sustainability when deciding what, if any, ground improvement method to employ. Historically, geotechnical engineers were primarily concerned with (1) providing an adequate factor of safety against failure of soil; (2) controlling settlements and movements of the ground; and (3) cost. Environmental and sustainability considerations are an important part of the decision process. Considerations of noise, historically or architecturally important structures, archeological finds, and inconvenience to the public are essential considerations when employing ground improvement. At the time of this writing (2021), it is clear that future projects will need to explicitly consider sustainability and legacy effects in the decision process.

1.4 IMPORTANCE OF CONSTRUCTION

There is a common thread that weaves through this chapter and this book: the design and performance of ground improvement is inextricably linked to construction. One cannot "design" a ground improvement program without a full understanding of the construction means and methods. In fact, credit for the development of ground improvement techniques lies largely with innovative contractors. Many of the experts in the field of ground improvement are or were contractors.

1.5 PROBLEMS

1.1 Are ground improvement techniques more sustainable than traditional alternatives such as deep foundations? Justify your answer.

1.2 Choose a ground improvement technique and prepare a 10-slide presentation appropriate for secondary school students to increase their interest in the fields of science, technology, engineering, or mathematics.

1.3 How are improvements in soil strength and stiffness fundamentally different?

1.4 Compare and contrast consolidation and compaction.

1.5 Using principles of sustainability, compare the use of ground improvement with more traditional deep foundation methods.

1.6 Ground improvement problems are largely those of soil-structure interaction. Explain.

1.7 The water content and degree of saturation will significantly impact the efficacy of certain ground improvement techniques. Relate your experiences on the beach building sandcastles to the effect of water content and degree of saturation.

1.8 Specialty contractors are more likely than geotechnical consultants to develop new and improved techniques in ground improvement. Why would this be the case?

REFERENCES

Ahmed, A. and Issa, U.H. (2014). Stability of soft clay soil stabilised with recycled gypsum in a wet environment. *Soils and Foundations*, 54(3), 405–416.

ASCE. (1968). *Specialty conference on placement and improvement of soil to support structures.* Reston, VA: American Society of Civil Engineers, Soil Mechanics and Foundations Division.

ASCE. (1978). *Soil improvement-history, capabilities, and outlook.* J.K. Mitchell (Ed.). New York: American Society of Civil Engineers, 182 pp.

Berechman, J. (2003). Transportation—economic aspects of Roman highway development: The case of Via Appia. *Transportation Research Part A: Policy and Practice*, 37(5), 453–478.

Burland, J.B., Jamiolkowski, M.B. and Viggiani, C. (2009). Leaning Tower of Pisa: Behaviour after stabilization operations. *ISSMGE International Journal of Geoengineering Case Histories*, 1(3), 156–169.

Dasaka, S.M., Dave, T.N., Gade, V.K. and Chauhan, V.B. (2014). Seismic earth pressure reduction on gravity retaining walls using EPS mm. In *Proceedings of 8th international conference on physical modelling in geotechnical engineering* (pp. 1025–1030), Perth, Australia.

Gray, D.H. and Ohashi, H. (1983). Mechanics of fiber reinforcement in sand. *Journal of Geotechnical Engineering*, 109(3), 335–353.

Hausmann, M.R. (1990). *Engineering principal of ground modification.* McGraw-Hill Publishing Company, 631 p.

Horvath, J.S. (2010). Emerging trends in failures involving EPS-block geofoam fills. *Journal of Performance of Constructed Facilities*, 24(4), 365–372.

Kirsch, K. and Kirsch, F. (2016). *Ground improvement by deep vibratory methods.* CRC press.

Lamont-Black, J., Hall, J.A., Glendinning, S., White, C.P. and Jones, C.J. (2012). Stabilization of a railway embankment using electrokinetic geosynthetics. *Geological Society, London, Engineering Geology Special Publications*, 26(1), 125–139.

Malekzadeh, M. and Sivakugan, N. (2017). Experimental study on intermittent electroconsolidation of singly and doubly drained dredged sediments. *International Journal of Geotechnical Engineering*, 11(1), 32–37.

Moseley, M.P. and Kirsch, K. (Eds.). (2004). *Ground improvement.* New York: Taylor and Francis.

Park, T. and Tan, S.A. (2005). Enhanced performance of reinforced soil walls by the inclusion of short fiber. *Geotextiles and Geomembranes*, 23(4), 348–361.

Powers, J.P., Corwin, A.B., Schmall, P.C. and Kaeck, W.E. (2007). *Construction dewatering and groundwater control: New methods and applications.* Somerset, NJ: John Wiley & Sons.

Schaefer, V.R., Berg, R.R., Collin, J.G., Christopher, B.R., DiMaggio, J.A., Filz, G.M., Bruce, D.A., Ayala, D. and Berg, R.R. (2016). *Geotechnical engineering circular no. 13 ground modification methods-reference manual volume II* (No. FHWA-NHI-16-028). National Highway Institute (US).

Chapter 2

Geotechnical fundamentals

It is expected that most readers of this book will have had at least one course in soil mechanics. It is also well known that deep learning can only be achieved by repeated retrieval and use of information previously learned. The fundamentals of soil behavior presented in this chapter are those most relevant and necessary to the understanding of ground improvement engineering. Several introductory soil mechanics textbooks present soil mechanics principles and it is not the purpose of this chapter to repeat that which is presented in greater detail elsewhere. Rather, this chapter is included to highlight those fundamentals that will need to be understood to understand the principles of ground improvement engineering presented in this book.

2.1 DEFINITIONS

While life is an open book, knowing, really knowing, fundamental definitions (or defined ratios) is essential to the successful understanding of ground improvement engineering. Water content, dry density, dry unit weight, specific gravity, total density, total unit weight, saturation, void ratio, and porosity are all defined ratios, the equations for which cannot be derived. Relationships between various defined ratios, such as the relationship between total unit weight, dry unit weight, and water content can be derived although some of the more commonly needed relationships should be committed to memory. Also, symbols vary from source to source, so this chapter serves to define variables that will be used here and throughout the book.

Soil, or ground, is a three-phase material having solid, liquid, and gas phases. For the purposes of ground improvement engineering, the liquid phase is water and the gas phase is atmospheric air. In totally dry soil, there is no water and in saturated soil, there is no air. These three phases can be characterized on a mass or weight basis and a volumetric basis as shown in Figure 2.1 where:

M_s = mass of solids $\qquad$ W_s = weight of solids
M_w = mass of water $\qquad$ W_w = weight of water
M_t = total mass $\qquad$ W_t = total weight
V_s = volume of solids
V_w = volume of water
V_a = volume of air
V_T = total volume
V_v = volume of voids

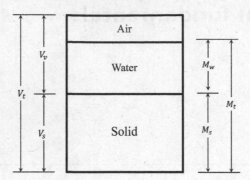

Figure 2.1 Phase diagram.

The masses of water and solids must sum to be equal to the total mass and the volumes of air, water, and solids must sum to be equal to the total volume. The volumes of air and water must sum to be equal to the volume of voids (the space between the solid particles).

2.1.1 Water content

The water content, also termed moisture content, has a substantial impact on the behavior of soils and is defined as the ratio of the mass of the water to the mass of the solids. In equation form, the water content, w, in percent, is:

$$w = \left(\frac{M_w}{M_s}\right)(100) \tag{2.1}$$

where terms are defined as shown in Figure 2.1. Widely used in geotechnical engineering, this gravimetric definition can give rise to water content values greater than 100%. It is noted that there are other gravimetric and volumetric definitions used in other disciplines including environmental engineering and geology.

In ground improvement engineering, water content is an incredibly important parameter. For example, high water content materials are stabilized using dry soil mixing, whereas low water content materials may call for stabilization using wet soil mixing. Since the water/cement ratio has a major impact on the strength of cemented materials, design mixtures must account for the in situ water content. The in situ water content is that water content of the soil in place prior to any disturbance during ground improvement.

Variation in water content with depth can also provide insight into stratigraphic variations. For example, clays may be found to have similar clay mineralogy, grain size distribution, and plasticity but different moisture contents. Those with the higher natural moisture content can be expected to have lower strength and greater compressibility than those with lower moisture contents.

The unit weight of water ($\gamma_w = 9.81\frac{kN}{m^3}$) is the density of water ($\rho_w = 1.0\frac{g}{cm^3} = 1.0\frac{Mg}{m^3}$) multiplied by the acceleration of gravity and can be assumed to be constant for the vast majority of ground improvement engineering problems. For problems in imperial units, the unit weight of freshwater can be assumed to be $\gamma_w = 62.4\frac{lb}{ft^3}$. The unit weight of saltwater is about $64\frac{lb}{ft^3}$.

2.1.2 Density, unit weight, density of solids, and specific gravity

Total density (ρ), total unit weight (γ), dry density (ρ_d), dry unit weight (γ_d), density of solids (ρ_s), and specific gravity of solids (G_s) are all defined ratios that, in some way, describe the degree of compactness of a soil. It is essential to be clear and precise in terminology when using these parameters in ground improvement engineering. The definitions are as follows:

$$\text{Total density, } \rho = \frac{M_T}{V_T} \tag{2.2}$$

$$\text{Total unit weight, } \gamma = \frac{g M_T}{V_T} = \frac{W_T}{V_T} \tag{2.3}$$

where g = acceleration due to gravity

$$\text{Dry density, } \rho_d = \frac{M_s}{V_T} \tag{2.4}$$

$$\text{Dry unit weight, } \gamma_d = \frac{g M_s}{V_T} = \frac{W_s}{V_T} \tag{2.5}$$

$$\text{Density of solids, } \rho_s = \frac{M_s}{V_s} \tag{2.6}$$

$$\text{Specific Gravity, } G_s = \frac{W_s}{\gamma_w V_s} \tag{2.7}$$

where the terms are shown in Figure 2.1. Note that these parameters are defined ratios. They are not derivable but are an essential part of the vocabulary of ground improvement engineering. The density of solids in g/cm^3 is mathematically identical to the specific gravity of solids.

Other material property definitions that provide insight into the density of soils are the void ratio, e, and porosity, n. The material properties are defined volumetrically as:

$$\text{Void ratio, } e = \frac{V_v}{V_s} \tag{2.8}$$

$$\text{Porosity, } n = \frac{V_v}{V_T} \tag{2.9}$$

Note that the limits of porosity are between 0 and 1.0, whereas void ratio can exceed 1.0.

In addition to water content, the amount of water in the soil can be examined through the volumetrically defined term *degree of saturation* which relates to the extent to which the voids are filled with water. The degree of saturation in percent is defined as:

$$S = \left(\frac{V_w}{V_v} \right)(100) \tag{2.10}$$

Example problem Ex.2.1: Phase diagram

The total density of dense, well-graded sand and gravel and the water content has been measured to be 2.20 g/cm³ (or 2.2 Mg/m³) and 10.0%, respectively. The density of the solids is given as 2.67 g/cm³. Determine the following properties:

1. dry density,
2. void ratio,
3. porosity, and
4. degree of saturation.

The first step is always to sketch the phase diagram to determine what is known. Use Figure 2.1 to do this. Since only defined ratios are given, it is not possible to enter any numerical values for the variables in the phase diagram. At this point, the simplest approach is to assume a unit value for one of the variables. For example, in this case, assume the M_s is 1.00 g (to three significant figures). Now using the values known for our defined ratio, the quantities in the phase diagram are calculated.

Using Equation 2.1 for the definition of the water content ($w = \dfrac{M_w}{M_s} *100$) and now knowing $w = 10.0\%$ and $M_s = 1.00$, M_w is calculated as 0.10. Now, knowing the M_w and M_s, add these together to determine the total mass, M_T, which is 1.10 g.

Now that the gravimetric side of the phase diagram is complete, the volumetric quantities can be computed using the density of solids to compute the volume of solids, the density of water to compute the volume of water, the total density to total volume, and the three volumes just computed to compute the volume of air as follows:

Using Equation 2.6 for the density of solids, $\rho_s = \dfrac{M_s}{V_s}$ and knowing $\rho_s = 2.67$ g/cm³ and $M_s = 1.00$ g, the $V_s = 0.382$ cm³ is computed. Similarly, knowing the density of water $\rho_w = 1.00$ g/cm³ and the $M_w = 0.10$ g, the $V_w = 0.102$ cm³ is computed. Continuing, using the total density $\rho_T = 2.20$ g/cm³ and the $M_T = 1.10$ g, the $V_T = 0.500$ cm³ is found. Lastly, the total volume is equal to the sum of the volume of solids, water, and air. Hence, the volume of air $V_a = V_T - V_w - V_s = 0.018$ cm³ and the volume of voids $V_v = V_T - V_s = 0.106$ cm³ can be computed.

Now the phase diagram has been filled out with quantities, the parameters requested can be calculated using the equations for each defined ratio provided in Equations 2.4, 2.8, 2.9, and 2.10, respectively.

Dry density, $\rho_d = \dfrac{M_s}{V_T} = \dfrac{1.0}{0.5} = 2.0$ g/cm³

Void ratio, $e = \dfrac{V_v}{V_s} = \dfrac{0.118}{0.382} = 0.309$

Porosity, $n = \dfrac{V_v}{V_T} = \dfrac{0.118}{0.5} = 0.236$

Degree of saturation (%), $S = \left(\dfrac{V_w}{V_v}\right)(100) = \left(\dfrac{0.1}{0.118}\right)(100) = 84.7\%$

2.2 WATER FLOW IN SOIL

Flowing water in soil causes several concerns for geotechnical engineers. Water changes the effective stress (Section 2.3) and, thus, the strength of the soil. Flowing water exerts pressure on buried objects and flowing water causes seepage forces that may cause erosion.

2.2.1 Darcy's law and one-dimensional flow

Darcy's law for one-dimensional flow of water through soils describes the rate of water flowing through the soil as follows:

$$q = kiA \qquad (2.11)$$

where
  q = flow rate of water
  k = hydraulic conductivity (or the coefficient of permeability)
  i = hydraulic gradient
  A = cross-sectional area of interest

The Darcy velocity (v) can be calculated by dividing the above equation by A

$$v = ki \qquad (2.12)$$

The Darcy velocity is a bulk parameter that uses the entire cross-sectional area of flow. Since the water cannot flow through solids, the water flows only through interconnected pores (or effective porosity) and thus the seepage velocity, v_s, is greater than the Darcy velocity.

$$V_s = k\frac{i}{n_e} \qquad (2.13)$$

where
  n_e = effective porosity

While total porosity can be readily calculated from using a phase diagram, determining the effective porosity is rather more difficult. For clean sands and gravels the effective porosity is essentially 100% of the total porosity. In contrast, for clay and rock, the effective porosity may be as low as 10% of the total porosity.

2.2.2 Flownets and two-dimensional flow

In many cases, ground improvement requires an understanding of the two-dimensional groundwater flow. The LaPlace equation can be used to represent the groundwater flow for the following assumptions:

1. S = 100%,
2. $k_x = k_y$ (isotropic with respect to hydraulic conductivity),
3. Darcy's law is valid, and
4. e = constant (incompressible formation).

For these assumptions two dimensional groundwater flow can be represented by the LaPlace equation as follows:

$$\frac{\partial^2 h}{\partial x^2} + \frac{\partial^2 h}{\partial z^2} = 0 \qquad (2.14)$$

Notice the solution is independent of the hydraulic conductivity.

Equation 2.14 can be solved using numerical approximation techniques, by electrical analog, or by graphical means termed a flownet (Harr 2012). A flownet is a map of flowing water as it dissipates energy with distance while flowing through the soil. Flownets are used to calculate:

a. porewater pressure so effective stress may be calculated,
b. uplift pressures on buried structures,
c. seepage forces on soil used to predict inter erosion or piping potential, and
d. the quantity of water flowing through soil by coupling knowledge of hydraulic conductivity to the flownet.

Energy, for groundwater, can be characterized as a total hydraulic head which is the sum of the elevation head and the pressure head. Given the slow flow velocities in the subsurface, the velocity head can be ignored. As water flows through the subsurface, it loses energy. Hence, the further the water flows, the less energy it has, and the lower the total hydraulic head will be. A map of this energy distribution is called a flownet which is a graphical solution to the LaPlace equation in two dimensions for groundwater flow. Figure 2.2 is a two-dimensional flownet of water flowing under a concrete dam.

The lines in the direction of flow are called flow lines, while those perpendicular to flow lines are called equipotential lines. *Potential*, here, means total hydraulic head energy potential. An equipotential line has the same energy at any point on the line, regardless of elevation. Flownets may be drawn by hand or by using software. As is always the case for computer-generated solutions, the results should be checked by the engineer using a hand-drawn flownet. McCarthy (2002) describes the manual drawing of flownets. Flownets are used to calculate three important quantities: the flowrate of water through the soil (employing both flownet results and hydraulic conductivity), the porewater pressure at points of interest (using the flownet and elevation data), and the exit gradient (using the flownet and length of flow path information).

2.2.3 Quantity of water flowing through soil

The quantity of water, Q, flowing through the soil in two-dimensional flow, is given by

$$Q = kh\left(\frac{n_f}{n_d}\right) \tag{2.15}$$

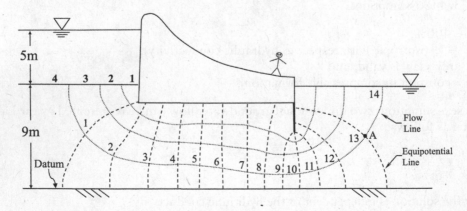

Figure 2.2 Flownet for water flowing under a concrete dam.

which is derived from Darcy's law incorporating the properties of a flownet. Here

Q = flowrate
k = coefficient of hydraulic conductivity
Δh = total head loss in the system
n_f = the number of flow channels
n_d = the number of equipotential drops

The coefficient of hydraulic conductivity can be estimated from a variety of field and lab tests or from correlations (Kulhawy and Mayne 1990). The total head (energy) loss in the system is the elevation difference between the headwater and tailwater (5 m in Figure 2.2). The number of flow channels, n_f, is the number of spaces between flowlines. In Figure 2.2, this is four. Note that, in Figure 2.2, the top and bottom impermeable boundaries are also flow lines. The number of equipotential drops, n_d, is the number of squares in a flow channel. The number of equipotential drops is the same in every flow channel, and for Figure 2.2 it is equal to 14.

For two-dimensional flownets, the units of Q are $(L^3/T)/L$; for example, cubic meters per second per meter of dam length normal to the flownet (i.e. measured into the page).

Example Problem Ex.2.2: Flow quantity

Using the flownet in Figure 2.2, calculate the flow in cubic meters per day beneath the dam per meter of dam length. The hydraulic conductivity of the dam foundation soils is 1×10^{-4} cm/s.

$$Q = k\Delta h \, (n_f/n_d)$$

and
Δh = 4m
n_f = 4
n_d = 14
k = 1×10^{-4} cm/s = 1×10^{-6} m/s
Substituting and calculating:

$$Q = 1 \times 10^{-6} \text{ m/s}(4\text{m})(4/14)(86000 \text{ s/day}) = \underline{0.1 \text{ m}^3/\text{day}}$$

2.2.4 Porewater pressure with water flowing through soil

In a non-flowing system, the porewater pressure is due to the static weight of the column of water above the point of interest. That is:

$$u = \gamma_w h_w \qquad (2.16)$$

where
u = porewater pressure
γ_w = unit weight of water
h_w = height of the water column above the point of interest

The water pressure can also be expressed in terms of head since the unit weight of the water can be considered a constant under most circumstances. For water under flowing conditions, Bernoulli's principle states that the total hydraulic head is the sum of the elevation head, pressure head, and velocity head as follows:

$$h_t = h_e + h_p + h_v \qquad\qquad (2.17)$$

h_t = total hydraulic head (units are in length)
h_e = elevation head, z
h_p = pressure head, u/γ_w, where γ_w is the unit weight of water
h_v = velocity head, $v^2/2g$, where g is the acceleration due to gravity

The equation has three terms summing for the total hydraulic head (or just total head), each representing energy: the *elevation* head (h_e), the *pressure* head (h_p), and the *velocity* head (h_v). Here, head refers to energy and has units of length. The head equation can be used for the laminar flow of water through soils. Laminar flow in soils has such a low velocity that the velocity head term, h_v, is ignored.

To calculate the porewater pressure at a point of interest, A, in Figure 2.1,

1. Choose a datum – any line perpendicular to lines of gravitational force. It is conveniently drawn at the base of the flownet.
2. Determine the vertical distance from the datum to the point of interest (z_A) to establish the elevation head (h_e) and the distance from the datum to the upstream water level to establish the upstream total hydraulic head (h_t).
3. Determine the total head loss across the dam (the elevation difference between the headwater and tailwater).
4. Divide the total head loss by n_d, to get the head loss per equipotential drop on the flownet.
5. Count the number of squares between the upstream soil surface and point A and multiply them by the head loss per drop as determined in step (d). This is the total energy lost between the upstream soil surface and point A (h_{lA}.).
6. Calculate the total head at A as $h_{tA} = h_t - h_{lA}$.
7. Use the Bernoulli equation written in terms of head, rearranging terms to calculate pressure head at A, h_{pA}.
8. To convert the pressure head at A to the porewater pressure at A (u_A), use the unit weight of water: $u_A = h_{pA} * \gamma_w$

Example problem Ex.2.3: Pore pressure in a flowing water regime

Using the flownet in Figure 2.2, calculate the porewater pressure at point A.

$h_t = h_e + h_p$ that is, the total hydraulic head is the sum of the pressure head and the elevation head.

Assuming the data as shown in Figure 2.2, upstream the total hydraulic head is $h_t = 5m + 9$ m = 14 m.

Downstream, the total hydraulic head is 9 m (elevation head only).

Since the n_d = 14, the total head loss per drop is 5 m/14 = 0.36 m/drop.

Point A is located downstream at equipotential drop number 13 indicating 13 equipotential drops.

Using 0.36 m/drop gives a total hydraulic head loss of 4.6 m (=13 * 0.36) at point A. Subtracting the head loss from the original total hydraulic head gives h_{tA} = 14 m − 4.6 m = 9.4 m of remaining total hydraulic head at point A.

The elevation head of point A is 4.5 m, its distance above the datum (i.e. h_{eA} = 4.5 m).

Finally, rearranging terms in Bernoulli's principle in terms of head yields the pressure head is the difference, that is, $h_{pA} = h_{tA} - h_{eA} = 9.4 - 4.5 = \underline{4.9\ m}$.

In terms of pore pressure at A, $u_A = h_{eA} * \gamma_w = 4.9$ m $* 1.0$ Mg/m³ $= 5.9$ MPa.

Notice the pressure head is 0.4 m greater than would be calculated using hydrostatic (no flow) conditions. This is as a result of the remaining energy in the water that originated upstream at a higher hydraulic head.

2.2.5 Uplift pressures

The uplift force on a structure can be calculated using the water pressures under the structure. For static groundwater conditions, the porewater pressures can be calculated assuming hydrostatic conditions. For conditions where water is flowing, the pressures can be calculated using the pressure distribution determined from the flownet. For example, for the dam shown in Figure 2.2, the porewater pressure can be calculated at various locations on the base of the dam using the methods shown in Example 2.3. The values of porewater pressure are then multiplied by a tributary area for that pressure and summed. This upward force is compared to the downward acting weight of the dam to establish a factor of safety with respect to uplift.

2.2.6 Seepage force

Flowing water exerts a seepage force on the soil particles as it flows through. The seepage force is calculated from

$$j = i\gamma_w \tag{2.18}$$

where

   j = seepage force, with units of F/L³
   i = hydraulic gradient, L/L

If the flowing water is exiting the soil into either the atmosphere or into free water, then the seepage force may be sufficient to displace the soil and lead to erosion that is termed piping. Piping can lead to failures in dams and levees.

The hydraulic gradient for the two-dimensional flow can be calculated from a flownet. With respect to piping, the critical location (largest hydraulic gradient) is at the smallest square along the equipotential line where the water exits the subsurface flow regime. This is the location where the seepage force is greatest. The seepage force is found using Equation 2.18. The hydraulic gradient is calculated thus

a. choose the smallest exit square,
b. divide the total head loss by n_d, to get the head loss in this square (recall that the head loss in every square is the same),
c. divide that head loss by the average distance *in the direction of flow* in that square, (This is the distance from the middle of the lower edge of the square to the middle of the upper edge of the square, in the direction of flow as indicated by the curved line.)
d. multiply (c) by the unit weight of water to get the seepage force.

The seepage force is a force per unit volume. The factor of safety against erosion (here, called piping) is the ratio between the soil buoyant unit weight at the exit location and the seepage force, j. Values of soil buoyant unit weights are often close to 10 kN/m³ which is approximately the unit weight of water. The FS$_{piping}$ approaches unity (pending instability) when the seepage force is about 10 kN/m³. This seepage force occurs when the exit hydraulic gradient is about unity. Hence, this exit gradient is called the *critical gradient*. That is, when the exit hydraulic gradient approaches unity, one can expect soil erosion (piping) to occur. It is interesting to note that once piping begins, the hydraulic gradient increases (because the flow length decreases). Increasing the hydraulic gradient increases the seepage force, accelerating the piping. This is a progressive failure. Once piping starts, it accelerates and is very hard to stop; time to call the authorities and evacuate downstream personnel. The US Army (1968) describes a lock and dam structure that failed in uplift, had seepage problems, and experienced seepage-induced erosion.

2.2.7 Capillary rise of groundwater

Groundwater may be found above the water table in what is termed the vadose zone. Groundwater rises from the groundwater table by capillarity, a surface tension phenomenon. This unsaturated soil has a negative porewater pressure, as the groundwater is in tension. The study of soil behavior in this zone is termed unsaturated soil mechanics (Lu and Likos 2004).

In cohesionless soils, the height of capillary rise is theoretically limited to approximately 11 m above the groundwater table. At this height, the porewater pressure (tension) reaches the vapor pressure of water at sea level, and the water cavitates. In cohesionless soils, the height of capillary rise may exceed 11 m because the space between the soil particles can be too small for cavitation to occur, there being insufficient space for water vapor to form.

Example problem Ex.2.4: Hydraulic conductivity

A falling head test is performed on a sandy soil using the apparatus illustrated on the right side of Figure Ex.2.4. The initial water level in the standpipe and the water level in the discharge tank is 80 cm and 15 cm, respectively, above the base of the permeameter. The cross-sectional area of the standpipe, a, is 5 cm^2, and the length, L, and cross-sectional area of the soil specimen, A, are 25 cm and 45 cm^2, respectively. Darcy's law, Equation 2.11, assumes a constant gradient (or a constant head over a fixed length of flow). It can be rewritten for a falling head as follows:

$$k = (aL/At)\ln(H_o/H_1) \qquad\qquad (ex2.4)$$

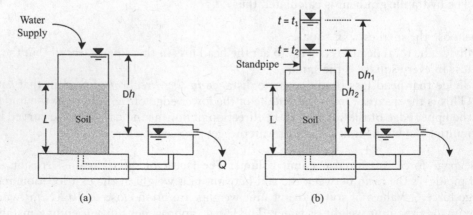

Figure Ex.2.4 Permeameter schematic for example 2.4.

1. Compute the hydraulic conductivity of the specimen (cm/s) if the water level in the standpipe drops 40 cm in 36 seconds.

The initial head, H_0, can be computed as $H_0 = 80 - 15 = 65$ cm.
The final head, H_1, can be computed as $H_0 = 40 - 15 = 25$ cm.
The elapsed time, t, for the test is 36 seconds.
Substituting the experimental measurements into equation ex 2.4 gives:

$$k = \left(\frac{aL}{At}\right)\ln\left(H_o/H_1\right) = \left(\frac{5*25}{45*36}\right)\ln\left(65/25\right) = \underline{7.3 \times 10^{-2}} \text{ cm/s}$$

2.3 EFFECTIVE STRESS

2.3.1 Effective stress equation

The stress between soil particles in a soil mass governs the soil behavior – strength and compressibility. Soils are particulate media; unconnected particles make up the soil mass. Because of this, soils do not behave the same as solid media. Soil particles interact with each other based on the stresses between them. In 1926, Terzaghi (1925) postulated soil behavior could be explained by what is termed *effective* stress, defined

$$\sigma' = \sigma - u \tag{2.19}$$

where
  σ' = effective stress
  σ = total stress
  u = porewater pressure

Effective stress in a soil mass can only be calculated (using the above formula). It cannot be measured. Total stress in a soil mass is calculated based on the weight of materials above the point in question. Porewater pressure is the water pressure in the soil pores at the point in question. For static water, this is the product of the unit weight of water and the distance from the point in question to the phreatic surface. For moving groundwater, a flownet may be used to calculate the porewater pressure. Holtz et al. (2010) provide an in-depth discussion, including a derivation of the equation.

2.3.2 Importance of effective stress

The effective stress between frictional soil particles governs how readily they slide past one another when loaded. That is, the soil strength is related to effective stress. The greater the effective stress, the stronger the soil. This is discussed further in section 2.4. Similarly, the compressibility of soils is related to the effective stress as discussed in Chapter 5.

The greater the difference between total stress and porewater pressure, the greater the effective stress, and vice versa. Hence, when porewater pressure drops, with no change in total stress, the soil strength increases. This is why geotechnical engineers like to drain soils – it decreases porewater pressure, increasing effective stress. And, conversely, when a site floods, or otherwise experiences an increase in porewater pressure, soils weaken which can cause landslides, bearing capacity failures and excessive settlement. In cases of an extreme increase in porewater pressure, such as settlement due to earthquakes, the effective stress may approach zero, resulting in a liquefied soil with no engineering strength.

Example problem Ex.2.5: Effective stress

What is the effective stress at a point in the soil four meters below the ground surface given the following conditions?

a. Initial conditions where $\gamma = 18$ kN/m^3 with the water table eight meters below the ground surface?
b. Later, conditions change when a dam is constructed nearby. The water table rises to eight meters above the ground surface. This increases the unit weight of soil to its saturated unit weight, 21 kN/m^3.

Solution:

a. The total stress is due to the weight of all the material above the point in question. Here, it is soil. The total stress above the point in question is:

$$\sigma = \gamma z = (18 \text{ kN/m}^3)(4 \text{ m}) = 72 \text{ kPa}$$

Since the point in question is above the water table, the pore pressure, u, is taken to be zero. Hence, the effective stress is

$$\sigma' = \sigma - u = 72 \text{ kPa} - 0 = 72 \text{ kPa}$$

b. The total stress at the point in question is due to the weight of soil and the weight of water above the point in question. Here, there are four meters of saturated soil and eight meters of water. The total stress at the point in question is

$$\sigma = \sigma_{\text{soil}} + \sigma_{\text{water}}$$

$$= \gamma_{\text{soil}} z_{\text{soil}} + \gamma_w z_w$$

$$= (21 \text{ kN/m}^3)(4 \text{ m}) + (9.81 \text{ kN/m}^3)(8 \text{ m})$$

$$= 84 \text{ kN/m}^3 + 78.5 \text{ kN/m}^3 = 163 \text{ kPa}$$

the pore pressure at the point in question is

$$u = \gamma_w z_w = (9.81 \text{ kN/m}^3)(12 \text{ m}) = 118 \text{ kPa}$$

the effective stress is

$$\sigma' = \sigma - u = 163 \text{ kN/m}^2 - 118 \text{ kPa} = 44.8 \text{ kPa}$$

2.4 SHEAR STRENGTH

Soils are called upon to resist a variety of applied loads including those from structures, gravitational forces producing self-weight, dynamic loads from seismic events, and mechanical loads. The term *shear strength*, often simply termed *strength*, is a measure of the ability of a soil to withstand applied shear and normal stresses. The shear strength is then the maximum applied shear stress a soil can resist without failing. Shear strength of the soil is a complicated and sophisticated topic and influenced by a myriad of factors including water content, rate of loading, stress path, material type, effective stresses, and drainage conditions. Further, there are various models and variations on models used to describe the shear strength of soils including Mohr-Coulomb, critical state theory, MIT-E3, and Cam-clay. For this introductory chapter, and consistent with most ground improvement applications, the Mohr-Coulomb model will be used.

2.4.1 The concept of soil strength

Soil strength is the ability of the soil to resist applied shear and normal stresses. When soils fail in shear, a failure surface develops, and soil particles move past one another. This requires energy, as there are frictional and surface forces between the soil particles that must be overcome. The magnitude of frictional and surface forces depends on the material properties (e.g. grain size, shape, mineralogy) and the stress between the particles. For geotechnical engineers, the appropriate stress is the effective stress. The effective stress is a function of total stress and porewater pressure. As the effective stress changes, such as from a raising or lowering of the groundwater table, the soil strength changes. This is a very important concept: soil strength is not constant. It depends on porewater pressure which can vary in both time and space. Hence, a dry soil slope may be stable until the slope saturates from, say, rain or flooding. With saturation, the effective stress decreases causing a decrease in the soil strength, and on occasion, leads to slope failure. It is common to read of landslides occurring after heavy rainfall (Iverson 2000).

Similarly, since effective stress depends upon both total stress and porewater pressure, soil strength is affected by the total stress on the soil. Hence, a given dry soil deposit with uniform density with depth, will not have a constant strength. In this case, the total stress increases with depth and, since there is no porewater pressure, the effective stress also increases with depth. Hence, the strength will increase with depth. This explains, in part, why shallow building foundations are (or should be) embedded in the ground – the deeper soil is stronger.

2.4.2 Laboratory evaluation of shear strength

Shear strength may be estimated using laboratory tests or field tests (Section 2.6). The direct shear and the triaxial shear are common laboratory tests.

2.4.2.1 Direct shear testing

The direct shear test is most commonly used for cohesionless soils and for the residual strength of sands. In this test, a normal stress is first applied to the soil and a shear stress is then applied such that the failure plane is forced horizontally through the soil. Figure 2.3 presents a schematic of the test.

The test procedure is as follows:

1. Place a soil sample in the testing box at the expected field density.
2. Place an arbitrary vertical load on the sample (normal force).
3. If desired, flood the sample.
4. Apply a horizontal force (shear force) to the sample, until the sample fails, continuously measuring the force and displacement to do so; if desired, measure vertical

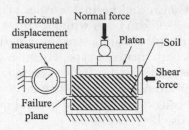

Figure 2.3 Direct shear test apparatus.

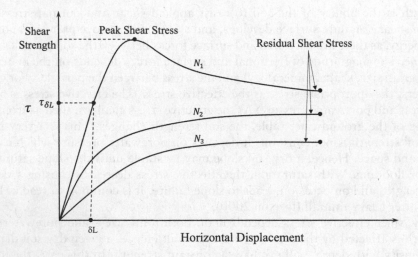

Figure 2.4 Direct shear initial data plot ($\sigma_1 > \sigma_2 > \sigma_3$).

displacement. Note that the horizontal displacement cannot be converted directly to strain but rather is presented as deformation.

5. Using the forces and area of the sample, calculate shear and normal stresses.

Figure 2.4 shows the data collected from the test, reduced and plotted. Here, the collected horizontal and vertical force data have been converted to stress by dividing the applied force by the area of the sample. Note the area of the sample changes with horizontal deformation.

This procedure is repeated three times, each time using a new sample prepared as in the previous case, but increasing the vertical force, N, on the sample. Hence, three curves shown in Figure 2.4 correspond to the three normal loads where $N_1 > N_2 > N_3$. Staged tests can also be conducted by stopping the application of shear load as soon as the peak force is reached, recentering the shear box, applying a new, higher normal load, and reapplying shear load.

Failure must be defined before the shear strength parameters are calculated for the assumed failure condition. Three possible criteria are shown in Figure 2.4 as follows:

1. peak shear stress,
2. residual shear stress, or
3. shear stress at a predetermined strain or deformation.

Dense soils and/or higher stresses tend to produce stress-deformation curves with a distinct peak stress followed by a post-peak drop such as that for N_1 in Figure 2.4. For such curves, the peak stress is easily identified. For those tests that do not produce a peak stress value followed by a post-peak drop, the peak value occurs at the end of the test and corresponds to the residual stress value. The curves for normal forces N_2 and N_3 are representative of such conditions.

The residual shear stress is that which occurs either after the peak stress (curve for N_1) or coincident with the peak stress (curves for N_2 and N_3). The residual value is used for design conditions where high strain is expected or for conditions where there is no stress-strain compatibility between various soil types. For example, consider two soils in a layered system soil (a) which reaches peak stress at 2% strain and soil (b) at 12% strain. In this case, it may make sense to use the peak stress from soil (b) and the residual stress from soil (a) so the analysis demonstrates stress-strain compatibility. Residual strengths are also used for slope stability analyses where geologic information provides evidence of prior instability.

Some geotechnical applications are deformation (strain) limited. Structures sensitive to deformation include brittle buildings, tall structures, brittle pipelines, and some large machines. For direct shear tests, the laboratory data are plotted as stress-deformation plots. Then, the stress at an acceptable deformation is defined as the failure stress, to be used to determine the strain-limited strength parameters. To illustrate this δL, an example of the limiting deformation is identified in Figure 2.4. Corresponding to this limiting deformation is a shear strength, $\tau_{\delta L}$, which is the strength at the limiting deformation and less than the peak strength that occurs at a higher deformation.

The data generated as in the direct shear test as presented in Figure 2.4 can be used to determine the *effective shear strength parameters*. The strength parameters are c′ (effective cohesion) and ϕ' (effective angle of internal friction) from the Mohr-Coulomb model shown in Figure 2.5 and explained below. These parameters are used to characterize the strength of the soil.

After the failure criterion is selected, above, the corresponding failure shear stress is noted for each of the three direct shear tests. The normal force, N, is converted to a normal stress by dividing by the cross-sectional area of the sample, remembering that the cross-sectional area changes during the test. Divide N by the area corresponding to the deformation where the failure criteria were chosen. This results in three pairs of normal stress (σ) and shear stress (τ). Since direct shear tests are almost entirely conducted in drained shear, the excess porewater pressure is zero and the total stresses are equal to the effective stresses.

As shown in Figure 2.5, the effective normal and shear stresses at failure, σ'_f, are plotted for each test in stress space. The data points for each of the three tests correspond to the three points in the figure. With good laboratory data, a straight line may be drawn through the data. The slope of the straight line is deemed the *effective angle of internal friction*, ϕ'. The intercept on the shear stress axis is deemed the *effective cohesion*, c′, sometimes called

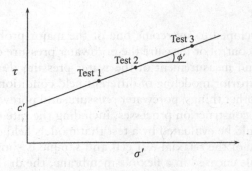

Figure 2.5 Direct shear reduced data plot.

the cohesion intercept. The line is called the *failure envelope*, because there are no pairs of stresses that can exist above this line. Together, c' and ϕ' form the effective strength parameters of the soil.

With these effective strength parameters, the shear stress at failure, τ_f, also known as the shear strength, at any σ', may be calculated:

$$\tau_f = c' + \sigma' \tan\phi' \tag{2.20}$$

This is the equation of the straight line in Figure 2.5. Note that the larger σ' is, the greater the shear strength. Put another way, soils increase in shear strength with depth because σ' increases with depth.

Example problem Ex.2.6: Direct shear tests

Given: Two direct shear tests are run on the same dry sandy soil in a circular direct shear box having an area 25 cm². At failure, the sample failure surface was 25 cm². For the first test, the normal force on the sample was 1.5 kN, while the shear force at failure was 1.5 kN. For the second test, the normal force on the sample was 3.0 kN, while the shear force at failure was 1.0 kN. No third test was run (but it would have been a good idea!). What are the total strength parameters for this soil?

Solution:

Calculate the normal stress on the sample for each test:

    Test one, σ = F/A = 1.5 kN/((5.0 cm)(5.0 cm)(1/10000)) = 600 kPa
    Test two, σ = F/A = 3.0 kN/((5.0 cm)(5.0 cm)(1/10000)) = 1200 kPa

Calculate the shear stress at failure for each test:

    Test one, τ = T/A = 1.0 kN/((5 cm)(5 cm.)(1/10000)) = 400 kPa
    Test two, τ = T/A = 2.0 kN/((5 cm)(5 cm.)(1/10000)) = 800 kPa

Plot these (σ, τ) data pairs (400 kPa, 600 kPa) and (800 kPa, 1200 kPa)

Draw the failure envelope through these two points. Once done, determine the slope and intercept of the resulting line. If drawn and or calculated correctly, c = 0 and ϕ = 33.7°. Remember, the axes of the (σ, τ) graph must be scaled equally. Since the pore pressure is zero, the total and effective stresses are the same; hence, the total and effective angles of internal friction are the same, as are the total and effective cohesion (Figure Ex.2.6).

2.4.2.2 Triaxial testing

Triaxial testing was developed to overcome one of the major problems with direct shear testing – the inability to control or measure the porewater pressure during the test. Triaxial testing allows control and measurement of the water pressure during testing. Moreover, triaxial testing allows superior modeling of different field conditions. Since the strength of soil depends on (among other things) porewater pressure, and porewater pressure is affected by (among other things) construction processes, including the rate of loading, the effective strength parameters should be evaluated by a test that models field conditions.

A simplified schematic of the triaxial test cell and sample is shown in Figure 2.6. Note the cylindrical soil sample encased in a flexible membrane, the drainage valve to allow for drained or undrained porewater conditions, a sensor to measure porewater pressure, and a

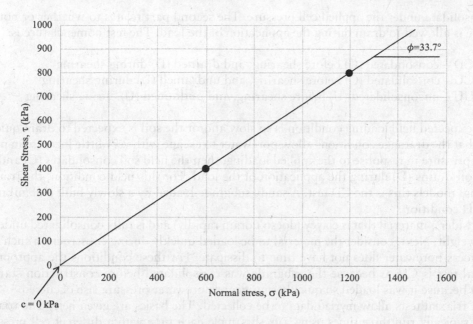

Figure Ex.2.6 Direct shear results for example 2.6.

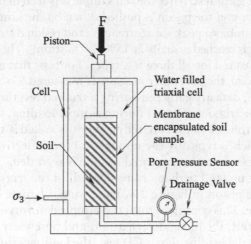

Figure 2.6 Schematic of triaxial test.

loading piston to apply an axial load. The cell is water filled and air pressurized to provide a confining pressure (σ_3) before testing begins. If the drainage valve is open at this time the confining pressure is also the three-dimensional consolidating pressure.

There are three common triaxial tests, each modeling a different field loading condition. Lade (2016), and Bishop and Henkel (1957) provide in-depth descriptions of these tests. The setup and conduct of triaxial tests are given in Holtz et al. (2010), McCarthy (2002), and many introductory geotechnical texts. The three tests are deemed CD, CU, and UU. The three tests, done on saturated soils, are distinguished by whether or not the sample is allowed to drain in response to the application of the cell pressure (consolidated, C, or unconsolidated, U), and the drainage conditions during loading (drained, D, or undrained, U). Notice each test name has two parts. The first part relates to whether not the material is permitted to consolidate under the applied cell pressure. The second part relates to whether or not the sample is allowed to drain during the application of the load. The test nomenclature is:

1. CD – consolidated (C) before shearing, and drained (D) during shearing
2. CU – consolidated (C) before shearing, and undrained (U) during shearing
3. UU – unconsolidated (U) before shearing, and undrained (U) during shearing

If the expected field loading condition is so slow and/or the soil is expected to drain quickly such that the drainage conditions allow porewater to escape with very little increase in pore-water pressure in response to the applied loading, then the field soil consolidates (C) and the field soil drains (D) during the application of the load. For this field condition, the triaxial test that models this is the CD test. A sandy subgrade loaded by a slowly built embankment is a CD condition.

Consider a material that is clayey (doesn't drain rapidly) and is fully consolidated under its own weight. Next, consider the material to be loaded quickly during construction such that the excess porewater does not have time to dissipate. For these conditions the appropriate triaxial test is CU. C because the subgrade was consolidated before construction started, and U because it was loaded so quickly significant porewater pressure rise occurred.

The triaxial tests allow myriad data to be collected. The basics are given here. The triaxial test is typically run three times using a fresh sample each time with a different cell pressure. Each triaxial test is run at a different confining (cell) pressure, σ_3, akin to the three different normal forces in the direct shear test. After the soil sample is placed in the triaxial cell and either permitted to consolidate or not, the piston is pushed down on the sample. This continues until the sample deforms and exhibits a peak shear strength (and residual strength if desired) or until the limit of the equipment is reached (usually at 15%–20% strain). The applied piston force and axial deformation are measured for all three test types. For tests that are undrained during the application of the piston load, the porewater pressure is measured.

Referring to Figure 2.6, data are gathered during a triaxial test throughout the test during sample preparation in the triaxial cell and during sample loading. During sample preparation in the triaxial cell, the cell is pressurized to what is called a total confining pressure (σ_3), or cell pressure, which is typically the minor total (not effective) principal stress. This is recorded. The pore pressure in the saturated sample is recorded.

After the sample is set up (and perhaps consolidated) in the triaxial cell, sample loading is done by depressing the piston in Figure 2.6, increasing the force (thus, stress) on the top and bottom of the sample. This continues until the sample deforms enough such that the test is terminated. For a CU test, the force, deformation, and porewater pressure are monitored continuously throughout the loading. For a CD test, the loading rate is slow such that excess pore pressure does not develop. For a UU test, only load and deformation are monitored. As in the direct shear test, the force is applied until failure occurs as per the same failure criteria described for the direct shear test. The total stress on the sides and top of the sample is the cell pressure (σ_3) plus the additional stress caused by depressing the piston, $\Delta\sigma_1$. At failure, these stresses sum up as follows:

$$\sigma_{1f} = \sigma_{3f} + \Delta\sigma_{1f} \tag{2.21}$$

For a CU test with pore pressure measurements, the major principal effective stress on the sample at failure is

$$\sigma'_{1f} = \sigma_{1f} - u_f \tag{2.22}$$

where subscript f denotes a stress at failure

Other data are sometimes recorded. More sophisticated triaxial testing may record the amount of water that leaves the saturated sample during sample preparation and consolidation as well as during loading in a CD test to determine the total volume change. Samples can be instrumented to determine diameter change.

Typically, three triaxial tests are run for a given soil resulting in three sets of either total and/or effective failure stresses, $(\sigma_{1f}, \sigma_{3f})$ and/or $(\sigma'_{1f}, \sigma'_{3f})$. The difference between the minor and major principal stresses is called the deviator stress, $\Delta\sigma$

$$\Delta\sigma = \sigma_1 - \sigma_3 \tag{2.23}$$

This is the difference between the pressure on the top of the soil sample and the triaxial cell pressure.

The principle effective stresses at failure $(\sigma'_{1f}, \sigma_{3f})$ are plotted in (σ', τ) stress space. Figure 2.7 shows the Mohr's circles for three triaxial tests. With good laboratory data, a straight line is drawn tangent to the circles, intercepting the shear stress axis. This is the failure envelope. The slope of the failure envelope is ϕ', the effective angle of internal friction, and the intercept on the shear stress axis is c', the effective cohesion. These are the strength parameters used in geotechnical design to predict the soil strength, τ_f

$$\tau_f = c' + \sigma'\tan\phi' \tag{2.24}$$

For the UU test, the sample is set up in the triaxial cell and the confining pressure is applied without allowing consolidation (i.e. unconsolidated). It is then immediately sheared in axial compression without allowing drainage (i.e. undrained). For this test, typically three samples are tested at three different cell pressures and the Mohr's circles are plotted as described above. However, since no drainage is permitted (and assuming S = 100% and incompressible water), each test results in the same strength. When plotted, the result is $\phi = 0$ and the y-intercept is the undrained shear strength, S_u, that is, radius of the Mohr's circle. A special case of the UU is the unconfined compression test where the confining pressure is zero (hence the name, unconfined). In this case, the radius of the Mohr's circle is the undrained shear strength and the diameter is the unconfined compressive strength (UCS) or, q_u. Mathematically, S_u is one-half of q_u.

2.4.3 Shear strength summary

The purpose of shear strength testing is to evaluate the shear strength parameters (c', ϕ') so the strength of the soil in the field can be estimated to see if it exceeds the shear stresses the proposed project will apply.

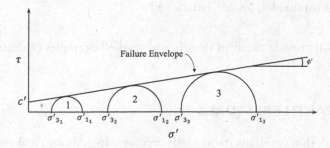

Figure 2.7 Mohr's circles and interpretations for triaxial test.

The direct shear test is straightforward and easier (and less expensive) than the triaxial test. However, the direct shear test does not readily allow the evaluation of effective strength parameters.

The triaxial shear test is more complicated than the direct shear test. Triaxial testing requires expensive equipment and a well-trained operator. Moreover, the test requires the engineer to think more about the field conditions before engaging the test. The choice of CD, CU, or UU test is critical. The major advantages of the triaxial test are its ability to produce *effective* strength parameters and better model field conditions.

Example problem Ex.2.7 triaxial shear

Given: Two triaxial tests are run on identical clayey soils. The confining pressure in the first test is 20 kPa and had a deviator stress at a failure of 26 kPa. The confining pressure in the second test is 40 kPa and had a deviator stress at a failure of 40 kPa.

Find: What are the total strength parameters? (hint: a graphical solution with Mohr's circles will solve this)

Solution:

Use the definition of deviator stress to find the major principal stress at failure.

$$\Delta\sigma = \sigma_1 - \sigma_3$$

For the first test, the deviator stress

$$\Delta\sigma = \sigma_1 - \sigma_3$$

$$26 \text{ kPa} = \sigma_1 - 20 \text{ kPa}$$

$$\text{rearranging}: \sigma_1 = 26 \text{ kPa} + 20 \text{ kPa}$$

$$= 46 \text{ kPa}$$

Similarly, for the second test

$$\Delta\sigma = \sigma_1 - \sigma_3$$

$$40 \text{ kPa} = \sigma_1 - 40 \text{ kPa}$$

$$\text{rearranging}: \sigma_1 = 40 \text{ kPa} + 40 \text{ kPa}$$

$$= 80 \text{ kPa}$$

Plot these (σ_1, σ_3) pairs, (20, 46) and (40, 80), and draw the failure envelope, tangent to the Mohr's circles (Figure Ex.2.7).

Measuring from the plot, $\phi = 15°$ and $c = 5$ kPa

Holtz et al. (2010) provide excellent details and worked examples of these, and other, soil strength tests.

2.5 LATERAL EARTH PRESSURES

Soil has weight and, thus, exerts vertical earth pressure. In addition, soil exerts lateral earth pressure. For example, the soil behind a foundation wall exerts a lateral pressure on the

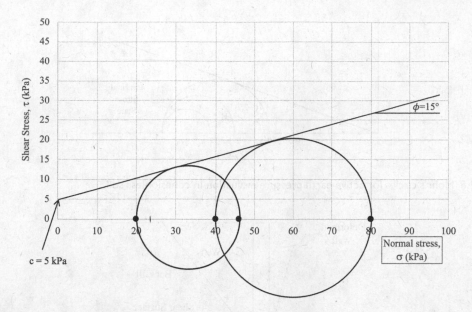

Figure Ex.2.7 Triaxial shear Mohr's circles for example 2.7.

wall. This pressure has a different magnitude than vertical earth pressure. It may be larger, equal to, or smaller than the vertical earth pressure.

The amount of lateral movement of a wall (foundation, or otherwise) is a major factor in determining the lateral earth pressure on the wall. Consider a retaining wall that is holding an excavation open. If the wall moves away from the backfill (into an excavation), the soil pressure on the wall becomes less than if the wall hadn't moved (active case). If the wall moves into the backfill, the soil pressure is greater than if the wall hadn't moved (passive case). Finally, if the backfilled wall doesn't move at all (at-rest case), the pressure is between the above cases.

2.5.1 Active earth pressure

The soil pressure on the wall that occurs when the wall moves away from the backfill is called the *active earth pressure*. After a very small movement, the earth pressure decreases to a lower limit state just prior to when the soil enters a failure state. The active earth pressure is determined from equations derived from Mohr's circle.

Consider the Mohr's circle and failure envelope shown in Figure 2.8, Mohr's circles for active earth pressure evaluation for cohesionless soil. Circle A represents the original, at rest, stresses in the soil before the wall moves away from the backfill. The vertical pressure is represented by σ'_1, and the horizontal pressure is represented by σ'_3. As the wall moves away from the backfill, the lateral pressure is relieved (σ'_3 becomes smaller) and the soil mobilizes some of its shear strength. σ'_1 remains the same because the soil weight causing σ'_1 is not changing (Circle B). Finally, when the limit of the shear strength of the soil is reached (Circle C), the soil is said to be at its active earth pressure condition. Any further reduction in σ'_3 would cause the soil to fail.

Perhaps it is worthy of note that what is described in the above paragraph is lateral unloading. As shown, soil can fail by lateral unloading.

Figure 2.9 depicts a retaining wall as it moves away from the soil and mobilizing the shear strength of the soil. This results in the soil being in an active state of stress. The figure can

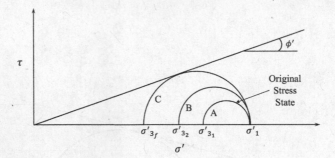

Figure 2.8 Mohr's circles for active earth pressure evaluation in cohesionless soil.

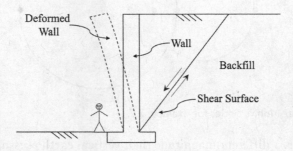

Figure 2.9 Wall moving into the active state (away from backfill).

be used to show one model which can explain why the lateral earth pressure decreases. For a wall that moves sufficiently to induce failure of the soil, a failure surface forms, and a wedge of soil tries to slide down. The soil movement is resisted by the retaining wall and by the internal strength of the soil. That is, the friction (and possibly cohesion) of the soil along the failure surface resists some of the weight of the sliding wedge; the retaining wall holds the rest of the weight. Hence only part of the weight of the sliding wedge presses against the retaining wall. This is because the soil has some internal strength, here manifested as friction (and possibly cohesion). If the soil had no internal strength, the vertical and lateral earth pressures would be equal, as it is in fluids, which have little internal strength.

As the wall continues to move away from the backfill, more and more of the sliding wedge's weight is carried by the soil until a limit soil strength (failure) is reached. Circles B and C represent the stresses in the soil as the wall continues to move. Eventually, the Mohr's circle touches the failure envelope (circle C), becoming the failure circle. The minor principal stress associated with the failure circle, σ'_{3f}, is the minimum pressure (horizontal) the soil can apply to the wall. It is desirable to allow the wall to move sufficiently to reduce the lateral pressure to σ'_{3f} as it allows for a less robust, and therefore less expensive, wall design. Almost all free-standing walls are flexible enough to move to allow this active lateral earth pressure to develop.

This value, σ'_{3f} can be calculated with this formula, based on the geometry of Figure 2.8, as:

$$\sigma'_{3f} = \left(\sigma'_{1f}\right)\tan^2\left(45 - \frac{\phi'}{2}\right) \tag{2.25}$$

for ϕ' in degrees. The Rankine active lateral earth pressure coefficient, K_A, is defined as:

$K_A = (\sigma'_{3f}/\sigma'_{1f})$, which, from the Mohr's circle geometry, is the same as $\tan^2(45 - \phi'/2)$

and is in the range of 0.3 for many soils.

Equation 2.25 is often written

$$\sigma'_{3f} = K_A \sigma'_1 \qquad (2.26)$$

Embedded in the Rankine lateral earth pressure coefficients are these assumptions:

1. The backfill is horizontal.
2. The wall is vertical and has no friction with the soil.
3. The failure surface is planar.
4. The soil is cohesionless.
5. The soil is dry.

Tanyu et al. (2008) and Kulhawy and Mayne (1990) provide information on lateral earth pressure coefficients that account for divergences from these conditions.

2.5.2 Passive earth pressure

Passive earth pressure develops when an external force pushes the retaining wall into the soil. Scenarios include ice pressure on a waterfront wall, bridge expansion on an abutment wall, and ship impact. As the wall moves into the backfill, the soil pressure increases up to a limit, the soil failure state. Figure 2.10 shows how the Mohr's circles change as the lateral pressure (σ'_3) increases for cohesionless soils. It is akin to the changes in Mohr's circles in the active case, only this time σ'_3 is increasing until the Mohr's circle touches the failure surface (circle D), at which point the soil fails, and the passive pressure, σ'_3, is maximized at σ'_{3f}. Passive earth pressure is about ten times larger than active earth pressure, requiring a very robust, expensive retaining wall. Passive applications of retaining walls are rare. Notice the rotation in principal stresses. For the at-rest case, the major principal stress is vertical, and the minor principal stress is horizontal. For the passive case, the major principal stress is horizontal, and the minor principal stress is vertical. Unlike the active case, the movement required to develop passive pressure is very large. σ'_{3f} can be calculated with this formula, based on the geometry of Figure 2.10 as follows:

$$\sigma'_{3f} = \left(\sigma'_{1f}\right)\tan^2\left(45 + \frac{\phi'}{2}\right) \qquad (2.27)$$

for ϕ' in degrees.

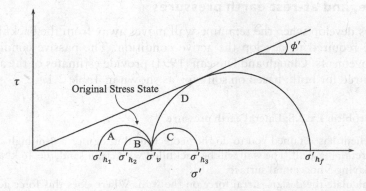

Figure 2.10 Mohr's circles for passive earth pressure evaluation in cohesionless soil.

The Rankine passive lateral earth pressure coefficient, K_P, is defined as:

$$K_p = \left(\frac{\sigma'_{3f}}{\sigma'_{1f}}\right) = \tan^2\left(45 + \frac{\phi'}{2}\right)$$ (2.28)

and is in the range of 3 for many soils.

Equation 2.27 is often written

$$\sigma'_{3f} = \sigma'_1 K_p$$ (2.29)

K_P, here, is subject to the same limitations on K_A, above, as both are Rankine lateral earth pressure coefficients.

2.5.3 At-rest (K_0) earth pressure

When a retaining wall is so constrained that it can't move into or out of the backfill, the lateral earth pressure is between the active and passive cases. This pressure is called the *at-rest (K_0)* earth pressure. Its magnitude is between the active and passive pressures. Unlike the active and passive pressures, it is not unique. The magnitude of the at-rest earth pressure is empirically determined. While there are many formulas, the Jaky (1944) formula, for cohesionless soils, is often cited as:

$$K_0 = \frac{(1 - \sin\phi')\left(1 + \left(\frac{2}{3}\right)\right)(\sin\phi')}{(1 + \sin\phi')}$$ (2.30)

which is often simplified to

$$K_0 = (1 - \sin\phi')$$ (2.31)

where ϕ' is in degrees.

K_0, between K_A and K_P, is often about 0.5 for cohesionless soils.

Box culverts, rigid buried pipes, very massive walls, and very rigidly braced basement walls are examples of walls subject to at-rest lateral earth pressures.

2.5.4 Amount of movement to develop active, passive, and at-rest earth pressures

Active pressures develop when the retaining wall moves away from the backfill. Very small movements are required to develop the active condition. The passive condition requires much larger movements. Clough and Duncan (1991) provide estimates of the amount of the movement required for both, based on soil type as shown in Table 2.1.

> **Example problem Ex.2.8: Lateral earth pressure**
>
> **Given:** A client has retained you to do the geotechnical design of a 4-m high concrete, cantilever retaining wall. The wall will be backfilled with clean sand, $\gamma = 16$ kN/m³, and $\phi' = 28°$, having a horizontal surface.
>
> **Find:** Calculate the design lateral force on the wall. Where does this force act on the wall?

Table 2.1 Wall movement to mobilize active and passive earth pressures (after Clough and Duncan 1991)

Type of Backfill	Values of δ/H^a	
	Active	Passive
Dense sand	0.001	0.01
Medium dense sand	0.002	0.02
Loose sand	0.004	0.04
Compacted silt	0.002	0.02
Compacted lean clay	0.01[b]	0.05[b]
Compacted fat clay	0.01[b]	0.05[b]

[a] δ is the amount of movement at the top of the wall and H is the height of the wall. δ/H is the ratio of the amount of movement as measured at the top of the wall and the wall height needed to mobilize active or passive earth pressures. Movement can be by rotation or translation of the wall.

[b] While these movements will mobilize active or passive pressures, clay soils will creep with time resulting in an increase in earth pressure for the active case or a decrease in pressure for the passive case and subsequent additional movement

Solution:

The first step is to determine the state of the backfill soil: active, at-rest, or passive. Passive is ruled out because there are no applied external forces pushing the wall into the backfill. While the wall is concrete and seems rigid, it is a cantilever wall and judged geotechnically flexible enough to deform the necessary amount needed to develop the active condition.

Calculate the active lateral earth pressure coefficient, K_A

$$K_A = \tan^2\left(45 - \phi'/2\right) = \tan^2\left(45 - 28/2\right) = 0.36$$

For the active case, the horizontal soil pressure on the wall, at a given depth is

$$\sigma'_{3f} = (\sigma'_{1f}) K_A$$

where σ'_{1f} is the vertical soil pressure at the given depth = $(\gamma)(\text{depth})$

The horizontal soil pressure distribution on the wall is given by this equation. A brief examination of this equation shows the lateral earth pressure is zero at the top of the wall ($z = 0$) and is maximum at the base of the wall ($z = 4$ m). Since this is a first-order equation, the pressure distribution is linear; triangular, in fact.

At the base of the wall, the lateral earth pressure is

$$\sigma'_{3f} = (\sigma'_{1f}) K_A = \gamma\ z\ K_A$$

$$= \left(\gamma = 16\ \text{kN/m}^3\right)\left(4\ \text{m}\right)(0.36) = 21.6\ \text{kPa}.$$

The lateral earth *force* on the wall is the area of this triangular pressure diagram

$$F = (\tfrac{1}{2})(21.6)(4) = \underline{43.2\ \text{kPa/m of the wall.}}$$

The force acts through the centroid of the triangular pressure distribution. The centroid is $^1/_3$ the distance from the base to the top of the diagram

Location is $(4\ \text{m})/(3) = \underline{1.33\ \text{m above the base of the wall.}}$

2.6 FIELD INVESTIGATIONS

In order to select, design, and implement the appropriate ground improvement technique, a thorough understanding of the subsurface conditions is required. Field investigations are therefore needed to reveal the subsurface conditions and to retrieve samples for laboratory testing. Field investigations should begin using information already available. Existing data can include air photos, previous site investigations, regional and local geology studies, and local contractor and engineer experiences in the vicinity of the site. The study of this available information can lead to a preliminary model of the site and subsurface conditions as well as to a more informed project-specific subsurface investigation.

Field investigations can include indirect, remote sensing methods (also known as geophysical methods) as well as direct methods such as drilling, sampling, and in situ testing. Geophysical methods use measurements of one property to infer another. For example, seismic methods measure arrival times of seismic waves which can then be interpreted to be, for example, the depth to bedrock. As a result, geophysical methods require verification known as ground truth. Other geophysical methods include resistivity, electromagnetic conductivity, ground-penetrating radar, magnetics, and gravimetrics. Geophysical methods are not discussed herein but an introduction to the various methods in the context of site investigations is available elsewhere (LaGrega et al. 2010)

Direct methods, such as drilling, sampling, and in situ testing involves means and methods to penetrate the subsurface, retrieve samples, and conduct tests in place to supplement or in lieu of laboratory tests. Samples retrieved from the subsurface are considered either disturbed or undisturbed. Disturbed samples are useful for laboratory tests when the in-place soil structure and density are not a factor in the outcome of the test. Examples include water content, specific gravity, grain size distribution, and plasticity. Undisturbed samples are useful for laboratory tests where the in-place soil density and structure are factors in the outcome of the test. Examples include shear strength and consolidation tests. Common in situ tests used in ground improvement engineering are the standard penetration test (SPT), (ASTM D1586–18 (2018)), and the Cone Penetration Test (CPT) (ASTM D5778 - 20 (2020)).

2.6.1 Drilling methods

Test borings are drilled into the subsurface as a means to access the materials encountered for identification, sampling, and in situ testing. A test boring method requires three separate and distinct components (functions):

1. soil at the bottom of the borehole needs to be excavated,
2. the cuttings need to be carried to the surface, and
3. the borehole must remain stable (not collapse into itself).

Keeping these components in mind, test borings in the soil are commonly drilled with hollow stem augers or with rotary mud drilling methods. Hollow stem augers consist of an auger flight welded to a pipe and with a cutting bit attached to the bottom as shown in Figure 2.11. The cutting bit performs the excavation function by loosening the in situ soil. The auger performs the function of carrying the cuttings to the surface. The pipe (i.e. the hollow stem) is typically 100 mm in diameter and performs the function of maintaining borehole stability while at the same time maintaining access to the bottom of the borehole for sampling and/or in situ testing.

Rotary mud drilling is a common alternative to hollow stem auger drilling. In this method, a cutting bit is attached to a hollow drilling rod through which drilling mud is circulated.

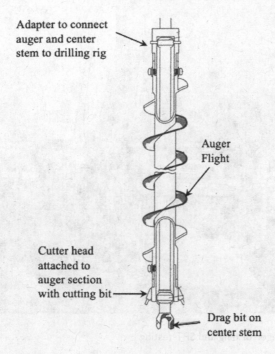

Adapter to connect
auger and center
stem to drilling rig

Auger
Flight

Cutter head
attached to
auger section
with cutting bit

Drag bit on
center stem

Figure 2.11 Hollow stem auger.

The mud goes down the center of the rod, out the end, and up the borehole on the outside of the rod. The cutting bit performs the excavation function while the drilling mud performs the remaining two functions of carrying cuttings to the surface and maintaining borehole stability.

In both hollow stem auger and rotary mud drilling borings, a drilling rig is required to rotate and lift all of the tools and equipment needed for the drilling, sampling, and in situ testing methods employed. Figure 2.12 is a track-mounted drilling rig showing the drilling and testing equipment including additional hollow stem augers lying on the ground in the background. Notice the drilling rod going into the ground through the hollow stem auger. The drilling rod is connected to a sampler at the bottom of the rod. A cylindrical hammer at the top drives the sampler into the soil as part of the Standard Penetration Test (discussed in the next section).

2.6.2 Sampling methods

One of the principal reasons to drill a test boring is to retrieve samples from various depths in the subsurface. One way to do this is to examine the cuttings as they come to the surface. These are highly disturbed samples. There is no way to know for sure from what depth the samples came. Further, if drilling with rotary mud, the samples are contaminated by the drilling mud. Nonetheless, examining the cuttings as they come to the surface does not delay the drilling nor does it require any specialized equipment. Given the particle size and shape information that can be gleaned by examining the cuttings, it is well worth the effort.

The most common way to retrieve a sample from a given depth is from the split-barrel sampler used in the Standard Penetration Test. This sampler has an outside diameter of 51 mm and an inside diameter of 35mm. It is driven into the ground with a hammer weighing 63.5 kg lifted and then dropped from a height of 0.76 m. An SPT sample is considered

Figure 2.12 Hollow stem auger drilling and SPT testing.

Figure 2.13 SPT split barrel sampler opened to show soil.

disturbed and useful for soil identification, grain size distribution, water content, and plasticity determinations, but not strength or compressibility. Figure 2.12 shows a safety hammer lifted and dropped using a rope and cathead system. Figure 2.13 shows the split barrel sampler open with the soil core exposed for examination and sampling for laboratory testing and preservation. Additional details on the SPT are included in section 2.3.6.1.

In order to conduct triaxial strength, consolidation, and permeability tests in the laboratory, an undisturbed sample is needed. Undisturbed samples are not truly undisturbed but rather minimally disturbed. That is, the sampling, handling, and removal from the in situ stress state all cause some degree of disturbance. The most common means of securing an undisturbed sample is through the use of a thin-walled sampler, commonly called a Shelby tube (presumably derived from one of the leading manufacturers in Shelby Township, Michigan, USA). Figure

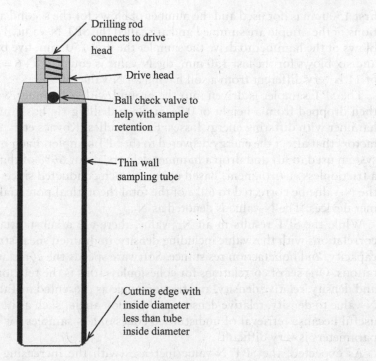

Drilling rod connects to drive head

Drive head

Ball check valve to help with sample retention

Thin wall sampling tube

Cutting edge with inside diameter less than tube inside diameter

Figure 2.14 Thin-walled sampler schematic.

2.14 is a schematic of a thin-walled sampler. This sampler is designed to be smoothly pushed into the ground rather than driven into the ground like the split barrel sampler. The sample from the thin-walled sampler is first extruded from the tube and then subsequently trimmed to the appropriate length and diameter for consolidation and/or shear strength testing.

2.6.3 In situ test methods

In situ tests are designed to ascertain the engineering properties of the soils and rock encountered in the subsurface. Notice in situ tests do not necessarily measure the desired property but may use other measurements to infer or correlate with the desired engineering property. It's also useful to note that the very process of testing is intrusive and may alter the desired properties compared with those properties that existed prior to the intrusive testing. For in situ geotechnical testing there are numerous methods available including the SPT, the cone penetration test (CPT), the vane shear test (VST), the presuremeter test (PMT), the dilatometer test (DMT), and the Iowa borehole shear test (BST). Only the SPT and CPT will be discussed in detail since these are the two most commonly used in situ methods associated with ground improvement projects.

2.6.3.1 SPT

The SPT hammer is used to drive a sampler to retrieve disturbed samples as described in section 2.6.2. Retrieving a sample is, in and of itself, sufficient reason to conduct an SPT. What makes the SPT an in situ test, however, is that the energy required to drive the sampler is recorded in such a way as to enable differentiation between loose and dense granular soils (or soft and stiff cohesive soils). Procedurally, the number of blows of the hammer for each of the three 150-mm penetrations of the sampler is recorded. The number of blows for the first 150 mm is not used and the number of blows for the second and third 150-mm penetrations of the sample are summed and are called the SPT N-value. For example, if it takes four blows of the hammer to drive the sampler the first 150 mm, five blows for the next 150 mm, and six blows for the last 150 mm, the N-value is equal to $5 + 6 = 11$. A soil with an N-value of 11 is very different from a soil having an N-value of 50.

The SPT sampler is driven into the ground with a hammer weighing 63.5 kg lifted and then dropped from a height of 0.76 m. Each drilling rig has a method to lift and drop the hammer with differing energy losses. Early studies (Kovacs et al. 1977) showed a number of factors that affect the energy delivered to the SPT sampler. Each blow for a rope and cathead system used to lift and drop a hammer delivers about 66% of that for a hammer free-fall in a frictionless environment. Based upon research conducted since 1977, ASTM recommends the N-value be corrected to 60% of the total theoretical potential energy for rope and hammer devices. The N-value is denoted as N_{60}.

While the SPT results in an N_{60} value, there are a substantial number of soil property correlations with that value including density, undrained shear strength, settlement, bearing capacity, and liquefaction resistance. Software speeds the correlations and assists in presentations. One set of correlations for cohesionless soils is the relationship between the N-value and density, relative density, and friction angle as presented in Table 2.2. Correlations of the N-value to density, relative density, and friction angle, such as in Table 2.2, are particularly useful because retrieval of undisturbed cohesionless samples for laboratory testing of these parameters is very difficult.

As expected, the SPT N-value increases with the increasing relative density of sands. Also, as would be predicted by the Mohr-Coulomb failure criteria (eq. 2.24), the N-value is influenced by the applied lateral stress on the sampler, which dissipates energy. A correction to the N-value is needed in order to filter out the effect of differing lateral pressures on the SPT sampler, in order to legitimately compare N-values at different depths. Here, the lateral in situ stresses are usually accounted for in terms of overburden pressure. The need to correct, or normalize, the N-value for overburden pressure was first clearly demonstrated by Gibbs and Holtz (1957). Since then, numerous studies have published correction factors (Marcuson and Bieganousky 1977, Liao and Whitman 1986, Skempton 1986).

A value of vertical effective stress of 95.6 kPa (1 ton/ft²) has been adopted as a standard pressure at which no correction to the N-values is needed. The N-values taken at lesser or greater vertical stresses require correction. The corrected N-value is widely used in foundation engineering and in liquefaction assessments. The correction is as follows:

$$N_{cor} = C_N N_F \tag{2.32}$$

where

N_{cor} = corrected N-value
C_N = correction factor (Figure 2.15)
N_F = N-value from the field

Table 2.2 Correlations using SPT N-values for sand properties

SPT N-value	Density	Relative Density (%)	Friction angle, $\phi°$
< 4	Very loose	< 20	< 30
4–10	Loose	20–40	30–35
10–30	Medium dense	40–60	35–40
30–50	Dense	60–80	40–45
>50	Very dense	> 80	> 45

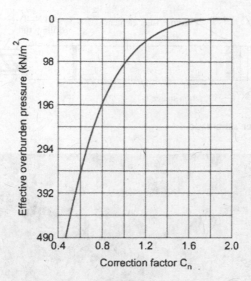

Figure 2.15 Correction values for SPT N-values.

Table 2.3 Correlations using SPT N-values for clay properties

SPT N-value	Consistency	Unconfined compressive strength q_u (kPa)
0–2	Very soft	< 25
2–4	Soft	25–50
4–8	Medium (firm)	50–100
8–15	Stiff	100–200
15–30	Very stiff	200–400
> 30	Hard	> 400

Undisturbed sampling and laboratory testing of clay soils is considerably easier for clays than sands but are expensive and time-consuming. As a result, correlations of the N-value to the undrained strength of clays are helpful. Table 2.3 is used to correlate N-value in clay with soil consistency and unconfined compressive strength, q_u.

The SPT is useful for retrieving a disturbed sample for identification and testing. The N-value can be useful for preliminary design or final design for smaller projects (with appropriate factors of safety). For ground improvement projects, the data from the SPT is useful for project planning and evaluation of alternative ground improvement techniques. Perhaps as important is the use of the SPT as a means to measure ground improvement by comparing before and after N-values.

2.6.3.2 CPT

The CPT is also widely used in ground improvement projects. In this test, a cone is pushed into the ground and tip resistance (stress), q_c, is measured as the cone penetrates the subsurface. Depending upon the project and the design of the cone, side friction (stress), f_s, and porewater pressure, u, are also measured. The ratio between the friction load and the tip load is termed the friction ratio and is key to using the CPT results to classify soils. A schematic of the standard cone penetrometer is shown in Figure 2.16 and a photograph in Figure 2.17.

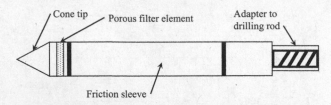

Figure 2.16 CPT penetrometer schematic.

Figure 2.17 CPT penetrometer.

Since the CPT cone is pushed into the ground, and loads are normally sensed electronically, the data can be acquired in a near continuous manner. Just as with the N-value, the CPT is most useful when the results are correlated with other geotechnical parameters (Robertson et al. 1986). For example, the CPT results can be used for soil classification, undrained shear strength, bearing capacity of shallow foundations, pile bearing and friction capacity, liquefaction resistance, and the SPT N-value (Robertson et al. 1983).

Figure 2.18 is a common correlation: classifying a soil using the tip resistance and friction ratio values from the CPT.

As with the SPT, the value of the CPT is in the use of the data for correlations to other parameters. And for ground improvement projects, again like the SPT, the data from the CPT is useful in project planning and evaluation of alternative ground improvement techniques. Perhaps as important, is the use of the CPT as a means to measure ground improvement by comparing before and after CPT results.

2.7 PROBLEMS

2.1 The zero air voids (ZAV) curve is plotted on a w vs. γ_d plot. The ZAV curve, calculated, is the saturated unit weight (zero air) of a given soil over a range of water contents.

a) Calculate and plot the zero air voids dry density curve for a material with a specific gravity of solids of 2.67 for a range of water content values between 5% and 40%. Repeat the calculation and plot for a degree of saturation of 90%.

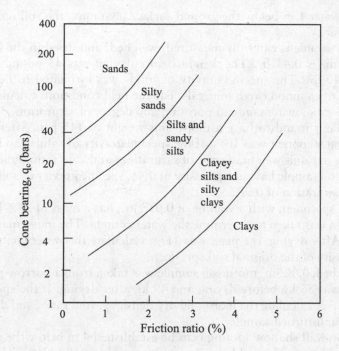

Figure 2.18 Soil classification using the CPT.

Table Pr.2.2 Sieve and Atterberg limits results for classification problem

Sieve Size	% finer by weight		
	Soil 1	Soil 2	Soil 3
No. 4	100	100	100
No. 10	85	95	100
No. 40	45	85	100
No. 100	11	75	98
No. 200	7	55	95
Atterberg limits			
Liquid limit	–	21	128
Plastic limit	–	14	37
Plasticity index	Non-plastic	7	91

b) Over the course of 25 field moisture-density tests, three data points were found to plot above the ZAV line. Should these three values be discarded as erroneous when computing the average moisture content and density for the project?

2.2 The data for three sieves and Atterberg limits tests are shown in the table below. Using the unified soil classification system classify and describe the soils (Table Pr.2.2).

2.3 At a depth of 5 m below the ground surface, for a soil having a saturated unit weight of 22 kN/m³, calculate the total stress, porewater pressure, and effective stress for the following groundwater conditions.

a. Groundwater at the ground surface

b. Groundwater 1 m above the ground surface

    c. Groundwater 1 m below the ground surface, assuming the soil remains saturated by capillarity.

2.4 A moist soil sample is carefully measured, weighed, and dried in the laboratory. The sample volume is 0.43 ft³. The weight before drying was 45 pounds. After drying, it weighed 40 lbs. The specific gravity of solids was estimated to be 2.67, the specific gravity of common earth minerals. For the field condition, calculate the dry unit weight, void ratio, water content, porosity, and degree of saturation.

2.5 A soil sample is found to have a moist unit weight of 145 pcf. After drying, it was found the water content was 10%. If the specific gravity of solids can be taken as 2.7, calculate the dry unit weight, degree of saturation, and void ratio. Find the moist unit weight of a soil sample having a porosity of 0.45, specific gravity of solids of 2.65, and a degree of saturation of 0.982.

2.6 A moist soil specimen, with a volume of 0.013 m³, has a mass of 26.5 kg. A sample of this specimen was taken to determine the water content. The moist mass of the sample was 135 g. After drying, the mass was 117g. Calculate the water content and the wet and dry density of the original soil specimen.

2.7 An undisturbed, 0.028m³ moist soil sample was taken from a borrow pit. The mass of the sample was 56 kg before drying and 49 kg after drying. If the specific gravity of solids was 2.72, calculate the moist and dry densities, void ratio, and degree of saturation of the undisturbed sample.

2.8 A foundation wall shallow footing is to be established 4 m below the ground surface, to avoid frost heave. The soil has a moist unit weight of 18 kN/m³. The water table is 5 m below the ground surface.

    a. Estimate the effective stress at the base of the footing.

    b. Years after footing installation and satisfactory performance of the footing, a long-term water leak from an adjacent swimming pool raises the water table in the vicinity of the footing to 3 m below the ground surface. If γ_{sat} is 23 kN/m³, estimate the effective stress at the base of the footing with these new ground conditions.

    c. Compare (a) and (b). Will the water leak affect the performance of the footing? How?

2.9 A pumped storage project uses an earthen berm to create a reservoir (lookup operation of pumped storage projects). How does the principle of effective stress play into the geotechnical design of the berm?

2.10 Several countries are subject to a monsoon season, characterized by heavy, sustained rain. Landslides sometimes occur in hillsides that were stable before the monsoon season. Explain, in geotechnical terms, what causes the instability.

2.11 Three direct shear tests are run on the same dry sand. Here's the data:

Test 1
    Normal stress: 2,100 psf
    Shear stress at failure: 970 psf

Test 2
    Normal stress: 3,799 psf
    Shear stress at failure: 1,700 psf

Test 3
    Normal stress: 4,500 psf
    Shear stress at failure: 2,080 psf

The soil sample is taken from the field, where the water table is 5 feet below the ground surface. The dry unit weight of the soil is 110 pcf. The saturated unit weight of the soil is 120 pcf.

Find the strength parameters for this soil.

Find the shear strength of this soil at a depth of 15 feet below the ground surface.

2.12 In a direct shear test on a dry, medium-dense sand, the normal stress on the failure plane at failure was 140 kPa at a shearing stress of 81.1 kPa.
  a. What is the effective angle of internal friction?
  b. Draw the Mohr's circle. What are the major and minor principal stresses at failure?

2.13 A direct shear was run on a medium-dense sand. The shear stress at failure was 58.6 kPa. If the effective angle of internal friction is 38°, what is the normal stress on the failure plane? What are the major and minor principal stresses at failure? (hint: draw the Mohr failure envelope, then the Mohr's circle).

2.14 Two direct shear tests are run on a two-inch square sample of the same sandy soil. At failure, the sample failure surface was two inches square. For the first test, the normal force on the sample was 400 lbs, while the shear force at failure was 200 lbs. For the second test, the normal force on the sample was 200 lbs, while the shear force at failure was 100 lbs. What are the total strength parameters for this soil?

2.15 A dry sand is tested in direct shear. Two tests are run. For the first test, the normal stress is 4,800 psf, and the shear stress at failure is 3,100 psf. Assuming c = 0 (dry sand), what is the internal angle of friction? For the second test on the same sand, the normal stress is 3,500 psf. What shear stress is expected to fail the sample?

2.16 Two identical clayey samples are tested in direct shear. One has a normal stress on it of 95kPa and fails at a shear stress of 71 kPa. The second test fails at a shear stress of 104 kPa under a normal stress of 150 kPa. What are the strength parameters for this soil?

2.17 A staged direct shear test on clean sand was carried out to failure with the following results (Table Pr.2.17).
  Determine:
  a. Angle of internal friction
  b. Major and minor principal stresses at failure for each stage

2.18 The sand described in problem 2.17 was tested in a consolidated drained (CD) shear triaxial test employing a cell pressure of 45 kN/m³. Predict the major principal stress at failure in this test.

2.19 The sand described in problem 2.17 was tested in a consolidated undrained (CU) shear triaxial test employing an effective consolidation pressure of 45 kN/m². The pore pressure at failure was 6 kN/m² but the axial stress load cell malfunctioned. Predict the major principal stress at failure in this test.

2.20 A sandy soil is tested in drained triaxial shear. The confining pressure is 21 kPa. The major principal stress at failure is 61 kPa. What are the strength parameters?

2.21 A triaxial CD test is run on sand. The confining pressure is 70 kPa. If the effective friction angle is 36°, what is the deviator stress at failure (note: deviator stress is the difference between sigma 1 and sigma 3)?

2.22 A CD triaxial test is run on a normally consolidated sand. The deviator stress at failure is 152 kPa, when the cell pressure was 70 kPa. What is the effective angle of internal friction?

Table Pr.2.17 Staged direct shear test information

Stage	Normal Stress (kN/m³)	Peak Shear Stress (kN/m³)
1	39	21
2	45	32
3	60	42

2.23 Two consolidated, undrained tests are run on identical soil samples. Pore pressures are measured during the tests. Here's the data

Test 1

Confining pressure: 200 kPa

Deviator stress at failure: 150 kPa

Pore pressure at failure: 140 kPa

Test 2

Confining pressure: 400 kPa

Deviator stress at failure: 300 kPa

Pore pressure at failure: 280 kPa

a. Plot the total stress failure envelope. What are the total stress strength parameters?

b. Plot the effective stress failure envelope. What are the effective stress strength parameters?

2.24 You have designed the wall given in the example lateral earth pressure problem, assuming $\phi' = 28°$. During construction, you discover the contractor did not compact the backfill, save for the last two feet of backfill. Do you have cause for concern? Explain your concerns in geotechnical terms.

2.25 A retaining wall you designed is constructed according to your design. However, you later find out that the owner installed a municipal water supply line in the backfill, running parallel to the wall. Are you concerned about this change? If so, what are your concerns (explain in geotechnical terms)?

2.26 A client has retained you to do the geotechnical design of a 6-m high cantilever, concrete retaining wall with a level backfill. You choose a free-draining backfill having a high angle of internal friction, 35°. Estimate the lateral *force* on the wall. Later, you find out that the owner substituted an incredibly massive concrete wall, much stiffer than the cantilever one you designed. Estimate the new lateral earth force on the wall. By what percentage is it different from your design?

REFERENCES

ASTM D5778–20. (2020). *Standard test method for electronic friction cone and piezocone penetration testing of soils*. West Conshohocken, PA: American Society of Testing and Materials.

ASTM Standard D1586/D1586M. (2018). *Standard test method for Standard Penetration Test (SPT) and split-barrel sampling of soils*. West Conshohocken, PA: American Society of Testing and Materials.

Bishop, A.W. and Henkel, D.J. (1957). *The measurement of soil properties in the triaxial test*. London: Edward Arnold publisher, 190 pages.

Clough, G.W. and Duncan, J.M. (1991). Earth pressures. In Hsai Yang Fang (Ed.), *Foundation engineering handbook* (pp. 223–235). Boston, MA: Springer.

Gibbs, H.J. and Holtz, W. G. (1957). Research on determining the density of sands by spoon penetration testing. In *Proceedings of the 4th International Conference on Soil Mechanics and Foundation Engineering*, London, UK, 1, 35–39.

Harr, M.E. (2012). *Groundwater and seepage*. North Chelmsford, MA: Courier Corporation.

Holtz, R.D, Kovacs, W.D., and Sheahan, T.C. (2010). *An introduction to geotechnical engineering*. Upper Saddle River, NJ: Prentice Hall.

Iverson, R.M. (2000). Landslide triggering by rain infiltration. *Water Resources Research*, 36(7), 1897–1910.

Jaky, J. (1944). The coefficient of earth pressure at rest. *Journal of the Society of Hungarian Architects and Engineers*, 22, 355–358.

Kovacs, W.D., Evans, J.C. and Griffith, A.H. (1977). Towards a more standardized SPT. In *Proceedings of the 9th International Conference on Soil Mechanics and Foundation Engineering*, Tokyo, v. 2, pp. 269–276.

Kulhawy, F.H. and Mayne, P.W. (1990). *Manual on estimating soil properties for foundation design* (No. EPRI-EL-6800). Palo Alto, CA: Electric Power Research Institute; Ithaca, NY: Cornell University. Geotechnical Engineering Group.

Lade, P.V. (2016). *Triaxial testing of soils.* Somerset, NJ: John Wiley & Sons.

LaGrega, M.D., Buckingham, P.L. and Evans, J.C. (2010). *Hazardous waste management.* Long Grove, IL: Waveland Press.

Liao, S.S. and Whitman, R.V. (1986). Overburden correction factors for SPT in sand. *Journal of Geotechnical Engineering, 112*(3), 373–377.

Lu, N. and Likos, W.J. (2004). *Unsaturated soil mechanics.* Somerset, NJ: Wiley.

Marcuson, W.F., III and Bieganousky, W.A. (1977). SPT and relative density in coarse sands. *Journal of the Geotechnical Engineering Division, 103*(11), 1295–1309.

McCarthy, D.F. (2002). *Essentials of soil mechanics and foundations* (6th ed.). Upper Saddle River, NJ: Prentice Hall, 730pp.

Robertson, P.K., Campanella, R.G., Gillespie, D. and Greig, J. (1986, June). Use of piezometer cone data. In *Use of in situ tests in geotechnical engineering* (pp. 1263–1280). New York, NY: ASCE.

Robertson, P.K., Campanella, R.G. and Wightman, A. (1983). SPT-CPT correlations. *Journal of Geotechnical Engineering, 109*(11), 1449–1459.

Skempton, A.W. (1986). Standard penetration test procedures and the effects in sands of overburden pressure, relative density, particle size, ageing and overconsolidation. *Geotechnique, 36*(3), 425–447.

Tanyu, B.F., Sabatini, P.J. and Berg, R.R. (2008). *Earth retaining structures.* Washington, DC: US Department of Transportation, National Highway Institute, Federal Highway Administration.

Terzaghi, K. (1925). *Erdbaumechanik auf bodenphysikalischer Grundlage.* Leipzig: F. Deuticke, 399 pages.

US Army. (1968). *Report on replacement--lock & dam 26, Mississippi River, Alton, IL.* St. Louis, MO: US Army Corps of Engineers; Washington, DC: US Department of the Army, 43pp plus appendices.

Chapter 3

Fundamentals of geosynthetics in ground improvement

3.1 INTRODUCTION

Geosynthetics are man-made (synthetic) materials embedded in or on the ground (geo) to improve soil behavior. Some natural materials, formed into geosynthetic shapes, are called geosynthetics because they are used for the same uses as man-made geosynthetics. Many geosynthetics are planar and are produced in rolls. Others are three dimensional. Geosynthetics are a proven technology, having been used by civil engineers for over sixty years. All are manufactured indoors, giving good quality control.

There are many civil engineering geosynthetics. The uses of geotextiles, geogrids, geocells, and geofibers are discussed here in detail. Other geosynthetics include geofoam, geomembranes, geopipe, geonets, and various geocomposites (combinations of materials, including geosynthetics). These are addressed in Koerner (2012) and other texts.

3.1.1 Geotextiles

Geotextiles are a type of cloth used in geotechnical applications. Most are polymeric, made of repeated patterns of multiple fibers. Geotextiles are made from fibers, which are sometimes combined into yarns that are entangled with one another to form the geotextile. The yarns or fibers may be entangled by weaving or by a nonwoven process. While there are many types of weaving, the simple basket weave of fibers or yarns is most common for geotextiles. Nonwoven geotextiles are most often created by needle punching fibers until they form a strong mat, which has the appearance of felt. Figure 3.1 shows woven and nonwoven geotextiles. Some nonwoven geotextiles are made by heat bonding, which produces a very smooth, shiny surface. Figure 3.2 shows a heat bonded nonwoven.

Geotextile manufacture takes a variety of forms. Woven geotextiles are produced on looms, where the warp and weft fibers are interwoven. Nonwoven geotextiles are produced from staples (short fibers) or continuous fibers. Here, a thick batt is stabbed repeatedly with barbed needles, which entangle the fibers, making cloth in a process called needle punching. Alternatively, a thinner batt may be passed between hot rollers, slightly melting the fibers together, making heat-bonded, nonwoven, cloth. Many geotextiles are made of polypropylene, polyethylene, or polyester, with various additives to improve performance. Woven geotextiles may have different warp and weft fibers or yarn, allowing different strengths in different directions. Since strength costs money and the same strength is not always needed in both directions, wovens may be less expensive than nonwovens in one-directional strength applications. Nonwovens, in the plane of the fabric, are largely isotropic. Because of the controlled manufacturing environment, excellent quality, by civil engineering standards, is possible.

Figure 3.1 Photo of woven and nonwoven geotextiles.

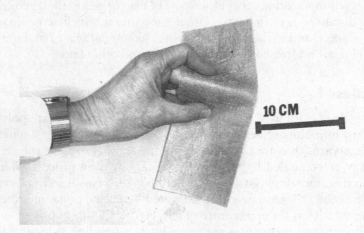

Figure 3.2 Photo of heat bonded nonwoven geotextile (note sheen on geotextile).

Geotextiles serve *functions*. When a design is considered, the needed function(s) are determined before a geotextile is selected that can perform those functions to complete the design. Geotextiles serve these primary functions:

1. Filtration – allowing water to pass through while retaining soil,
2. Reinforcement – making soil stronger,
3. Separation – isolating different soil types,
4. Erosion and sediment control – holding surface soil in place and/or catching it once it's moved,
5. Drainage – expediting water removal from soil,
6. Waterproofing (when impregnated with asphalt);
7. Cushioning – protection from damage, and
8. Insulation from heat/cold.

Geotextiles are used for these basic civil engineering designs:

1. Filtration - soil drains and filters in slopes, in dams, road structures, behind retaining walls;
2. Reinforcement - strengthening foundations, strengthening fill slopes, retaining walls, roads;
3. Separation - separating clayey soils from cohesionless ones (especially in roadways) and separating filter soils from the soil being drained;
4. Erosion and sediment control - erosion control materials, silt fences;
5. Drainage - removing water from slopes, walls, and dams; road drainage, vertical drains during preloading usually as part of a geocomposite; and
6. Waterproofing - under asphalt concrete overlays, pond liners, foundation waterproofing.

The first three uses are addressed in this book. For example, Figure 3.3 shows a picture of a geocomposite wall drain being installed on a basement wall. A geotextile filter overlays an egg-carton-shaped plastic sheet drain.

3.1.2 Geogrids

Geogrids are plastic sheets with apertures much larger than those in geotextiles. Figure 3.4 shows various geogrids. There are two common manufacturing processes. One is to punch holes in a plastic sheet and stretch the sheet in one, two, or three directions. The other is to weave strips, in a grid pattern, and connect at the strip overlaps (junctions). The manufacturing process allows geogrids to have different strengths in different directions. Because geogrids are manufactured in a factory setting, quality control is very good.

The primary geogrid functions are:

1. Reinforcement – making soil stronger, and
2. Containment – may be used to wrap large particles into a mattress.

Figure 3.3 Photo of geocomposite wall drain being installed (Koerner 2012).

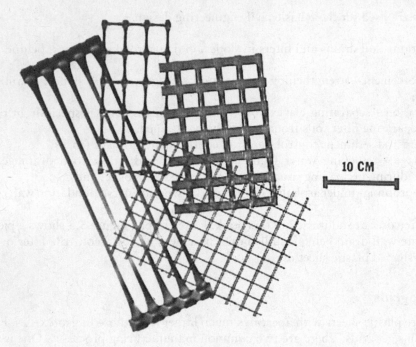

10 CM

Figure 3.4 Photo of geogrids.

Geogrids are used for these basic civil engineering designs:

1. When in sheet form as reinforcement – strengthening foundations, strengthening fill slopes, retaining wall construction, road construction; and
2. When wrapping soil in a mattress – foundation and embankment construction and strengthening.

The first use is addressed in this book.

3.1.3 Geocells

Geocells (geocellular confinement systems) are basically plastic honeycombs (Figure 3.5). They are manufactured from plastic sheets or geotextiles, welded together at the joints. Unlike geotextiles and geogrids, they have a significant height – some as much as 300 mm. Geocells are delivered flat (compressed), expanded on site, and filled with soil. The result is an extremely strong mat.

The geocell's functions are primarily reinforcement and erosion control.

The first function, reinforcement, is addressed in this book.

Geocells are used for these basic civil engineering designs:

1. Strengthening fill slopes,
2. Strengthening building foundations,
3. Reinforcing roadbeds,
4. Stacking to form walls, and
5. Erosion control.

Figure 3.5 Geocell photo showing as-shipped (flat) geocell in the foreground with expanded geocells as placed by workers (courtesy of Presto Geosystems 2014).

The first three designs are addressed in this book.

3.1.4 Geofibers

Geofibers are polymeric fibers mixed with the soil to improve it. Short fibers (staples) and continuous fibers are used. All are manufactured indoors, giving good quality control. Short fibers are delivered in bales, while continuous fibers are delivered on spools.

The geofibers' functions are primarily reinforcement and erosion control.

The first, reinforcement, one is addressed in this book.

Geofibers are used for these basic civil engineering designs:

1. Strengthening building foundations,
2. Strengthening roadbeds,
3. Building steepened slopes,
4. Compaction aids, and
5. Erosion control.

The first four are discussed in this book.

3.1.5 Historical notes

Geosynthetics have been used for over sixty years. Koerner and Welsh (1980), Rankilor (1981), and Veldhuijzen van Zanten (1986) provide excellent summaries of the early work. Richardson and Koerner (1990) summarize the current usage of the various geosynthetics, in the chapter introductions in the book *Design Primer: Geotextiles and Related Materials*.

Some credit Robert J. Barrett as the "father of geotextiles" for his pioneering work with geotextile filters, and his early designs and applications (Holtz and Christopher 1990). Koerner (1986) and John (1987) also provided early books on geosynthetic use.

The use of geosynthetics has significantly changed geotechnical engineering because geosynthetics allow designs and construction procedures that would have previously been impossible, very difficult, or impractically costly. At its essence, geosynthetics are a ground improvement technology that can make soils stronger, less compressible, and/or more impermeable.

3.2 PROPERTIES OF GEOSYNTHETICS

Geosynthetics are engineering materials with engineering properties. A civil engineer specifying concrete might specify properties such as durability, permeability, and strength. Similarly, civil engineers specifying geosynthetics must specify the properties of the geosynthetic if the geosynthetic is to serve the required function.

The American Society of Testing and Materials (ASTM) and the International Organization for Standardization (ISO) are two major standards bodies that publish test procedures to determine the engineering properties of geosynthetics. ASTM standards are used in this text.

3.2.1 Tensile strengths

To serve as soil reinforcement, geotextiles must have tensile strength. There are several tensile tests, modeling different field conditions. The engineer is responsible for choosing the best test to model field conditions. For reinforcement, the choice is normally the wide-width tensile test (ASTM D4595 2011). In this test, an eight inch (20 cm) square sample is cut and pulled apart in tension. The entire width of the sample is held by the tensile machine jaws. The force (maximum, or at a given strain) is divided by the geotextile width, and reported as force per unit width. Figure 3.6 shows a wide-width test underway.

Grab Tensile tests (ASTM D5034 2009) result in a different tensile strength than wide-width testing because the test is run differently. Here, a four inch wide (10 cm) sample is gripped by one inch (2.5 cm) wide square jaws and pulled apart. The failure mode is different, resulting in a different strength. The results of the test are reported in units of force. This test originated as a quality control test in the textile industry and was adapted to geotextiles. Figure 3.7 shows the Grab Tensile test.

Geogrid strengths are evaluated by tensile testing. Various sample sizes are allowed. ASTM D6637 (2011) gives the details. The number of grid intersections (nodes) tested, and the number of sections tested affect the result and must be reported.

Other tensile strength tests exist but are rarely used compared with those discussed above.

3.2.2 Permittivity (used in drainage)

Geotextiles are used to carry water from soil. The geotextiles must be pervious enough to allow water to pass through, perpendicular to the plane of the geotextile, while limiting soil particles from passing into or through the geotextile. ASTM D4491 (2009) describes the test, where water (or another liquid) is passed through the geotextile. The volume of water that flows through the geotextile per unit time (direction of flow perpendicular to the plane of the geotextile), divided by the geotextile area, divided by the geotextile thickness is the permittivity (Ψ):

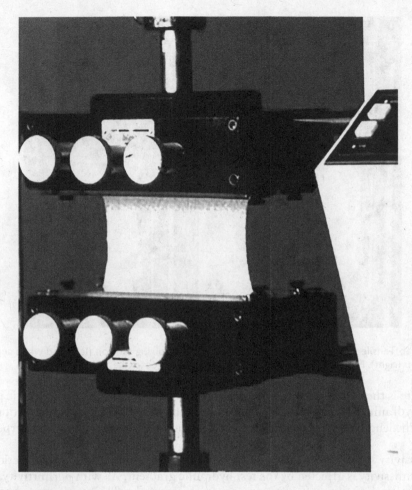

Figure 3.6 Wide-width tensile test in progress.

$$\Psi = \frac{\dfrac{\text{volume of water}}{(\text{time})(\text{area})}}{\text{thickness}} \tag{3.1}$$

Permittivity is reported in units of 1/Time. The fluid used, and its temperature, affect permittivity, because of the fluid's viscosity, which changes with temperature. Laboratory conditions should model field conditions.

3.2.3 Transmissivity (used in drainage)

Some geotextiles, and some geocomposites, are used to carry water in the plane of the geosynthetic. These geosynthetics are used as drains. Transmissivity, Θ, is defined as the planar coefficient of permeability of the geotextile times the thickness. Transmissivity is calculated, from test results, as the in-plane flow rate divided by the hydraulic gradient used during the test, divided by the width of the test specimen:

$$\theta = \left(\frac{\text{flow rate}}{(\text{hydraulic gradient})(\text{width of geosynthetic})} \right) \tag{3.2}$$

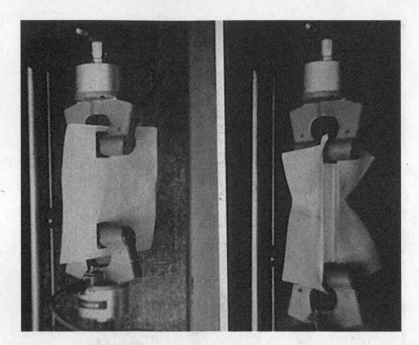

Figure 3.7 Grab Tensile test showing the geotextile prior to applying tension (left) and while applying tension (right). Note the non-planar deformation of the geotextile.

The flow rate is the volume of water passing through the plane of the geosynthetic in a given time. The hydraulic gradient is the head loss across the geosynthetic, in the direction of flow, divided by the length of the geosynthetic. The geosynthetic width is measured perpendicular to flow.

Transmissivity is often reported with units of Length^2/Time (L^2/T), a contraction of (L^3/T)/L. Transmissivity is affected by the test hydraulic gradient. As with permittivity, the fluid used, and its temperature, affect transmissivity because the fluid's viscosity changes with temperature. Again, laboratory conditions should model field conditions.

3.2.4 Pore size determination (used in filtration)

Nonwoven and woven geotextiles are used as filters to keep soil particles from moving through them while allowing water to pass through. The geotextile's effectiveness depends on the sizes of the geotextile's inter-fiber spaces. Apparent opening size (AOS) is the common measure of the spaces in nonwoven geotextiles. ASTM D4751 (2012) gives the procedures for determining the AOS. The procedure involves sieving different diameter glass beads through the geotextile until a certain percentage of a certain size bead is retained. Hence, it is only a *measure* of the spaces. The AOS is expressed as a length – the diameter of the certain size bead. AOS is also reported as a US Sieve number, with a hole size corresponding to that length (diameter). This same size is also designated O_{95}, referring to the opening size in the geotextile where 95% of the certain size glass beads are retained on a geotextile after sieving.

The capillary flow test (ASTM D6767), also known as a bubble point test, can be used to evaluate the pore size distribution in nonwoven geotextiles. This test provides a distribution of sizes, to compare with the pore size distribution of the soil being filtered. There are no standard design procedures (2012) that use the pore size distribution.

While the AOS test can be used for woven geotextiles, percent open area (POA) is the more common method of characterizing the sizes of spaces in woven geotextiles. One method to determine AOS is to place the geotextile on an overhead projector. Light penetrates the holes. Manually, the area allowing light through is measured and compared to the total area. The POA is defined as:

$$POA = \frac{\text{total area allowing light through}}{\text{total area of geotextile sample}}(100) \tag{3.3}$$

Alternatively, electronically scanned geotextiles can be analyzed with software to measure hole sizes and calculate the AOS.

3.2.5 Interface friction (used in mechanically stabilized earth and steepened slope design)

Mechanically stabilized retaining walls, and other geotechnical structures, consist of geo-synthetics and soil. Interface friction, also called an angle of external friction, is the amount of friction between the soil and the geosynthetic used in design. The interface friction speci-fication, ASTM D5321 (2012), uses a variation of ASTM D3080 (2011), the direct shear test for soils. When testing geosynthetics, a horizontal sample geosynthetic is vertically pressed against the candidate soil with a normal force. With this pressure intact, the soil sample is slid horizontally across the geosynthetic. The shear force is recorded, and converted to the interface friction angle, δ, thus

$$\delta = \arctan\frac{(\text{shear force/area})}{(\text{normal force/area})} \tag{3.4}$$

Interface friction is reported in degrees. Figure 3.8 shows an alternative interface shear apparatus termed a tilt table (after Narejo 2003).

3.2.6 Survivability and durability

Geosynthetics must withstand installation, and, once in place, must retain their properties for the project lifetime. The American Association of State Highway and Transportation Officials (AASHTO) specification M288 (AASHTO 2017) provides baseline geosynthetic

Figure 3.8 Photos of tilt table (left) and tilt table apparatus showing soil on the bed (right).

properties needed to withstand installation conditions for projects. AASHTO M288 references ASTM standards. The relevant standards for applications in this text are:

1. ultraviolet light degradation, ASTM D4355 (2007) and D7238 (2012)
2. trapezoidal tearing, ASTM D4533 (2011)
3. puncture, ASTM D6241 (2009)
4. mass/area, ASTM D5261 (2010)

All polymeric geosynthetics break down in ultraviolet light, present in sunlight. In order to preserve the engineering properties of the geosynthetic, this must be prevented. Actinic resistance is typically enhanced by additives, primarily carbon black or titanium oxide. A common practice is to limit ultraviolet light exposure to less than two weeks. Covering geosynthetics with any amount of soil eliminates ultraviolet light-induced degradation. Hence, it's important to keep geosynthetics out of sunlight before installation. ASTM tests D4355 (2007) and D7238 (2012) evaluate the actinic resistance of geosynthetics. These tests do not particularly simulate field performance but give relative resistance.

While a geosynthetic may have adequate strength to resist service loads, it may not have adequate durability to resist installation. The trapezoidal tearing test, ASTM D4533 (2011), is a measure of resistance, not to service criteria, but to an installation condition. Here, a small cut is made in a rectangular geotextile sample, which is then pulled apart such that the tear propagates from the cut. The maximum force to tear the geotextile is recorded as the trapezoidal tear strength. Figure 3.9 (left) shows a test specimen, Figure 3.9 (center) shows the specimen in the test apparatus, and Figure 3.9 (right) shows the specimen after failure.

Geotextiles may be exposed to puncture forces during installation. The relative resistance of geosynthetics to these forces is evaluated with ASTM D6241 (2009). Here, a metal two inch (50 mm) diameter piston is pushed through a restrained geosynthetic. The force required to puncture the geosynthetic is recorded. Figure 3.10 shows the test setup.

The mass/area of a geosynthetic suggests several properties e.g. strength, durability, and elasticity. Typically, the greater the mass/area (amount of polymer), the greater the resistance

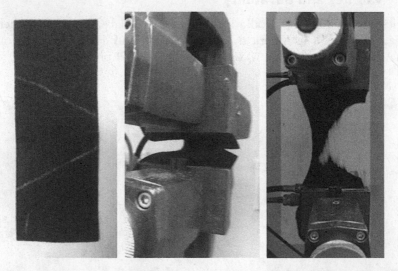

Figure 3.9 Trapezoidal tear test with specimen marked for testing (left), trapezoidal tear test with specimen in grips (center), and trapezoidal tear test showing specimen at failure (right). (courtesy of Golder Associates, Inc.)

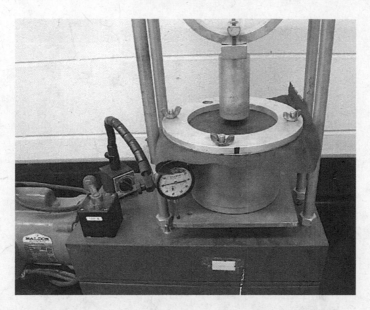

Figure 3.10 Puncture apparatus setup.

to installation damage. AASHTO M288 uses a mass/area criterion, determined from ASTM D5261 (2010), as a general criterion to help resist installation forces.

In addition to the above specifications, which satisfy AASHTO M288, the geosynthetic may be exposed to liquids, biological hazards, high temperatures, or abrasion during installation or service. ASTM provides specifications for these durability concerns.

3.3 GEOTEXTILE FILTER DESIGN

3.3.1 Introduction

Geotextiles, in contact with soil, can be used as filters that allow water to pass through while retaining most soil particles. These filters are used in retaining wall design, slope drainage, highway drainage, dewatering, earth dams, and preloading projects.

Filters must meet the following criteria:

1. Adequate permeability, so water can pass through;
2. Adequate soil particle retention, to reduce soil particle penetration and transmission to an acceptable level; and
3. Must not clog or blind. Clogging occurs when soil particles get stuck inside the geotextile, reducing permeability. Blinding occurs when soil particles coat the outside of the geotextile, reducing permeability.

These are conflicting criteria – greater permeability leads to decreased retention. Increased soil particle retention leads to decreased permeability. Hence, filter design is a trade-off between these criteria. The proper filter has holes large enough to allow adequate passage of water, and small enough to retain sufficient soil particles. There are many methods for filter design. Luettich et al. (1992) present a comprehensive, well-accepted method for designing geotextile filters. The method considers the grain size distribution of the soil being filtered, and a representative geotextile hole size (O_{95}), or POA.

Figure 3.11 Geotextile bulging out from under displaced articulated rip rap (FHWA 1992).

Not all soils can be filtered with geotextiles. High plasticity soils and dispersive clays cannot be effectively filtered by geotextiles. Runoff with high concentrations of suspended solids, or precipitating chemicals should not be filtered with geotextiles. Gap graded soils, and those with high pH runoff shouldn't be filtered with geotextiles.

Interestingly, geotextiles do NOT filter the soil directly. Rather, the geotextile serves as a catalyst for a graded soil filter to form on the upstream side of the geotextile. This soil filter forms because soil arching takes place, the same phenomenon that makes a graded granular soil filter work.

Figure 3.11 shows the effects of a clogged geotextile filter. In this project, an oceanfront structure was subject to tidal groundwater fluctuations. When the tide ebbs, water flows out of the beach into the ocean. The water carried small soil particles that clogged the geotextile. The force of the water on the clogged geotextile disrupted and moved the articulated concrete blocks.

3.3.2 Design procedure

Luettich et al. (1992) propose an eight-step procedure for geotextile filter design. The steps are itemized and each step is explained in detail in the following text.

1. Characterize soil to be drained.
2. Define filter boundary conditions.
3. Determine O_{95} required for the geotextile.
4. Determine required geotextile permeability.
5. Check anti-clogging specifications.
6. Check survivability specifications.

7. Check durability specifications.
8. Check miscellaneous considerations.

Step 1. Characterize material to be drained. Obtain the grain size distribution of the soil (ASTM D6913 2009), the plasticity index (PI) (ASTM D4318 2010), and the relative density (ASTM D4354 2006). The engineer then decides if the application favors retention of fines or favors permeability. Retention means the geotextile is designed to retain more fines but will have lower permeability, while permeability means the geotextile will have a higher permeability but will not retain as many fines. If the drain material has a relatively low void volume, the application favors retention, so that the small amounts of fines passing through the filter do not seriously impede the drain capacity. A geonet drain, for example, has considerably lower void space than, say, a crushed rock drain, and, hence, could accept fewer fines before clogging. Hence, a geotextile filter over a geonet drain should favor retention. A larger void space drain (e.g. crushed rock) can accept a much larger quantity of fines before clogging. Hence, a geotextile filter over a crushed rock drain should favor permeability.

Step 2. Define the geotextile filter boundary conditions in terms of flow and soil stress on the filter. The filter may be subject to one-directional flow, as in a basement wall drain, where the water only flows into the drain. Two-directional flow, a much more rigorous condition, can occur in the case of, say, tidal flows, where the water flows both in and out of the filter.

The soil stress on a filter is a function of the depth of the filter below grade. If the filter is under high soil stress, there is a tendency for finer soil particles to be pushed through the filter. High stress conditions tend to push the geotextile into the drain pore spaces, reducing drain capacity. Higher stresses tend to push soil particles into the geotextile, leading to filter, and possibly, drain clogging.

Step 3. Determine O_{95} required for geotextile (retention criteria). Figures 3.12 and 3.13 present decision trees used to evaluate the required retention criteria for the geotextile filter, expressed as the geotextile's O_{95}, for one-directional and two-directional flow, respectively.

One-directional flow, Figure 3.12, is examined first. While the figure is largely self-explanatory, some points on figure use are given.

The notation "d_x" refers to the diameter of the soil particle smaller than x percent passing. The value d_x is taken from the soil's grain size distribution. Soils with "more than 20% clay," as defined by the d_{20}, require the double hydrometer ratio test, DHR, (ASTM D4221 2011) to evaluate dispersion potential. Dispersive soils cannot be filtered with a geotextile – it will clog. Soils with less than 10% fines or more than 90% gravel require the determination of a pseudo coefficient of uniformity, C'_u. C'_u is determined from d'_{100} and d'_0. These are determined by drawing a straight line on the soil's grain size distribution, extending to the 100% passing and 0% passing lines, as shown, for example, in Figure 3.14. The line on Figure 3.14 is drawn by following the decision tree for this path: Less than 10% fines → Application favors Retention → Stable Soil. The straight line is drawn through two points the d_{30} and d_{60} points on the grain size distribution curve and extended until it reaches the top and bottom axes, where, d'_{100} and d'_0 are read from the axes. These are not the soil's true d_{100} and d_0 but are pseudo values used in this calculation.

Similar lines are drawn for the other cases. With d'_{100} and d'_0 known, a pseudo coefficient of uniformity, C'_u is calculated using the equation in the figure. Depending on the value of C'_u and the soil relative density, O_{95} (or a range of O_{95}'s) is read from the figure. This is the retention specification.

Step 4. Determine the required geotextile permeability. The required geotextile permeability is based on the permeability of the soil being filtered. The geotextile permeability,

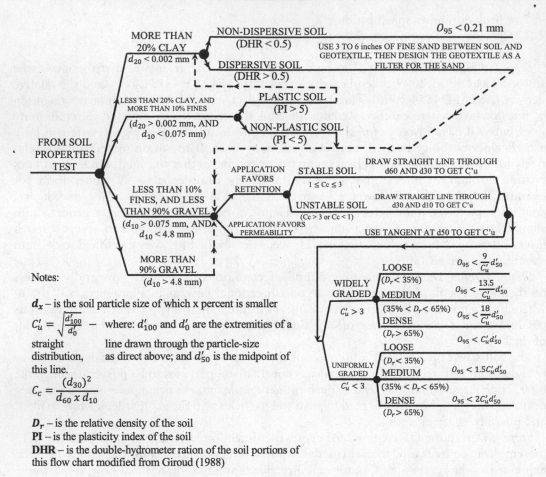

Notes:

d_x – is the soil particle size of which x percent is smaller

$C'_u = \sqrt{\dfrac{d'_{100}}{d'_0}}$ – where: d'_{100} and d'_0 are the extremities of a straight line drawn through the particle-size distribution, as direct above; and d'_{50} is the midpoint of this line.

$C_c = \dfrac{(d_{30})^2}{d_{60} \times d_{10}}$

D_r – is the relative density of the soil
PI – is the plasticity index of the soil
DHR – is the double-hydrometer ration of the soil portions of this flow chart modified from Giroud (1988)

Figure 3.12 Flow chart to determine O_{95} for one-dimensional flow through a geotextile (after Luettich et al. 1992).

$k_{\text{geotextile}}$, must be greater than the soil permeability, k_s, times the expected hydraulic gradient across the geotextile, i:

$$k_{\text{geotextile}} > (k_s)(i) \tag{3.5}$$

The engineer estimates the hydraulic gradient. A hydraulic gradient of one is typically used for gravity drains in civil engineering structures (walls, foundations, etc.). High head difference situations (dams, fine-grained soils) may generate gradients as large as three. Shoreline wave impact may generate gradients larger than ten. Giroud (1988) provides guidance.

The calculated $k_{\text{geotextile}}$ must be adjusted with reduction factors (RF) for non-quantifiable, but important, field effects: soil intrusion (IN) into the filter, geotextile creep (CR), *chemical clogging* (CC), biological clogging (BC), and installation damage (ID). RFs are chosen by the engineer, and multiplied by the $k_{\text{geotextile}}$ calculated above, to yield the adjusted geotextile permeability

$$k_{\text{geotextile adjusted}} > (k_{\text{geotextile}})(RF_{IN})(RF_{CR})(RF_{CC})(RF_{BC})(RF_{ID}) \tag{3.6}$$

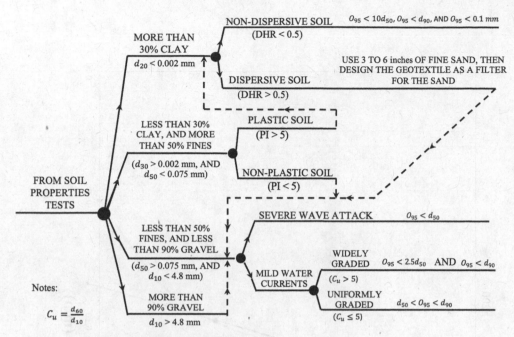

Figure 3.13 Flow chart to determine O_{95} for two-dimensional flow through a geotextile (after Luettich et al. 1992).

This is the permeability that is specified. Koerner (2012) suggests RFs. These are called *reduction* factors in the literature because, for non-filter applications, the property of interest is usually divided by the factors, not multiplied, as in this case. Geotextile permeability is rarely a constraint because geotextiles have very high permeability compared to most soils.

Step 5. Check anti-clogging requirements. Clogging means the soil has coated or filled the geotextile, significantly reducing the permeability. To reduce clogging potential, choose a geotextile that has the largest O_{95} that satisfies the retention criterion. For nonwoven geotextiles, use the largest porosity (n) geotextile available, but not less than 30%. Geotextile porosity is calculated from:

$$n = 1 - \frac{m}{\rho t} \tag{3.7}$$

where

   m = mass/area of the geotextile
   ρ = mass density of polymer (not the geotextile)
   t = thickness of geotextile

Or, if using a woven geotextile, select the largest percent open area (POA) available that satisfies the O_{95} criterion, but not less than 4%.

Step 6. Check survivability specifications. The geotextile should conform to the AASHTO M288 specification, discussed earlier.

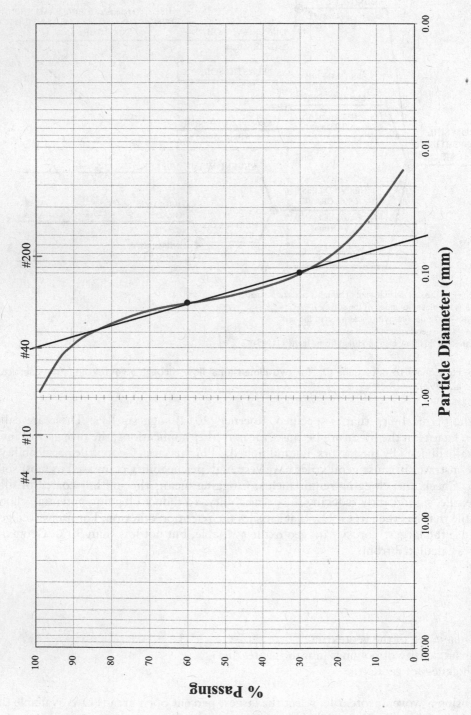

Figure 3.14 Grain size distribution plot with an interpretive straight line added.

Step 7. Check durability specifications. These are achieved, first, by meeting the AASHTO M288 ultraviolet light criterion. Other durability tests should be run if the engineer anticipates the geotextile will be exposed to abrasion, chemicals, or other abuses. ASTM has specifications.

Step 8. Specify installation criteria. Compaction of the soil at the face of the filter is critical. Loose soils tend to pipe, leading to filter clogging. Moreover, loose soils allow fines to migrate easily, which leads to filter clogging. When geotextiles are placed against a perforated pipe, the soil must be in intimate contact with the geotextile. Air spaces lead to very high hydraulic gradients, and large seepage forces, which move fine soil particles to the geotextile, causing clogging. Seams and overlaps must not allow soil penetration into the drain. The drain behind the geotextile filter must be graded so water will flow downhill.

Software for the above procedure is available (GeoFilter 2013).

Some perforated pipes have geotextile sleeves as a filter. This is not recommended, as the hydraulic gradient, and, thus, the seepage force, is very high at the pipe-geosynthetic interface. This leads to soil migration into the geotextile and possibly into the pipe.

There are other considerations. Nonwoven and woven geotextiles are used in filters. Selection criteria should be based on the properties of the geotextile, as determined above, rather than on the geotextile structure. The retention ability of geotextiles is related to the geotextile's thickness. Thicker filters, which reduce the hydraulic gradient across the geotextile, are better at retaining soil particles, despite the above criterion. If the soil being filtered is problematic (dispersive clay, gap graded, high fines content), the candidate geotextile should be *tested* using the very soil and using ASTM D5567 (2006) or ASTM D5101 (2006). More critical applications justify more testing.

Example Problem Ex.3.1: Geotextile filter design

A rural farm-to-market road, overlaid with asphalt concrete, is being rebuilt because it experienced severe breakup after only five years of service. The county engineer is paying much more attention to drainage. The new design includes elevating the road, crowning the road, and sloping ditches. On your advice, the engineer is installing a geosynthetic pavement underdrain – a geonet (Koerner 2005) with a geotextile filter. The geonet will be placed beneath the base course, whose grain size distribution is given in Figure 3.14. The base soil is granular and will be well compacted. Write the geotextile filter specifications.

Solution:

The eight-step Luettich et al. (1992) filter design procedure will be used. The symbols used in the calculations are given in Figure 3.12.

Step 1. Characterize soil to be drained: The soil grain size distribution is given in Figure 3.14. The soil will be well compacted ($D_r > 90\%$ expected). The soil has more than 10% fines, but they are nonplastic.

Step 2. Define filter boundary conditions: The geonet drain is in a low-stress location, subject to one-directional flow, downward.

Step 3. Determine O_{95} required for the geotextile: Use Figure 3.12 to determine O_{95}. Starting from the left, first go to the "Less than 20% clay..." node, since the soil d_{20} is > 0.002mm and d_{10} < 0.075mm. The soil is nonplastic. Follow the flow chart to the retention or permeability node. Since geonets have relatively little volume making them clog-susceptible if too many soil particles pass through the geotextile, choose retention.

Next calculate C_c: $C_c = \dfrac{d_{30}^2}{d_{10}d_{60}} = \dfrac{0.1^2}{(0.19)(0.038)} = 1.4$

Draw a straight line on the grain size distribution through d_{60} and d_{10} (as shown in Figure 3.14). The intersection of this line with the abscissa, and the 100% passing line, yields

$$d'_{100} = 0.4 \text{ mm} \quad \text{and} \quad d'_0 = 0.055 \text{ mm}$$

Next calculate C'_u: $C'_u = \sqrt{\dfrac{d'_{100}}{d'_0}} = \sqrt{\dfrac{0.4}{(0.055)}} = 2.7$

For $C'_u < 3$, and Dense soil ($D_r > 65\%$), giving: $O_{95} < 2C'_u\, d'_{50} < (2)(2.7)(0.15) < 0.8$ mm
This is the geotextile retention specification.

Step 4. Determine required geotextile permeability: The soil permeability can be estimated from any of a variety of approximations. Here, 0.0005 cm/sec is estimated, based on grain size distribution. The expected hydraulic gradient in the geotextile is 1, based on estimates provided by Luettich et al. (1992).

$$k_{\text{geotextile}} \geq (i)(k_{\text{soil}}) \geq (1)(0.0005 \text{ cm/sec}) \geq 0.0005 \text{ cm/sec}$$

RFs given by Koerner (2005) are applied to account for nonquantifiable but important considerations. Here, use

$$RF_{\text{soil clogging}} = 3$$
$$RF_{\text{creep}} = 1.75$$
$$RF_{\text{intrusion into the drain}} = 1.1$$
$$RF_{\text{chemical clogging}} = 1.1$$
$$RF_{\text{biological clogging}} = 1.15$$

The product of these RFs is 7.3. Thus, the adjusted

$$k_{\text{geotextile}} \geq (k_{\text{soil}})(\text{RFs}) \geq (0.0005 \text{ cm/sec})(7.3) \geq 0.004 \text{ cm/sec}$$

This is the geotextile permeability specification.

Step 5. Check anti-clogging specifications
For nonwoven geotextiles, use the largest porosity (n) geotextile available, but not less than 30%. Recall, geotextile porosity is calculated from

$$n = 1 - \frac{m}{\rho t}$$

Step 6. Check survivability specifications
The AASHTO M288 (2017) specifications for strength survivability for Class 1 installations of nonwoven geotextiles. These criteria are reflected in Table Ex.3.1, below. The M288 specifications for all applications are in Chapter 9.

Table Ex.3.1 Geotextile property specifications

Geotextile Property	Required
O_{95}	< 0.8 mm
$k_{\text{geotextile}}$	$\geq$ 0.004 cm/sec
Porosity	> 30%
(from AASHTO M288-17) Grab strength	$\geq$ 200 lbs
Tear strength	$\geq$ 80 lbs
Puncture strength	$\geq$ 430 lbs
UV resistance at 500 hours	$\geq$ 50% strength retention

Step 7. Check durability specifications
For this application (geotextile buried under a road, no destructive chemicals, no abrasion), no additional specifications are needed for durability.
Step 8. Check miscellaneous considerations
Installation criteria must be specified.
The complete geotextile filter specification is presented in Table Ex.3.1.

3.4 SUMMARY

Geosynthetics are man-made materials devised to improve engineering properties of soil systems – strength, compressibility, permeability, and related properties. Made of plastic, they are durable, have excellent uniformity, and are inexpensive. Geosynthetics are standard materials for geotechnical projects, having engineering properties that must be evaluated before specification. The use of geosynthetics has reduced the time needed for project completion, reduced material costs, and allowed geotechnical engineers to complete projects heretofore too expensive to consider.

3.5 PROBLEMS

3.1 What is a geosynthetic?
3.2 What are the two basic structures of geotextiles? Describe each.
3.3 A long retaining wall will use geotextiles to strengthen the soil behind the wall. Should the geotextile have high strength in one direction only? Two directions? Why?
3.4 Describe the failure criteria for the wide-width test of geotextiles.
3.5 Nonwoven and woven geotextiles are used for filters.
What are the advantages/disadvantages of each?
When might you prefer one over the other?
3.6 Describe three different ways of creating the junctions in the manufacture of geogrids.
3.7 a. What was the original purpose of geocells? b. How do geocells improve soil strength? c. Name three distinct applications of geocells.
3.8 A wide-width test of a geotextile resulted in the data shown in Figure Pr.3.8
What is the ultimate wide-width strength for each test?
What is the strain at ultimate failure for each test?
What is the 10% strain strength for each test?
3.9 What are the initial tangent moduli for the data in problem 3.8?
3.10 Why isn't the Grab Tensile strength value used in reinforcement design?
3.11 What is the difference between a geotextile and a geogrid?
3.12 Why was AASHTO M288 developed?
3.13 What do transmissivity and permittivity measure? How are they different?
3.14 A geotextile is tested for its ability to transmit water. Water is run perpendicularly through the geotextile with the following results:
1 cubic foot of water ran through the specimen in 4.4 minutes.
The geotextile was nonwoven.
The square specimen was 3.5 inches on a side.
The geotextile was 0.11 inches thick.
The hydraulic gradient across the test specimen was about 1.3.
O_{95} = 0.12 mm
What is the geotextile's permittivity?

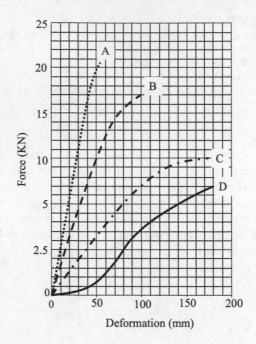

Figure Pr.3.8 Wide-width force-deformation plot for use with Problem 3.8.

3.15 In the context of geotextile filters, what's the difference between blinding and clogging? What effect does each have on geotextile filter performance?

3.16 What are the three major criteria every soil filter must meet? Describe each.

3.17 What is soil arching? How does it relate to geotextile or soil filters?

3.18 What geotextile properties are needed for a geotextile used for a filter? Be complete.

3.19 Would you specify a slit film woven geotextile for a filter? Justify your answer.

3.20 Name three civil engineering projects that require a drain and filter.

3.21 Design a geotextile to be used as part of a wall drain system. The geotextile will cover a gravel drain placed against the wall. The wall is 20 feet high, and retains soil with these properties:
$\gamma = 18.8$ kN/m^3
$n = 0.23$
$\phi = 31°$
$c = 1.44$ kPa
PI = 3
$D_r = 95\%$
and the grain size distribution is shown in Figure Pr.3.21.

3.22 Your client is constructing a four-lane highway, expected to carry over 2,000 semi-trailers/day. You've recommended gravel-filled, geotextile-wrapped edge drains, to improve drainage and, thus, lengthen the life of the road. The grain size distribution of the base soil, in which the drain will be placed, is given in Figure Pr.3.22. Specify the properties of a nonwoven geotextile filter.
$\gamma = 17.3$ kPa
$n = 0.23$
$\phi = 28°$
PI = 3
$D_r = 85\%$

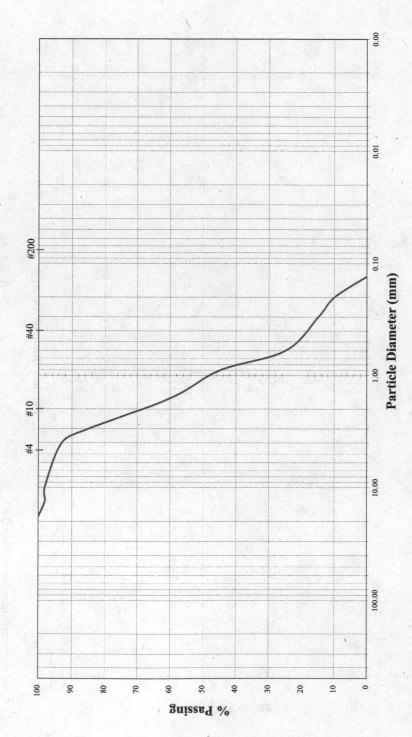

Figure Pr.3.21 Grain size distribution for problem 3.21.

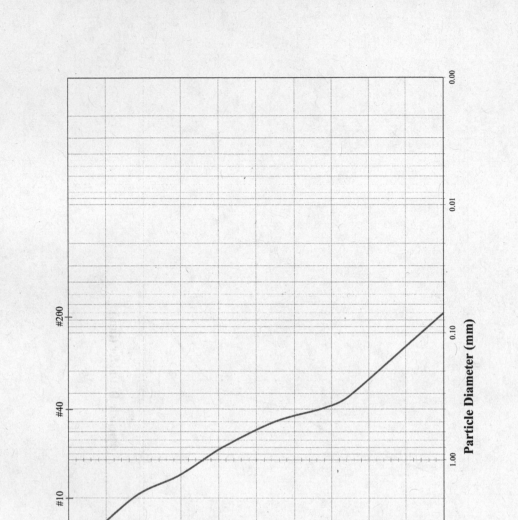

Figure Pr.3.22 Grain size distribution for problem 3.22.

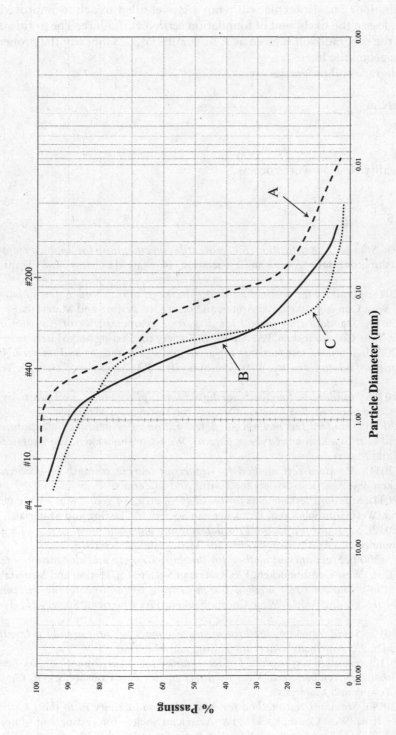

Figure Pr.3.23 Grain size distribution plot for problem 3.23.

3.23 A geotextile will be used as part of a foundation drain around a parking structure. Specify the properties of a nonwoven geotextile to filter the soil with the given grain size distribution. The geotextile will wrap a gravel-filled trench to improve drainage and, thus, lessen the likelihood of foundation settlement/failure. The grain size distribution of the subgrade soil is given as A on Figure Pr 3.23. Specify the properties of a nonwoven geotextile filter.

The subgrade soil properties are:

PI = 3

γ = 17.3 kN/m^3

D_r = 85%

ϕ = 28°

n = 0.23

3.24 What is reality? Use 50 words or less.

REFERENCES

AASHTO (2017). *Standard specification for geotextile specification for highway applications M288*. Washington, DC: American Association of State Highway and Transportation Officials.

ASTM D3080 (2011). *Standard test method for direct shear test of soils under consolidated drained conditions*. West Conshohocken, PA: American Society for Testing and Materials.

ASTM D4221 (2011). *Standard test method for dispersive characteristics of clay soil by double hydrometer*. West Conshohocken, PA: American Society for Testing and Materials.

ASTM D4354 (2006). *Standard test methods for minimum index density and unit weight of soils and calculation of relative density*. West Conshohocken, PA: American Society for Testing and Materials.

ASTM D4318 (2010). *Standard test methods for liquid limit, plastic limit, and plasticity index of soils*. West Conshohocken, PA: American Society for Testing and Materials.

ASTM D4355 (2007). *Standard test method for deterioration of geotextiles by exposure to light, moisture and heat in a xenon arc type apparatus*. West Conshohocken, PA: American Society for Testing and Materials.

ASTM D4533 (2011). *Standard test method for trapezoid tearing strength of geotextiles*. West Conshohocken, PA: American Society for Testing and Materials.

ASTM D4595 (2011). *Standard test method for tensile properties of geotextiles by the wide-width strip method*. West Conshohocken, PA: American Society for Testing and Materials.

ASTM D4751 (2012). *Standard test method for determining apparent opening size of a geotextile*. West Conshohocken, PA: American Society for Testing and Materials.

ASTM D5034-09 (2009). *Standard test method for breaking strength and elongation of textile fabrics (Grab Test)*. West Conshohocken, PA: American Society for Testing and Materials.

ASTM D5101 (2006). *Standard test method for measuring the soil-geotextile system clogging potential by the gradient ratio*. West Conshohocken, PA: American Society for Testing and Materials.

ASTM D5261 (2010). *Standard test method for measuring mass per unit area of geotextiles*. West Conshohocken, PA: American Society for Testing and Materials.

ASTM D5321 (2012). *Standard test method for determining the shear strength of soil-geosynthetic and geosynthetic-geosynthetic interfaces by direct shear*. West Conshohocken, PA: American Society for Testing and Materials.

ASTM D5567 (2006). *Standard test method for hydraulic conductivity ratio (hcr) testing of soil/geotextile systems*. West Conshohocken, PA: American Society for Testing and Materials.

ASTM D6241 (2009). *Standard test method for the static puncture strength of geotextiles and geotextile-related products using a 50 mm probe*. West Conshohocken, PA: American Society for Testing and Materials.

ASTM D6637 (2011). *Standard test method for determining tensile properties of geogrids by the single or multi-rib tensile method.* West Conshohocken, PA: American Society for Testing and Materials.

ASTM D6767 (2011). *Standard test method for pore size characteristics of geotextiles by capillary flow test.* West Conshohocken, PA: American Society for Testing and Materials.

ASTM D6913-04 (2009). *Standard test methods for particle-size distribution (gradation) of soils using sieve analysis.* West Conshohocken, PA: American Society for Testing and Materials.

ASTM D7238 (2012). *Standard test method for effect of exposure of unreinforced polyolefin geomembrane using fluorescent UV condensation apparatus.* West Conshohocken, PA: American Society for Testing and Materials.

ASTM D4491-99a (2009). *Standard test methods for water permeability of geotextiles by permittivity.* West Conshohocken, PA: American Society for Testing and Materials.

FHWA (1992). Geosynthetic design and construction guidelines: Reference manual. *Publication no. FHWA NHI-NHI-07-092*, NHI Course 132013. Washington, DC: National Highway Institute and Federal Highway Administration, U.S. Department of Transportation.

GeoFIlter (2013). GeoFilter software for geotextile filter design. http://www.tencate.com/amer/geosynthetics/design/default.aspx, June 11, 2013.

Giroud, J.P. (1988). Review of geotextile filter design criteria. In *Proceedings of the first Indian conference on reinforced soil and geotextiles, Bombay, India.* New Delhi: IBH Publishing, pp. 1–6.

Holtz, R.D. and Christopher, B.R. (1990). In remembrance of Robert J. Barrett (1924–1990), *Geotechnical News*, 8(3), 41.

John, N.W.M. (1987). *Geotextiles.* New York: Chapman and Hall.

Koerner, R.M. and Welsh, J. P. (1980). *Construction and geotechnical engineering using synthetic fabrics.* New York: Wiley, 267 pp.

Koerner, R.M. (1986). *Designing with geosynthetics* (1st ed.). New York: Prentice Hall.

Koerner, R.M. (2005). *Designing with geosynthetics* (5th ed.). New York: Prentice Hall.

Koerner, R.M. (2012). *Designing with geosynthetics* (6th ed.). Vols. 1 and 2. West Conshohocken, Pennsylvania: Xlibris.

Luettich, S.M, Giroud, J.P., Bachus, R.C. (1992). Geotextile filter design guide. *Geotextiles and Geomembranes, 11*, 355–370.

Narejo, D. (2003). A simple tilt table device to measure index friction angle of geosynthetics. *Geotextiles and Geomembranes*, Amsterdam: Elsevier, *21*, 49–57.

Rankilor, P.R. (1981). *Membranes in ground engineering.* New York: Wiley, 377 pp.

Richardson, G.N. and Koerner, R.M. (1990). *A design primer: Geotextiles and related materials.* Roseville, MN: Industrial Fabrics Association International, 104 pp.

Veldhuijzen van Zanten, R. (Ed.). (1986). *Geotextiles and geomembranes in civil engineering.* New York: Wiley.

Chapter 4

Compaction

4.1 INTRODUCTION

Compaction is the densification of soil at a constant water content. This differs from consolidation which is the densification of soil at a changing water content. During compaction, air is expunged from a partially saturated soil in response to the imparted compaction energy (normally a dynamic load). In contrast, consolidation requires the movement of water out of the consolidating soil in response to an applied stress (normally a static load). Ground improvement from the densification of soil via compaction is performed to increase strength and decrease compressibility and permeability, and, in the case of granular materials, reduce liquefaction susceptibility. Compaction can be considered "shallow" or "deep" according to the following criteria. Shallow compaction is that which occurs beneath a surface-operated compactor such as a roller or plate compactor. Deep compaction is that which occurs in the region surrounding a vibrator that penetrates the ground surface such as a vibrating probe. This distinction between shallow and deep will be made to enable discussion of compaction mechanisms and equipment. Like most classification systems, there may be some overlap in applications. Equipment can vary from small plate compactors to large vibratory rollers to large vibratory probes.

The energy applied to soils during the compaction process is termed compactive effort (or compaction energy). Compaction energy is applied to soils in several different ways. Figure 4.1 (left) shows *kneading* compaction of clayey soil with a padfoot roller at a hazardous waste site in California. Figure 4.1 (right) shows *vibratory* compaction of a sandy and gravelly soil with a smooth drum vibratory roller for a dam.

In the laboratory, impact compaction such as the standard proctor test (ASTM 2012a) is the method most commonly employed to evaluate the compaction characteristics of a soil. The standard proctor mold and hammer are shown in Figure 4.2. The energy of compaction can be varied by varying the weight of the hammer, the drop height of the hammer, the number of blows per lift, the number of lifts, and the size of the mold.

The most common form of delivering compaction energy in the field to cohesive soils is through kneading and to cohesionless soils through vibration as illustrated in Figure 4.1. Impact compaction is not generally used for shallow compaction but impact compaction is employed for deep compaction (Section 4.5).

This chapter presents the underlying soil mechanics for soil improvement by compaction and presents the means and methods by which compaction is achieved in the field.

4.2 THEORETICAL UNDERPINNINGS OF COMPACTION

Fundamentally, compaction is the densification of soil in response to the expulsion of air from void space through the application of mechanical energy at constant water content.

Figure 4.1 Pad foot kneading compactor (left) and smooth drum vibratory compactor (right).

Figure 4.2 Laboratory impact compaction hammer and mold.

For a clayey (or cohesive) soil, the resulting compacted density depends upon soil type, compaction energy, and compaction water content. These interrelationships for a clayey soil are illustrated in Figure 4.3. Notice that, for any given compaction water content, the compacted dry density increases with increasing compaction energy. The water content corresponding to the peak of each curve is deemed the *optimum water content* (OMC). Notice the optimum moisture content decreases with increasing compaction energy. The zero air voids density (ZAVD) curve defines a unique relationship between water content and dry density for any given density of solids (specific gravity). As the name implies, when there are zero air voids, all voids are filled with water and S = 100%, making further compaction impossible. The line of optimums is roughly parallel with the *zero air voids density line* (ZAVD, S = 100%). Different soils will plot at different locations in this moisture-density space, but all will plot beneath the ZAVD. Compaction of soils with water contents less than the optimum water content is termed *dry side* compaction (or "dry of optimum") and compaction of soils with water contents larger than the optimum water content is termed *wet side* compaction (or "wet of optimum").

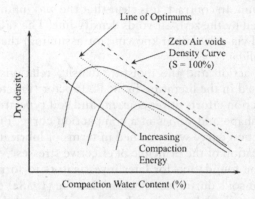

Figure 4.3 Compaction moisture-density relationships.

In the lab, two compaction energies are commonly employed (*standard proctor* and *modified proctor* (ASTM 2012b)) both utilizing impact compaction and equipment similar to that shown in Figure 4.2 (Germaine and Germaine 2009). The Standard Proctor test uses a 24.4 kN (5.5 lb.) hammer falling 300 mm (12 in.) on soil placed and compacted in three layers (lifts) with 25 blows per layer. A standard proctor-based specification is typically used for fills that will not carry loads. The modified proctor uses a 44.5 kN (10 lb) hammer falling 460 mm (18 in.) on soil placed in five layers with 25 blows per layer. A modified proctor-based specification is typically used for fills that will carry loads. In both cases, the size of the mold is 115 mm (4.6 in) high with 105 mm (4 in.) diameter. Hence the total compaction energy is 600 kN-m/m³ (12,400 ft-lb/ft³) and 2,799 kN-m/m³ (56,300 ft-lb/ft³) for the standard and modified tests, respectively. While standard and modified proctor tests are the most common, other standards have been developed for special conditions such as coarser-grained materials and kneading compaction effort. However, the above discussion forms the foundation for understanding the laboratory development of the compaction curve shown in Figure 4.3.

To relate laboratory values to values obtained in the field, *relative compaction* has been defined as a means to evaluate how a dry density measured in the field compares with that practicable for a given soil at a laboratory standardized compaction energy. Relative compaction is defined as:

$$RC(\%) = \frac{\rho_{d,field}}{\rho_{d,lab\,max}} \times 100 \tag{4.1}$$

where

RC (%) is the percent relative compaction

$\rho_{d,\,field}$ is the dry density of the soil in the field after compaction and

$\rho_{d,\,lab\,max}$ is the maximum dry density of the soil as determined in the laboratory.

An examination of the definition of relative compaction reveals it is possible to achieve values over 100% by applying greater compactive effort in the field than that applied in the laboratory. While relative compaction is important, and often specified, the compaction moisture content has a substantial impact on the resulting properties of the compacted soil, and is often specified, as will be discussed later.

It is useful to examine the shape of the moisture-density curves and the theories that explain the shape. There are multiple theories with considerable disagreement, particularly on the dry side of optimum. In contrast, it is clear that the maximum dry density on the wet side of optimum is limited by the zero air voids density line. The values associated with this upper limit are obtained via calculation knowing (or assuming) the density of solids, ρ_s (or the specific gravity of solids).

The concept of compaction and the moisture-density relationship (a.k.a. compaction curve) was first articulated in the literature by R. R. Proctor (Proctor 1933). Proctor noted that dry density, compaction effort, water content, and soil type are the four variables that determined the precise shape and values of a compaction curve. Proctor's explanation for the shape of the compaction curve was couched in terms of lubrication and put forth prior to the widespread application of the principle of effective stresses.

Proctor's work laid the foundation for later investigators to formulate explanations for the observed behavior of soils during compaction. Lambe (1958a) explained the moisture-density relationship on a colloidal chemistry basis. Lambe found that clay soils compacted dry of optimum exhibited a flocculated soil structure, whereas clay soils compacted wet of optimum exhibited a more dispersed structure. Two identical soils with identical dry densities exhibited dramatically different engineering properties (shear strength, permeability, and strength) attributable to the difference in soil structure. The difference in structure relates to the nature of the inter-particle force system. Clays, with their crystalline structure, have a net negative charge due to isomorphous substitutions which are balanced by cations on the dry clay surface. As water is added, these cations dissolve in the water creating a diffuse ion layer. During compaction at a water content dry of the soil's optimum water content, the clayey particles are more commonly oriented edge-to-face giving rise to the flocculated soil structure. As more water is added, the diffuse ion layer grows, affecting particle-to-particle interactions and resulting in a more dispersed (or oriented) structure.

The increasing density with increasing water content on the dry side of optimum can also be explained with porewater and pore air pressures (Hilf 1956; Hilf 1991). As the compaction water content increases and a greater percentage of pore space is filled with water, the sum of the capillary tension decreases rendering the soil easier to compact i.e. a greater density is achieved at any given compaction energy. This theory is consistent with the observed phenomena of a "tail" at very low water contents, where very dry soils are more readily compactable than soils having a higher water content but still quite dry compared to optimum (see Figure 4.4). At very low water contents, there is insufficient water to produce substantial capillary stresses to resist compaction.

Highlighted in the previous paragraphs are just three of the possible explanations provided explaining the shape of the moisture density relationship. Many others have contributed to our understanding of the nature of compaction of clayey soils including Hogentogler and Willis (1936), Seed and Chan (1959), and Olson (1963).

The compaction theory for cohesionless soils is different than for cohesive soils. While the foregoing discussion of theoretical underpinnings of compaction relates to cohesive soils, there are many times that it is desirable to compact sandy soils. Notably, cohesionless soils do not generally yield the pronounced moisture-density relationship shown in Figure 4.3. In fact, standard and modified compaction laboratory methods applicable to cohesive soils are not applicable to cohesionless soils. Sandy and gravelly soils (cohesionless soils) are a particulate medium with reasonably uniform dimensions in all three directions (at least compared to clays). Further, they tend to be free of electrical charges and thus free from the surface forces that dominate clay behavior. In the presence of water, capillarity plays an important role.

The engineering behavior of any given cohesionless soil is dependent upon density. Descriptors of density include "very loose" meaning the soil has a low density and high void ratio. At the other end of the spectrum, the soil may be "very dense" meaning the soil has

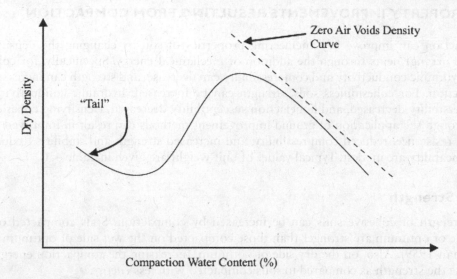

Figure 4.4 Moisture-density relationship with a dry side tail.

a high density and low void ratio. Laboratory tests have been devised to measure the lowest density state, e_{max}, and the highest density state, e_{min}. These laboratory tests rain dry soil into a mold to produce the loosest state or vibrate the dry soil to produce the densest state. Hence, the field density condition (in terms of void ratio) is compared to *both* maximum and minimum density states, rather than only the maximum density in the case of relative compaction. This comparison is termed *relative density*. Note the relative density is independent of water content. For cohesionless soils, relative density is defined as:

$$D_r(\%) = \frac{e_{max} - e}{e_{max} - e_{min}} \times 100 \tag{4.2}$$

where
D_r = the relative density, that is, the density relative to a soil's loosest and densest state
e_{max} = the maximum void ratio representing the loosest particle packing state
e_{min} = the minimum void ratio representing the densest particle packing state
e = the void ratio at the particle packing state for which the relative density is being calculated

The relative density equation can be rewritten in terms of the soil's field dry unit weight (γ_d) compared to its maximum and minimum dry unit weight as follows:

$$D_r = \frac{\gamma_{dmax} \left(\gamma_d - \gamma_{dmin} \right)}{\gamma_d \left(\gamma_{dmax} - \gamma_{dmin} \right)} \tag{4.3}$$

where
D_r = the relative density, that is, the density relative to a soil's loosest and densest state
γ_{dmax} = the maximum dry unit weight representing the densest particle packing state
γ_{dmin} = the minimum dry unit weight representing the loosest particle packing state
γ_d = the dry unit weight at the particle packing state for which the relative density is being calculated

4.3 PROPERTY IMPROVEMENTS RESULTING FROM COMPACTION

Compaction can improve key engineering properties of soil by changing the density and particle arrangements through the addition of mechanical energy. Specifically, for cohesive soils, hydraulic conductivity and compressibility can decrease, and strength can increase with compaction. For cohesionless soils, strength can be increased, hydraulic conductivity and compressibility decreased, and liquefaction susceptibility decreased. Densifying cohesionless soil through the application of ground improvement methods can result in improved lique-faction resistance, reduced compressibility, and increased strength and stability. Reductions in permeability are modest. Typical values of unit weight are given in Table 4.1.

4.3.1 Strength

The strength of cohesive soils can be increased by compaction. Soils compacted on the dry side of optimum are stronger than those compacted on the wet side of optimum (Seed and Chan 1959). Also, on the dry side of optimum, increasing the compaction energy will increase the strength as compared to soils compacted with less energy.

Compaction on the wet side of optimum produces strengths that are generally weaker than those same soils compacted on the dry side of optimum. Further, since the densification by compaction on the wet side of optimum is limited by the ZAVD, increasing the compaction energy will have little impact on the resulting soil strength.

4.3.2 Compressibility

The compressibility of soils can be decreased by compaction. For two identical samples at the same void ratio, soils compacted on the dry side of optimum are less compressible than those compacted on the wet side of optimum (Lambe 1958b) at low stress levels. One explanation for this is that the compressibility is controlled by the interparticle forces and the flocculated soil structure on the dry side of optimum is more resistant to rearrangement than the dispersed soil structure results from compaction on the wet side of optimum. However, it has also been shown (Lambe 1958b) that the reverse can be true. That is, soils compacted on the dry side of optimum are more compressible than those compacted on the wet side of optimum for samples subjected to high stress levels.

4.3.3 Hydraulic conductivity (permeability)

The hydraulic conductivity of soils can be decreased by compaction. For two identical samples at the same void ratio, soils compacted on the wet side of optimum are less permeable than

Table 4.1 Typical compacted unit weight values

Soil Classification (Unified soil classification system)	Range of values for unit weight (kN/m³)		
	Very loose	Compacted (standard)	Very dense
GW	17–19	20–22	22–23
GW-GM, GM, GW-GP, GP-GM	17–19	18–21	21–23
GP	17–18	18–20	21–22
SW	15–17	18–21	20–21
SW-SM, SP-SM, SM	13–16	18–20	19–21
SP	14–16	16–19	18–20

those compacted on the dry side of optimum (Lambe 1958b). Again, soil structure can be used to explain this observed behavior. Clayey soils compacted on the wet side of optimum have more dispersed structure than those compacted on the dry side resulting in a structure that has smaller, more uniformly distributed voids and, therefore, a lower hydraulic conductivity. Also, on the dry side of optimum, the air voids are contiguous; not so on the wet side of optimum.

4.3.4 Optimizing compacted soil properties

Consider the desirable soil properties for a homogeneous dike constructed to retain storm-water runoff including strength and hydraulic conductivity. The dike needs to be strong enough to be stable during its operational lifetime and at the same time be impermeable enough to retain stormwater. The preceding discussion of strength and permeability leads to the conclusion that compaction should be dry of optimum for the highest strength and wet of optimum for the lowest permeability. As in many engineering decisions, it is necessary to balance conflicting priorities.

Daniel and Benson (1990) developed a procedure to define an acceptable zone of compaction such that soils compacted within the acceptable zone, defined by water content and density limits, would produce strength and hydraulic conductivity values that meet the project requirements (that is assuming the project requirements are reasonable and such a zone does indeed exist). One side of the acceptable zone is defined by the ZAVD curve. As an example, for the proposed dike described earlier in this section, the approach is to define a range of compaction energies and water contents, perform permeability and strength tests on the compacted soils, and plot them along with the compaction curve and the project criteria to establish the acceptable zone.

4.4 SHALLOW COMPACTION

4.4.1 Field compaction equipment

Equipment used in the field for shallow compaction of soils can range from small plate compactors to large, self-propelled vibratory drum rollers. The key to successful ground improvement via compaction is the proper choice of compaction equipment and the proper water content. An overview of the compactor equipment choices is presented so when design considerations are presented, there is a connection between the theoretical and physical aspects of compaction.

Rammer (tamper, jumping jack) compactors are useful on any soil. They are operated by a construction worker walking behind the compactor and guiding it over the soil to be densified. These are motor driven (battery-electric, two-stroke, four-stroke, or diesel) with an eccentric flywheel that causes the compactor to "jump" and ram the soil as it comes down. These compactors work best with a thin lift (<100 mm) and with granular materials. They weigh from 70 kg to 100 kg and have plate sizes of about 0.1 m². Rammer compactors are particularly well suited for small areas such as backfilling a narrow trench.

For greater compactive effort, vibratory plate compactors can be used (Figure 4.5). These generally weigh from 100 kg to 250 kg, have a typical plate area of approximately 0.4 m², and are motor driven (four-stroke or diesel). Vibrating plate compactors, best for granular materials, may be suitable for some mixed materials having a small cohesive component. Like rammer compactors, they are operated by a construction worker walking behind the compactor. Lift thickness is increased over rammer compactors but should generally be less

Figure 4.5 Vibratory plate compactor.

Figure 4.6 Smooth drum vibratory compactor (left), and sheepsfoot compactor (right); (sheepsfoot roller photo courtesy Dimitrios Zekkos, Geoengineer.org.).

than 200 mm. As shown in Figure 4.5, plate compactors are often used adjacent to structures where heavier compactors might damage the structure.

Walk-behind compactors generally do not have the capacity to compact soils at a sufficient rate or density for large compaction projects. In these cases, self-propelled or pulled drum rollers are available in a wide range of sizes and configurations. The two self-propelled compactors shown in Figure 4.1 are a padfoot roller on the left and a smooth drum roller on the right. Not all large compactors are self-propelled. Two commonly used roller compactors are shown in Figure 4.6. On the left side of Figure 4.6 is a towed, smooth drum, vibratory roller working in a confined space. On the right side of Figure 4.6 is a towed sheepsfoot roller. It is common for larger smooth drum and padfoot compactors to be self-propelled whereas sheepsfoot rollers are typically pulled with a tracked dozer.

The compaction energy imparted to the soil increases with each pass of the roller. Different rollers impart different amounts of energy. Vibratory smooth drum rollers are commonly employed for compacting granular soils, whereas sheepsfoot rollers are best for highly plastic clays. Padfoot rollers are versatile for a wide range of soil types. For a smooth drum roller, the drum is in direct contact with the soil. The compaction stress is distributed downward from the top of the lift. The depth of compaction influence depends upon the weight, dimensions, and frequency of vibration of the roller. Larger and heavier rollers have a greater depth of influence than smaller, lighter rollers. Figure 4.7 shows a double-axis plot of both vertical dynamic pressure and soil type plotted against depth. The figure shows that, for a given depth, the energy needed for effective compaction increases as the material being compacted goes from sand to mixed sand, silt and clay, to clay. The space in the upper left-hand corner of Figure 4.7 represents low-energy vibratory compactors such as plate compactors. The figure shows these low-energy vibratory compactors are only effective with sandy materials and only to shallow depths. At the other limit, the figure shows that compactors that exert a large vertical dynamic pressure can only compact clays to limited depths compared to sands.

A sheepsfoot roller applies compaction stress to the soil from the tip of the spike (or foot) which is not initially at the top of the loose lift. Rather, the spike of the roller penetrates the loose lift deeply and compacts the lift from the bottom up. Thus, the sheepsfoot roller "walks out" of the lift as evidence of the effectiveness of compaction as schematically represented in Figure 4.8. Mechanistically, the padfoot roller works in a way somewhat between the smooth drum and sheepsfoot rollers. Both sheepsfoot and padfoot rollers function primarily as kneading compactors.

Pneumatic tired compactors are useful on many soils. They can be considered the universal compactor. Pneumatic tired compactors typically consist of an axle with several tires, side by side. Better compactors have two axles, with the tires on the front axle not aligned with the rear axle tires. This allows complete coverage with one pass of the compactor. Some pneumatic tired rollers are simply large, heavy machines with large rubber tires. The space between the rubber tires provides the kneading action needed to compact clayey soils, and the weight provides the action needed to compact cohesionless soils. The effect on cohesionless soils can be enhanced with vibration.

Grid rollers are used to compact rockfill. Grid rollers are heavy, large drums that, instead of having a smooth or sheepsfoot surface, have a large opening grid as the drum. The grid breaks the rocks, which begin to fill the spaces in the rockfill.

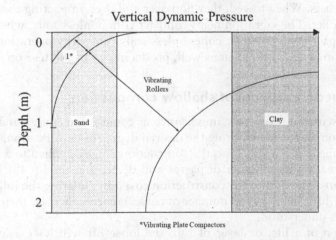

Figure 4.7 Effect of compaction energy and soil type on the effective depth of compaction.

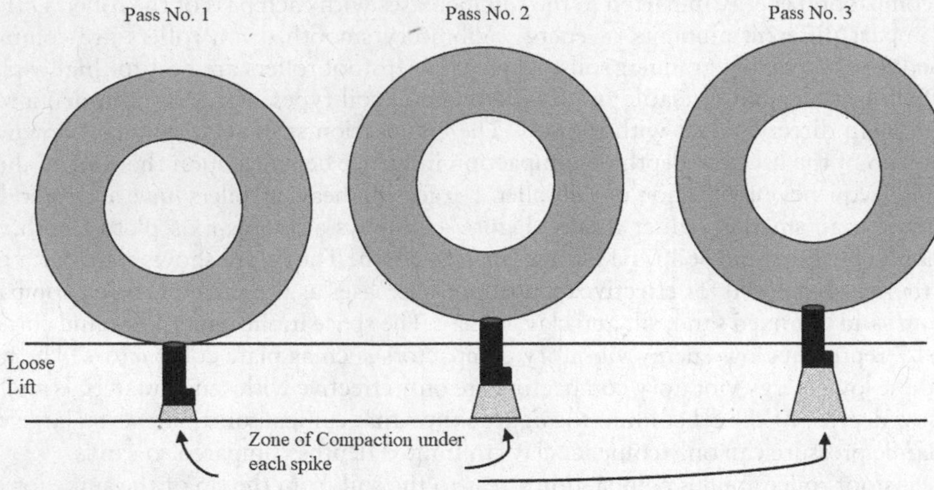

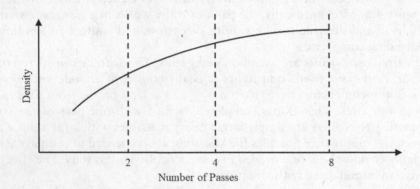

Figure 4.8 Sheepsfoot roller walks out of compacted clay layer.

Figure 4.9 Compaction density growth curve.

Impact rollers are best used on cohesionless soils. These rollers are triangular, or square, with radiused corners. When towed, they flop over and over, impacting the ground with the flat sides of the roller. The vibration helps compact cohesionless soils, where they are most effective. The impact helps compact cohesionless soils. The jerky operation of these rollers impacts the operators and equipment as well, producing wear and tear on each.

4.4.2 Construction aspects of shallow compaction

The foregoing discussion reveals the importance of compaction energy and the selection of the appropriate compactor in achieving the desired degree of densification. For a given type and weight of compaction equipment, the compaction energy applied to a given lift can be increased by increasing the number of passes and decreasing the lift thickness. However, both of these approaches add to the construction costs. By isolating the influence of each of these parameters (lift thickness and number of passes) on compaction, their combined influence can be better understood.

Upon deposition of a lift, or layer of soil, the loose lift will have a low starting density as shown in Figure 4.9. With each pass of the compactor, the density will increase.

However, the density increase (change in density) with each pass of the compactor decreases with the increasing number of passes. That is, there is diminishing effectiveness with the increased number of passes. Eventually, further passes with the compactor produce little benefit, illustrating the concept of diminishing return. The curve of increasing density with the increasing number of passes (Figure 4.9) typically levels off somewhere between four to eight passes. Only in rare cases are the desired results achieved in less than four passes. It is also rare that there is substantial benefit after more than eight passes.

The concept of depth of compaction influence shown in Figure 4.10 can be coupled with the notion that soil is placed in lifts as shown in Figure 4.10. Imagine the first lift of a cohesionless sandy soil is placed at its loose lift density. After compaction (e.g. four to eight passes), the density versus depth relationship is as shown in Figure 4.10. Notice only the lower portion of the sandy soil lifts is dense enough to meet the density requirement. For cohesionless soil, the lack of confinement at the top of the lift reduces the effectiveness of compaction. However, placing the next lift adds confinement to the previously placed and compacted lift. After compaction of this next lift, the density is increased in both the placed lift and the upper portion of the prior lift such that everywhere in the first lift, the minimum required compacted density is achieved. The process is repeated as a third lift is placed and compacted as shown in Figure 4.10. The optimum lift thickness is the thickest lift that achieves the required density throughout.

Based upon the forgoing discussion of the nature of compaction and compaction equipment, a summary of recommended compaction equipment for various soil types is presented in Table 4.2. The first two columns show the soil type as both a verbal descriptor and by the Unified Soil Classification System symbols. The third column rates the soil type for general compactability. The fourth column provides a recommendation for compactor type.

The foregoing discussion illustrates both the fundamentals and the widely applied means and methods for surface compaction and compaction of deep fills in lifts. However, there are other approaches useful for certain site-specific constraints. These techniques are used for the compaction of materials at deeper depths than possible with surface compaction

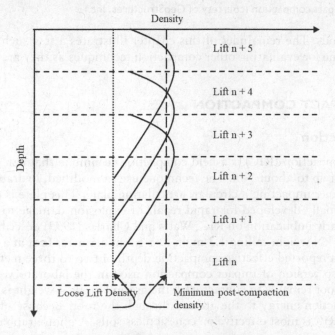

Figure 4.10 Schematic of compaction effectiveness with individual lifts.

Table 4.2 Compaction recommendations

Soil description	Unified soil classification	Compactability	Recommended compactor
Sand and sand-gravel mixtures (trace fines)	GW, SW, GP, SP	Good	Vibratory drum Pneumatic tire
Silty sand and sand-gravel mixtures with some silt	GM, SM	Good	Vibratory drum Pneumatic tire
Silty, clayey sand and sand gravel mixtures	GC, SC	Fair to good	Vibratory padfoot, vibratory sheepsfoot, pneumatic tire
Silt	ML	Poor to good	Pneumatic tire, vibratory padfoot
	MH	Poor to fair	Pneumatic tire, vibratory sheepsfoot, vibratory padfoot
Clay	CL	Fair to good	Pneumatic tire, vibratory padfoot
	CH	Poor to fair	Sheepsfoot, padfoot.
Organic soil	OL, OH, PT	Not recommended for structural fill	

Figure 4.11 Rapid impact compaction (courtesy of GeoStructures, Inc.).

means and methods. The remainder of this chapter illustrates a few such approaches and serves as a baseline for evaluating other compaction techniques as they are developed.

4.5 RAPID IMPACT COMPACTION

4.5.1 Introduction

Rapid impact compaction (RIC) is a field compaction technique that can compact soils in place to depths of up to about 7 m. The technique uses a modified, hydraulically operated, pile-driving hammer impacting a circular, articulating plate. This plate is often termed the foot. It was originally developed for rapid repair of explosion damage to military airfield runways. In an early publication on RIC, Watts and Charles (1993) describe the compactor as including a 70 kN weight falling 1.2 m on to a 1.5 m diameter foot at a rate of 40 blows per minute with a reported effective compaction depth of two to three meters. In principle, RIC is a scaled-up version of impact compaction used in the laboratory. In the RIC process, a patented foot remains in contact with the ground while a weight is rapidly dropped imparting compaction energy to the ground through the foot. Because of the large vibrations generated, RIC is most effective on cohesionless soils. A photograph of the equipment and the resulting impact crater is shown in Figure 4.11.

4.5.2 Applications

RIC has been shown to be viable for ground improvement of a wide variety of site conditions. In one study (Serridge and Synac 2006) collapsible loess, which exhibits low bearing and is susceptible to sudden collapse, was treated with RIC. The RIC program used a 70 kN hammer dropping 12.2 m onto a 1.5 m diameter steel plate. Interestingly, slightly different soil conditions in two areas of the site (termed areas 1 and 2 in this text), and trials with RIC on 1.5 m centers led to the decision to use the technique in area 1 but not area 2. Pre- and post-RIC testing showed improvement to depths up to 2.5 m in area 1. In contrast, area 2 did not show any immediate improvement attributed to excess pore pressures. Area 2 had a slightly higher plasticity and clay content than area 1. Area 1 was subsequently subjected to area-wide ground improvement using RIC on a 1.7 m grid with two passes. Shallow foundations were then used for the support of structures. In contrast, Area 2 was treated by excavating the top 2 m of soil and piles were used for the support of structures. This case study demonstrated the success of RIC in loess but also illustrated the need for site-specific trials to confirm design expectations. For this site, small differences in grain size distribution made large differences in RIC performance.

RIC has also been successfully used as a ground improvement technique for liquefaction mitigation as a case study by Kristiansen and Davis (2004) illustrates. In this case study for a multistory structure, one component of the structure was a fire hall that was deemed critical and needed to withstand a 1 in a 475-year earthquake with limited structural damage. The site was underlain by fill, and variable density granular deposits with SPT N_{60} values (ASTM D1586) less than 10 in the upper 2 m. After an evaluation of alternatives that included deep dynamic compaction and vibroflotation stone columns (both discussed later in this chapter), RIC was chosen. For this site, in situ testing demonstrated that RIC depth of influence varied in different site areas but generally ranged from 6 to 8.5 m.

4.5.3 Construction vibrations

As with many ground improvement and traditional deep foundation techniques, vibrations during construction need to be considered. Falkner et al. (2010) presented vibrations from RIC as a function of distance from the impact for a variety of soil types. Shown in Figure 4.12 is a log-log plot of maximum particle velocity as a function of distance for dense gravels, loose sandy gravels, gravelly silty sands, and sandy silts. Also shown on the plot is a horizontal line at 10 mm/s which is the standard maximum value allowed in Austria (source of the data) above which damage to structures may occur. Notice in Figure 4.12 that the loose sandy gravels show the most damping of particle velocities of the four soils shown. For these soils, the safe distance is approximately 11 m. At the other extreme, dense gravels are more efficient in transmitting energy and the resulting safe distance was found to be 34 m. The results for the other two soils lie within these limits. Also notice that the relationship between maximum particle velocity and distance is linear on a log-log plot. These results, good for design, again point to the need for field trials of RIC prior to full-scale production to confirm the site-specific particle velocities versus distance. It is also expected that the transmission of peak particle velocities would change as RIC proceeds and the site soils become increasingly compacted. Mitigation methods can include the use of shallow trenches excavated near vibration-sensitive structures to minimize the effect of vibrations from RIC upon these structures (Kristiansen and Davis 2004).

RIC is, indeed, rapid, hence, its attraction. RIC creates vibrations much larger than roller compaction. This vibration may make it unsuitable in urban, or developed environments where structures may be damaged, or humans annoyed.

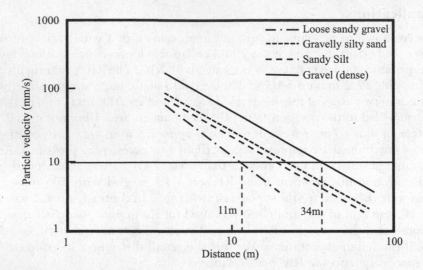

Figure 4.12 Particle velocities due to rapid impact compaction (after Faulkner et al. 2020).

4.6 DEEP DYNAMIC COMPACTION

4.6.1 Introduction

Deep dynamic compaction (DDC, or heavy tamping) is a scaled-up version of the standard proctor impact compaction. While it is reported that the technique has been used as far back as the Romans (Welsh 1986), the modern application of DDC was developed by Louis Menard in the 1970s (Menard and Broise 1975). This technique involves dropping a heavy weight of 20 kN to 500 kN from a height ranging from 10 m to 40 m. The weight is normally lifted with a crane and dropped in free fall by releasing the lifting cable. The resulting momentum of the falling weight provides the compaction energy for each drop. The location and number of drops are designed to achieve the desired compaction outcome. The process is shown schematically in Figure 4.13.

The technique is widely applicable to granular soils, rubble, and waste. DDC on finer-grained, cohesive soils is difficult as these soils generate excess porewater pressures upon impact reducing the effectiveness of compaction. Ground improvement can result in reduced compressibility, increased shear strength, reduced liquefaction susceptibility, and in cases of landfill waste, increased air space. Construction site photos are shown in Figure 4.14.

4.6.2 Design considerations for dynamic compaction

The factors that influence the effectiveness of dynamic compaction are hammer weight, the number of drops, drop height, drop spacing, and weight shape. Increasing any of these (except weight shape) increases the effectiveness. Large weights from large drop heights have compacted cohesionless soils to depths of 40 m.

There are empirical methods of evaluating the effect of these variables. These allow an estimate of the cost of dynamic compaction. Once a project has started, the number of drops and drop spacing are typically the only variables. The contractor chooses these and then evaluates the effectiveness (density increase at depth) using a probing tool, such as the standard penetration test (SPT) or the cone penetration test (CPT) which correlates to the density.

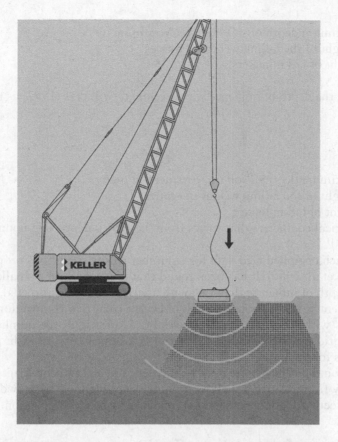

Figure 4.13 Deep dynamic compaction schematic (courtesy of Keller North America).

Figure 4.14 Deep dynamic compaction (courtesy of GeoStructures, Inc.).

One of the earliest and simplest methods to estimate the effective depth of compaction is an empirical relationship from Menard and Broise (1975) as follows:

$$D = 0.5\sqrt{WH}$$
(4.4)

where

    D is the maximum depth of soil improvement in meters
    W is the weight of the falling weight in tonnes
    H is the height of fall in meters

A variation on the depth of influence was developed by Lukas (1992) as follows:

$$D = n\sqrt{WH} \tag{4.5}$$

where

    D is the maximum depth of soil improvement in meters
    W is the weight of the falling weight in tonnes
    H is the height of fall in meters
    n is an empirical coefficient that varies from 0.3 to 1.0 depending upon site conditions

One study which compiled field data for a number of site conditions proposed an *n* value of 0.5 (Leonards et al. 1980). It has been found that a weight with a smaller base area will have a greater depth of influence than one with a larger base area (Bo et al. 2009).

Soil improvement peaks at about one-half of D. It is clear that compaction energy per unit volume increases with increasing hammer weight, drop height, the number of passes and with decreasing grid spacing. Typically, field trials are done to fine-tune the grid spacing and number of passes for the project.

Depending upon the needs of the project, the depth of soil improvement by surface compaction may vary from relatively shallow to quite deep. Figure 4.15 provides a visual comparison of the effective depth of compaction influence by roller compaction, RIC, and DDC.

4.6.3 Verification of compaction effectiveness

There are a number of ways to determine the effectiveness of DDC in the field. It is common to develop before and after pressuremeter test (PMT), CPT, and/or SPT profiles. An example

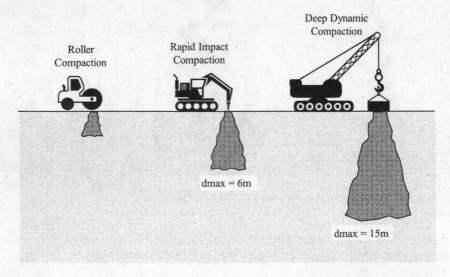

Figure 4.15 Comparison between shallow compaction, rapid impact compaction, and deep dynamic compaction.

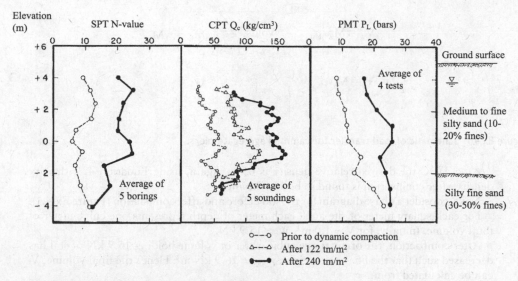

Figure 4.16 Use of SPT, CPT, and PMT to evaluate the effectiveness of deep dynamic compaction (from Mayne et al. 1984 used by permission ASCE).

of a project where all three techniques were used is shown in Figure 4.16. Using correlations to relative density, in situ test results can be used to compute before and after relative density values and to determine if the compaction produced the required density.

DDC produces craters and settlements observable at the surface and this offers an additional means by which the effectiveness of compaction can be evaluated. By careful surveying of pre- and post-compaction elevations, the total volume change can be determined. Using the volume change with a phase diagram and calculations as shown in Chapter 2, the increase in density (or decrease in the void ratio) can be computed.

Example Ex.4.1

In situ testing with the CPT has revealed an eight-meter thick layer of sand with an average relative density of 35%. To minimize liquefaction susceptibility and settlement under static load beneath the planned structure, calculations reveal it is necessary to densify the soil to a relative density of 65%. Deep dynamic compaction is being considered for this necessary ground improvement. Laboratory studies have found the $\gamma_{d\,min}$ = 14.9 kN/m^3 and $\gamma_{d\,max}$ = 18.0 kN/m^3.

QUESTION

One means of quality control is to measure the surface settlement. How much settlement due to DDC is predicted?

ANSWER

First, calculate the in situ density of the soil using equation 4.3.

$$D_r = \frac{\gamma_{dmax}\left(\gamma_d - \gamma_{dmin}\right)}{\gamma_d\left(\gamma_{dmax} - \gamma_{dmin}\right)} \tag{4.3}$$

For a relative density of 35% and the values of maximum and minimum density provided above it is found using Equation 4.3 that γ_d = 15.9 kN/m^3.

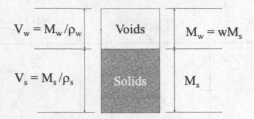

Figure Ex.4.1 Schematic of load transfer for rammed aggregate piers.

After DDC the desired relative density is 65%. Again, using Equation 4.3, the dry density after compaction is found to be 16.9 kN/m³.

Next, consider a phase diagram for the soil before and after compaction (Figure Ex.4.1).

For each square meter of site area, each meter of depth represents one cubic meter of total volume. Initially, for $V_t = 1.0$ m³, $W_s = 15.9$ kN.

After compaction, the original one cubic meter of volume holding 15.9 kN of soil has decreased such that the final unit weight, $\gamma_{dfinal} = 16.9$ kN/m³. Hence the final volume, V_f can be calculated from:

$$\frac{W_s}{V_f} = \gamma_{dfinal} \Rightarrow \frac{15.9}{V_f} = 16.9 \text{ kN/m}^3 \Rightarrow V_f = 0.94 \text{ m}^3$$

Thus, the change in volume per meter of soils is $\Delta V = V_i - V_f = 1.0 - 0.94 = 0.06$ m³

Since the area is constant the total change in height is (0.06 m/m)(8 m) = **0.48 m**.

4.6.4 Applications of deep dynamic compaction

DDC has been proven effective on a wide range of soil types including granular soil, loess, and municipal solid waste. The technique is most effective on free-draining materials and materials with large void volumes. DDC is least effective on poorly draining clayey materials.

An excellent case study on the use of DDC to densify a hydraulic fill has been provided by Bo et al. (2008). In this project, the hydraulic fill was clean sand used for land reclamation for future expansion of the Singapore Changi airport. The fill thickness was as much as 20 m. The CPT profiles conducted before and after DDC were effective in demonstrating the desired densification had been achieved. Interestingly, data on the depth of the impact crater were gathered and were normalized for a given energy and the average relationship is plotted as a function of the number of drops as shown in Figure 4.17. Note the resulting curve is the same shape as the density growth curve shown in Figure 4.9.

In one study Feng et al. (2015), DDC was effectively used to eliminate loess collapse potential. In the project, a maximum energy level of 12,000 kN-m was used to eliminate collapse potential to a depth of 10 m. Field testing included standard penetration tests and plate load tests to evaluate the final effect of the DDC. At the completion of the work, an allowable bearing value of 250 kPa was achieved for the site.

DDC has been successfully used to densify municipal solid waste (Van Impe and Bouazza 1997; Zekkos and Flanagan 2011; Zekkos et al. 2013). One of the principal goals of DDC on landfills is that by increasing the density of the waste, the storage capacity, or airspace, is increased. In one study (Van Impe and Bouazza 1997) it was shown that young landfills behaved differently than aged landfills when subjected to DDC. They found that the *n* value

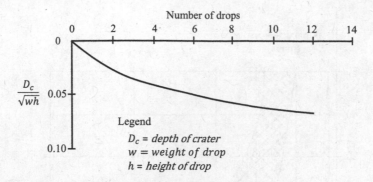

Figure 4.17 Normalized crater depth as a function of the number of drops.

of equation 4.5 had an upper limit of approximately 0.35 for aged landfills and a lower limit of approximately 0.65 for young landfills. The study also found that geophysics can be useful in evaluating the effectiveness of DDC on landfills. Specifically, the surface analysis of surface waves (SASW) method was used to determine shear wave and Raleigh wave velocities and convert these to stiffness moduli.

4.6.5 Construction vibrations

Not surprisingly, impacting the ground by dropping a large weight from a great distance produces vibrations that may damage adjacent structures. The construction vibrations produce particle velocities that decrease with distance. It is common to restrict peak particle velocities from construction vibrations to less than approximately 50 mm/s to avoid structural damage. Architectural damage can occur at peak particle velocities of approximately 10 mm/s. Equation 4.6 (after McCarthy 2002) can be used to estimate the safe distance from the point of impact during DDC as follows:

$$L = C\sqrt{E} \tag{4.6}$$

where
   L is the safe distance from the point of impact, meters
   C is a constant equal to 1.8 in SI units (5.0 in imperial units)
   E is the energy of impact in N-m (ft-lb in imperial units)

Rather than using an equation such as equation 4.6, graphical guidance is also available. Figure 4.18 shows peak particle velocities versus distance for two DDC weights and for vibrocompaction (see section 4.7). Notice for heavy weights used in DDC, architecturally damaging vibrations can occur as much as 45 m from the impact point of the weight.

4.7 DEEP VIBRATORY METHODS

4.7.1 Introduction to deep vibratory methods

Deep vibratory methods encompass a range of techniques and equipment centered on the use of a vibratory probe to densify deep, cohesionless soils or to create a void in deep cohesive materials to be filled with granular materials. Broadly, *vibrocompaction* is applied to

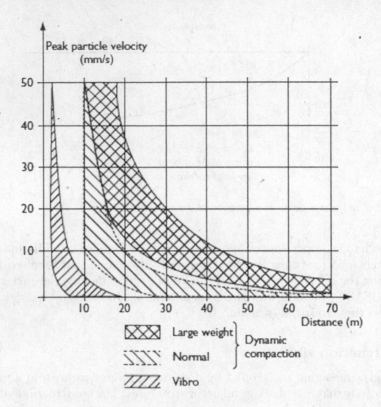

Figure 4.18 Particle velocities due to deep dynamic compaction.

deep densification of granular materials and vibroreplacement is applied to fine-grained and cohesive soils. In all cases, deep vibratory methods use a vibrating probe that is lowered into the ground, densifying cohesionless soils or creating a columnal void for replacement in cohesive soils. The development of deep vibratory compaction techniques and equipment was a key step in the evolution of many types of ground improvement.

The Keller Company in Germany originally developed the technique and the first successful (technically and economically) project for the deep compaction of sand was completed in 1937. (Kirsch and Kirsch 2016). There are a variety of "vibro" techniques, and many contractors have proprietary names for the particular version or equipment they use. This chapter will stress fundamentals and use commonly employed terminology for the methods discussed.

4.7.2 Vibrocompaction

Vibrocompaction is the deep vibratory method that is an excellent method for densifying deep deposits of cohesionless sands and gravels. Vibrocompaction is also known as vibroflotation. In so doing, liquefaction susceptibility is reduced, bearing capacity is increased, and settlement potential is decreased.

Vibrocompaction consists of a downhole vibrator held using a crane as shown schematically in Figure 4.19. As shown in this figure, the construction process employs a vibrating probe which is vibrated into the ground. The probe penetrates to the treatment depth under the actions of its own vibrations and self-weight. Penetration can be aided by the use of

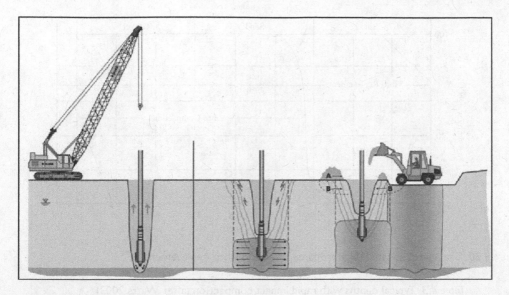

Figure 4.19 Schematic of vibrocompaction (courtesy of Keller North America).

water jetting or compressed air to assist the probe in penetrating to the desired depth. The vibrations not only cause the probe to penetrate the ground but also causes densification of the surrounding granular materials. As the soil grains are repositioned into a denser packing, void space is diminished. Above the water table, the air is expelled from the voids. Below the water table, water is expelled from the voids. The space created by the densification of the in situ materials is filled by additional granular materials as the vibrator is withdrawn. This entire process is illustrated as three steps in Figure 4.19. Variations on the technique include bottom-feed vibrators that permit the fill to be added from the bottom up rather than from the top down. For projects where the final grade is to be lower than the starting grade, additional fill material may not be needed.

The Palm Islands off the coast of Dubai is a well-known example of a project where vibrocompaction was used to compact loose sand fills (Wehr and Sondermann 2012). Without suitable means of deep compaction, the Palm Islands fills would have been unsuitably loose for their intended purpose, that is, the support of hotels and other structures planned for the islands. Vibrocompaction was used to compact the sand to a relative density of greater than 85% over a depth of 14 m. Deep deposits of loose sand represent the ideal project for the application of vibrocompaction.

The vibrations are largely horizontal. The combination of energy imparted from the vibrations and the overburden stress increased the relative density of the soil (McCabe et al. 2007). The increase in density comes from the granular particles rearranging into a denser configuration. Cohesive soils are not suitable for vibrocompaction as the energy of compaction is insufficient to overcome the cohesive strength and, if saturated, the low permeability prevents the rapid drainage needed for densification. Vibrocompaction of saturated loose sand may result in liquefaction.

The effectiveness of vibrocompaction depends on a number of factors. The most important factor is the soil (see Table 4.3). Vibrocompaction works best with relatively clean cohesionless sands and gravels with fines content less than 15% as shown in Figure 4.20. Mitchell (1981) proposes that the most desirable soils for vibrocompaction are those where the soil's fines content is limited to 18%. In addition to grain size, the effectiveness of

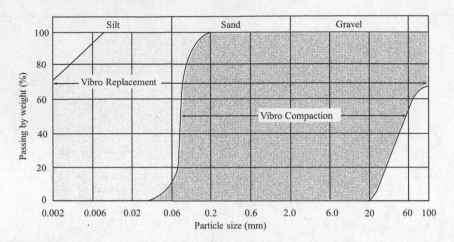

Figure 4.20 Grain sizes suitable for vibrocompaction and vibro-replacement.

Table 4.3 Typical depths with rapid impact compaction (after Watts 2003)

Ground description	Total applied energy (m-kN/m²)	Effective depth of compaction (m)
Loose building rubble	1500	4.0
Ash fill	1500	3.5
Clean granular fill	1500	4.0
Sandy silt	800	2.0
Silty sand	2000	3.0

vibrocompaction is influenced by the grain shape, distribution of grain sizes, initial density, specific gravity of particles, hydraulic conductivity, and depth due to increasing overburden stress levels with depth (Kirsch and Kirsch 2016). Finally, the position of the water table impacts vibrocompaction effectiveness. Note that Figure 4.20 also presents the appropriate grain size range for vibro-replacement discussed in Section 4.6.3.

Additional factors influencing the effectiveness of compaction include the frequency of vibration and the centrifugal force of the vibrator. The horizontal force transmitted to the ground typically varies from 150 kN to 700 kN. When unimpeded, such as when suspended in the air, the vibrator displacement is generally between 10 mm and 50 mm. Once inserted in the ground, these amplitudes are reduced by the confinement provided by the soil. Based upon field studies, the vibrators work best when the vibration frequency ranged from 20 Hz to 30 Hz (Sondermann and Wehr 2012).

As expected, the effectiveness of the vibrator in compacting the soil diminishes with distance from the probe. The result is that the effectiveness of vibrocompaction on a given site is also influenced by the probe spacing. Probe insertion points are typically 1.5 m to 3.0 m on center (Broms 1991). Common grid patterns are shown in Figure 4.21 While the triangular pattern is the most efficient for the treatment of the entire site, many times localized ground improvement is sufficient. Square patterns are commonly employed at column locations where square spread footings are planned. Linear patterns may be employed to improve the ground beneath strip footings.

Preliminary design of vibrocompaction can be done using design graphs such as that shown in Figure 4.22. From this figure, the area per compaction probe can be estimated for

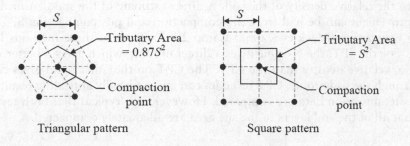

Figure 4.21 Grid patterns for vibrocompaction.

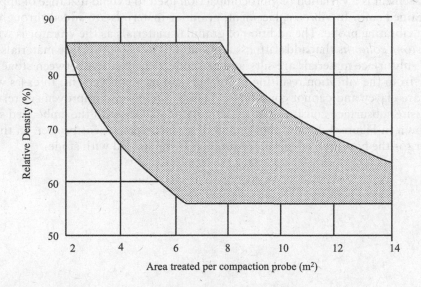

Figure 4.22 Preliminary design chart for Vibrocompaction.

a design relative density. Note that the design relative density decreases as the area per probe increases. Also, note that there is considerable latitude in the results from Figure 4.22. Most significant projects use a field test section to verify outcomes prior to beginning production vibrocompaction.

There are multiple measures of ground improvement resulting from vibrocompaction. Increased density increases shear strength and decreases compressibility allowing buildings to be supported on shallow foundations that might otherwise have required deep foundations. For areas where seismicity is a concern, densification reduces liquefaction susceptibility. By reducing liquefaction susceptibility, a number of possible failure modes are eliminated including loss of bearing capacity, excessive settlement, lateral spreading of slopes, and general loss of support for pavements, buildings, sidewalks, and landscaping.

As with most ground improvement methods, post-ground improvement verification of the effectiveness of vibrocompaction is typically undertaken. Just as with DDC, vibrocompaction resulting in soil densification results in the need to replace the lost void volume with fill soils. Both top and bottom feed vibrators use granular material to replace the volume lost during densification to maintain the site grade. By keeping track of fill used, calculations can be done to show improvement in density or relative density. In addition, both the SPT and CPT can be used for in situ evaluations of the site. Both of these in situ tests can be correlated to the relative density of the soil. A direct estimate of the soil densification from ground improvement can be had from pre-compaction and postcompaction SPT N-values or CPT tip resistances (Mackiewicz and Camp 2007). Both of these methods have their advantages. For the SPT, the N-value gives a direct number which gives a better numerical correlation to relative density than the CPT. The CPT on the other hand gives continuous readings through the soil profile. CPT test data can give a visually appealing result when the pre- and postcompaction data are compared. However, it is typical that both tests are run to ensure that all of the soil layers in the site area are adequately compacted.

4.7.3 Vibroreplacement

Vibroreplacement is a variation of vibrocompaction used to extend its range of applicability to finer-grained soils. In vibro-replacement, granular materials are added through the bottom of the vibrating probe. The addition of granular materials as the vibrator is withdrawn results in *stone columns* that add stiffness and stability to the subsurface materials. In cases where the subsurface materials are silty and or sandy, densification between stone columns will occur from the vibration resulting in further ground improvement. In cases where the materials are clayey and cannot consolidate quickly, the ground improvement results from the composite subsurface consisting of the original materials with the embedded stone columns. Shown in Figure 4.23 is a photograph of a vibroreplacement project. In this photo, the hopper for the bottom-feed vibratory probe is being loaded with stone.

Figure 4.23 Vibroreplacement project.

In principle, vibroreplacement is a ground improvement method where the improvement results from the composite behavior of the stiffer, stronger stone columns and the more compressible, weaker finer-grained soils through which they are installed. The stone columns are not structural members designed to transmit load from the structure deeper into the formation as a pile or drilled shaft would be designed. Vibrocompaction equipment can be used to construct stone columns in cohesive soils. The stone used in stone columns installed in cohesive soils generally have high friction angles and low compressibility compared to the cohesive soils in which they are installed resulting in a substantial ground improvement.

The design of vibro-replacement columns can be a complex process, leading to multiple methods of analysis in the literature. It is recommended that the underlying assumptions of the various methods of analysis be well-understood and applicable to the site conditions for which the analysis is being applied. One approach presented by Priebe (1995) allows for an illustration of the fundamentals of vibro-replacement design as well as a straightforward design procedure. In this method, it is assumed that:

1. The column is based upon a rigid layer.
2. The column material is incompressible.
3. The bulk density of both the column and soil can be neglected.
4. Soil displacement during construction produces an at-rest earth pressure condition such that $K_o = 1.0$.
5. The soil has a Poisson's ratio of $\mu_s = 0.33$.
6. The columns shear from the start while the surrounding soil behaves elastically.

The idea in Priebe's design methodology is that stone columns improve the performance of the ground as compared to the original, unimproved state. Hence an improvement factor, n_0, is introduced to permit pre-and post-ground improvement engineering evaluations. The improvement is calculated for a given grid area which may or may not be the total area of the site. Based upon the assumptions above, the basic improvement factor, n_0, can be calculated as follows:

$$n_0 = 1 + \frac{A_C}{A}\left[\frac{5 - A_C / A}{4K_{aC}(1 - A_C / A)}\right]$$

(4.7)

where

n_0 = basic improvement factor
A = total grid area
A_C = area of the columns within the total grid area
K_{aC} = the active earth pressure coefficient for the columns

The active earth pressure coefficient for the columns can be computed as follows:

$$K_{aC} = \tan^2\left(45 - \phi_C / 2\right)$$

(4.8)

where

K_{aC} = the active earth pressure coefficient for the columns
ϕ_C = friction angle of the column backfill material

Equations 4.7 and 4.8 can be used for design and alternative analyses to consider various column fill materials and area ratios to produce the needed basic improvement factor. These equations can also be used to produce a design chart (Priebe 1995) as modified and presented in Figure 4.24. Note the abscissa is the area ratio between the total grid area and the area of the columns within the grid area.

The above approach to the design of vibroreplacement columns illustrates the basic parameters controlling the design and allows for preliminary design calculations. Priebe (1995) extends this work to account for variation from the assumption that the columns are incompressible and to account for the impact of overburden pressure. Specific guidance is also available for the use of vibro-replacement to prevent earthquake-induced liquefaction (Priebe 1998). Schaefer et al. (2012) describe a comprehensive web-based system that provides additional guidance on the design and construction of vibroreplacement columns (as well as many other ground improvement techniques).

The design procedure above, and many others, do not account for the creep of cohesive soils in which the stone columns are installed. As a result, the improvement factors computed considering primary consolidation settlement (see Chapter 5) only tend to overpredict the improvement unless creep is accounted for in the design (Sexton et al. 2017). Similarly, the design procedure is not suitable to predict the time of consolidation of the clay reinforced with stone columns. It would be expected that the placement of free-draining stone columns would accelerate the time rate of consolidation. A simplified method to determine the time rate of consolidation has been developed (Han and Ye 2001) which accounts for the stress transfer from the soil to the stone column and the pore pressure dissipation in the clay. In these cases, and in general, as analyses become more sophisticated, so does the need for data to use in the analyses.

The design method just presented does not take into consideration the nature of the project to be built upon the improved ground. Embankments are one of the more common structures constructed over soft ground and supported by columns, such as those installed by the vibroreplacement technique. Due to the propensity for lateral spreading at the base of an embankment over soft ground, geosynthetics (see Chapters 3, 9, and 10)

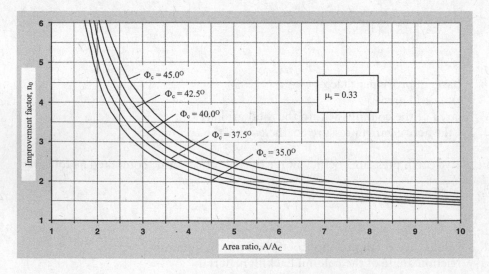

Figure 4.24 Vibro-replacement design chart.

are often used in conjunction with columns. The use of geosynthetics improves the capacity of the columns to support the embankment and adds additional considerations to the design process. Based upon the field, laboratory and analytical studies, the critical height for a column supported embankment on soft clay using geosynthetics is a linear function of the column diameter and spacing (Filz et al. 2012). The critical height is that height above which base differential settlements did not produce differential settlement at the top surface of the embankment.

The heave of soft ground during the construction of vibroreplacement columns is common and is due to the lateral displacement of the soil as the column is constructed. For most projects, this is not a problem and does not negatively impact the long-term foundation performance (Egan et al. 2008).

Vibroreplacement can be either a dry or wet process. The dry method employed to construct the columns uses the equipment shown in Figure 4.25. The equipment, suitable for subsurface conditions allowing for a stable drill hole, consists of an auger rig to drill a hole and a downhole vibrator to densify the stone placed into the hole. For the wet process, a hole is first jetted into the ground with a water jet through the vibratory probe. After reaching the desired depth and flushing fines from the hole, the stone is incrementally added and compacted by the vibratory probe.

Vibroreplacement is a tried and proven method of ground improvement. Numerous case studies found it successful in mitigating the potential for settlement (Mitchell and Huber 1985) and for liquefaction susceptibility reduction (Baez and Martin 1992; Boulanger et al. 1998; Mitchell et al. 1998; Ausilio and Conte 2007).

Figure 4.25 Schematic of aggregate piers constructed using vibration method (courtesy of Keller North America).

4.8 AGGREGATE PIERS

Aggregate piers are columns of aggregate that are installed by augering a hole typically 0.3 m to 1.0 m in diameter. Lifts of stone are then placed in the open hole and compacted. Compaction can be done using vibrators or with rams. Aggregate piers densified using vibration are known as vibrated piers. Aggregate piers densified using impact compaction are known as rammed piers. Rammed Aggregate Pier® elements are a registered trademark for impact compaction piers and are clearly differentiated from vibrated stone piers. The specific name, Vibro Piers™, is also trademarked in the United States. That said, a number of contractors have expertise in both types of aggregate piers as they are widely used worldwide. A schematic of the construction process using vibration to densify the aggregate is shown in Figure 4.25. In comparison, a schematic of the construction process using impact compaction to construct a rammed aggregate pier is shown in Figure 4.26. Also shown in Figure 4.26 is a photo of a rammed aggregate pier that has been excavated to examine the finished element.

Aggregate piers are typically installed to intermediate depths of 2 m to 8 m as they require the augered hole to remain open during the process. Vibro-replacement stone columns described in section 4.6 have no such requirement and can be installed to much deeper depths. Aggregate piers are installed to increase bearing capacity and soil modulus as well as reduce settlement and liquefaction susceptibility by creating a composite foundation material of in situ soils with stiff aggregate piers. The compaction energy causes the aggregate to bulge into the formation beyond the original bored column diameter. Aggregate piers seem most economic when the top layer of weaker soil is underlain by a stiffer, stronger layer (Handy 2001). Aggregate piers offer an economical alternative to overexcavation and replacement with compacted, load-bearing fill, and when the cost of construction is high.

Design considerations include diameter, depth, and spacing leading to a needed increase in bearing capacity, and reduction in the settlement. Phenomenologically, the compaction of the aggregate induces passive stresses in the surrounding material as shown in Figure 4.27. The impact of this is that ground improvement is achieved by both:

1. the replacement of in situ materials with a stronger, less compressible aggregate, and
2. the change in stress state from at-rest to passive in the vicinity of the aggregate pier.

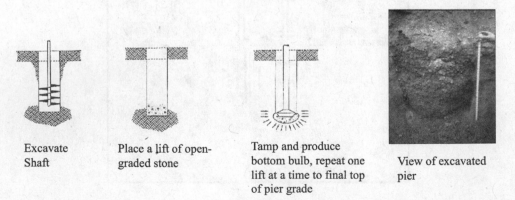

Excavate Shaft

Place a lift of open-graded stone

Tamp and produce bottom bulb, repeat one lift at a time to final top of pier grade

View of excavated pier

Figure 4.26 Schematic of rammed aggregate pier construction and exhumed pier (courtesy of GeoStructures, Inc.; www.geostructures.com).

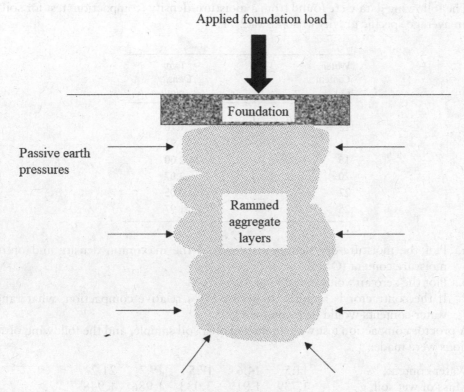

Applied foundation load

Foundation

Passive earth
pressures

Rammed
aggregate
layers

Figure 4.27 Schematic of load transfer for rammed aggregate piers.

Aggregate piers have been used on many projects worldwide. As with most ground improvement methods, case studies reveal the use of aggregate piers along with other ground improvement techniques. For example, an aggregate pier-supported embankment was constructed over soft ground using a load transfer platform composed of geosynthetics to aid load transfer from the embankment to the aggregate pier (Abdullah and Edil 2007). In another case, aggregate piers were used in a stratigraphy with peaty soils to support a railroad line (Carchedi et al. 2006). In addition to the ground improvement to support building and embankment loads, aggregate piers have been used to improve the stability of slopes (Kwang et al. 2002; O'Malley et al. 2004).

4.9 PROBLEMS

4.1 Relative density is defined in terms of void ratio in Equation 4.2. Derive an expression for relative density in terms of density.

4.2 Figure 4.9 presents a density growth curve for a typical compacted soil. Imagine compacting a rockfill where it is difficult to measure density, so, instead, you survey the top of the fill after loose lift placement and after each pass with the compactor. Sketch a plot of the relationship between the average surface elevation versus the number of passes.

4.3 For a compacted soil, given $G_s = 2.72$, $w = 18\%$, and $\gamma_d = 0.8\ \gamma_{zav}$, determine the dry unit weight of the compacted soil.

4.4 Calculate the zero-air-void unit weight for a soil (in kN/m^3) at $w = 5\%$, given $G_s = 2.68$.

4.5 The following data were found from a moisture-density (compaction) test for soil with an average specific gravity of solids of 2.71.

Water Content (%)	Total Density (g/ml)
10	1.51
13	1.70
16	1.91
18	2.00
20	2.07
22	2.05
25	1.97

    a. Plot the moisture-dry density curve. Find the maximum density and optimum moisture content (OMC).
    b. Plot the zero air voids density curve.
    c. If the contractor is required to secure 90% relative compaction, what range in water contents would be recommended?

4.6 A proctor compaction test was conducted on a soil sample, and the following observations were made:

Water content, %	11.5	14.6	17.5	19.7	21.2
Mass of wet soil, g	1,739	1,919	2,033	1,986	1,948

For a mold volume of 950 cm³ and a specific gravity of soil grains of 2.65, perform the necessary calculations and draw,
    a. compaction curve and
    b. 80% and 100% saturation lines.

4.7 A proctor compaction test has been performed on soil that has G_S = 2.68. The test results were as follows. The weight of the empty mold was 5.06 lb.

Test No.	Weight of Compacted Soil plus mold (lb.)	Moisture Content Test Results		
		Mass of Can (g)	Mass of Can + Moist Soil (g)	Mass of Can + Dry Soil (g)
1	8.73	22.13	207.51	202.30
2	9.07	25.26	239.69	225.27
3	9.40	19.74	253.90	230.64
4	9.46	23.36	250.93	219.74
5	9.22	20.28	301.47	250.95

    Plot the laboratory test results and the 80 and 100% saturation curves, then draw the proctor compaction curve. Determine the maximum dry unit weight (lb/ft³) and the optimum moisture content. Proctor molds are 1/30 ft³.

4.8 Redo Example Problem Ex.4.1 by computing the surface settlement for a target relative density of 75%.

4.9 For the soil in the table below, G_s = 2.65. Plot the proctor curve from this data. What is the OMC? What is the maximum dry unit weight? Use US customary units.

Total weight of compacted soil in 1/30 ft³ mold	Sample data for water content	
(lb)	Wet mass (g)	Dry mass (g)
3.69	353	311
3.95	344	298
4.06	383	327
4.17	376	316
4.19	345	284
4.12	416	338
4.07	381	307

a. Plot the water content versus dry unit weight curves and determine the maximum dry unit weight and optimum water content.

b. On the same graph, plot the zero air voids density curve and the curves for 90% and 80% saturation assuming $G_s = 2.65$.

4.10 A silty clay soil is planned for use in the core of an earth-fill dam. The laboratory compaction curve from problem 4.8 represents the moisture-density relationship for the soil. For a large project such as this, a method specification is required. Outline this method specification including at least five specific requirements.

4.11 Calculate the zero-air-void unit weight for a soil (in lb/ft³) at $w = 5\%$, 8%, 10%, 12%, and 15%, given $G_s = 2.68$.

4.12 For a slightly organic soil, $G_s = 2.54$. For this soil, calculate and plot the variation of γ_{zav} (in lb/ft³) against w (in percent) (with w varying from 5% to 20%).

4.13 For a compacted soil, given $G_s = 2.72$, $w = 18\%$, and $\gamma_d = 0.9\gamma_{zav}$, determine the dry unit weight of the compacted soil.

4.14 Using first principles (definitions, not derived expressions), derive the equation for the zero air voids density curve.

4.15 Vibrocompaction is planned to improve the properties of a 9 m thick deposit of loose sandy soils beneath a proposed casino for weightlifters in Atlantic City. It is necessary to increase the relative density to 85% for satisfactory building performance (it is a heavily loaded structure).

a. What probe spacing would you estimate for your preliminary design evaluation?

b. How near the adjacent casinos can you safely employ the vibrocompaction probes?

c. Laboratory data from relative density testing on this soil indicates the maximum dry density is 19 kN/m³ and the minimum dry density is 15 kN/m³. What is the minimum in situ dry density required after ground modification?

d. Consider deep dynamic densification as an alternative to vibrocompaction. Using a quantitative evaluation, will DDC achieve the required relative density?

e. What pre- during- and post-densification investigations would you recommend for this project to ensure successful performance of the new casino?

REFERENCES

Abdullah, C.H. and Edil, T.B. (2007). Behaviour of geogrid-reinforced load transfer platforms for embankment on rammed aggregate piers. *Geosynthetics International*, 14(3), 141–153.

ASTM (2012a). *ASTM D698 - 12e2, standard test methods for laboratory compaction characteristics of soil using standard effort (12 400 ft-lbf/ft³ (600 kN-m/m³))*. West Conshohocken, PA: American Society of Testing and Materials.

ASTM (2012b) *ASTM D1557 - 12e1, standard test methods for laboratory compaction characteristics of soil using modified effort (56,000 ft-lbf/ft³ (2,700 kN-m/m³))*. West Conshohocken, PA: American Society of Testing and Materials.

ASTM D1586/D1586M. (2018). *Standard test method for Standard Penetration Test (SPT) and split-barrel sampling of soils*. West Conshohocken, PA: American Society of Testing and Materials.

Ausilio, E. and Conte, E. (2007). Soil compaction by vibro-replacement: a case study. *Proceedings of the Institution of Civil Engineers: Ground Improvement, 11*(3), 117–126.

Baez, J.I. and Martin, G.R. (1992). Liquefaction observations during installation of stone columns using the vibro-replacement technique. *Geotechnical News, 10*(3), 41–44.

Bo, M.W., Na, Y.M., Arulrajah, A. and Chang, M.F. (2009). Densification of granular soil by dynamic compaction. *Proceedings of the Institution of Civil Engineers: Ground Improvement, 162*(3), 121–132.

Broms, B.B. (1991). Deep compaction of granular soils. In Hsai Yang Fang (Ed.) *Foundation engineering handbook* (pp. 814–832). Boston, MA: Springer.

Boulanger, R.W., Idriss, I.M., Stewart, D.P., Hashash, Y. and Schmidt, B. (1998). Drainage capacity of stone columns or gravel drains for mitigating liquefaction. In *Drainage capacity of stone columns or gravel drains for mitigating liquefaction* (pp. 678–690). New York, New York: American Society of Civil Engineers.

Carchedi, D.R., Monaghan, J. and Parra, J. (2006). Innovative stabilization of peat soils for railroad foundation using rammed aggregate piers. In Ali Porbaha, Shui-Long Shen, Joseph Wartman, and Jin-Chun Chai (Eds.) *Ground modification and seismic mitigation* (pp. 127–134). Reston, VA: ASCE.

Daniel, D.E. and Benson, C.H. (1990). Water content-density criteria for compacted soil liners. *Journal of Geotechnical Engineering, 116*(12), 1811–1830.

Egan, D., Scott, W. and McCabe, B.A. (2008, August). Installation effects of vibro replacement stone columns in soft clay. In *Proceedings of the 2nd international workshop on the geotechnics of soft soils*, Glasgow (pp. 23–30).

Falkner, F.J., Adam, C., Paulmichl, I., Adam, D. and Fürpass, J. (2010, June). Rapid impact compaction for middle-deep improvement of the ground–numerical and experimental investigation. In *The 14th Danube-European conference on geotechnical engineering "from research to design in European practice* (p. 10). Bratislava, Slovakia.

Feng, S.J., Du, F.L., Shi, Z.M., Shui, W.H. and Tan, K. (2015). Field study on the reinforcement of collapsible loess using dynamic compaction. *Engineering Geology, 185*, 105–115.

Filz, G., Sloan, J., McGuire, M.P., Collin, J. and Smith, M. (2012). Column-supported embankments: settlement and load transfer. In *Geotechnical engineering state of the art and practice: keynote lectures from GeoCongress 2012* (pp. 54–77). Oakland, California.

Germaine, J.T. and Germaine, A.V. (2009). *Geotechnical laboratory measurements for engineers*. Upper Saddle River, NJ: John Wiley & Sons.

Green, R.A. (2001). *Energy-based evaluation and remediation of liquefiable soils*. Ph.D. Dissertation, Virginia Polytechnic Institute and State University, 397 pp.

Han, J. and Ye, S.L. (2001). Simplified method for consolidation rate of stone column reinforced foundations. *Journal of Geotechnical and Geoenvironmental Engineering, 127*(7), 597–603.

Handy, R.L. (2001). Does lateral stress really influence settlement? *Journal of Geotechnical and Geoenvironmental Engineering, 127*(7), 623–626.

Hennebert, P., Lambert, S., Fouillen, F. and Charrasse, B. (2014). Assessing the environmental impact of shredded tires as embankment fill material. *Canadian Geotechnical Journal, 51*(5), 469–478.

Hilf, J.W. (1956). *An investigation of porewater pressure in compacted cohesive soils*. Washington, DC: Transportation Research Board.

Hilf, J.W. (1991). Compacted fill. In Hsai Yang Fang (Ed.) *Foundation engineering handbook* (pp. 249–316). Boston, MA: Springer.

Hogentogler, C.A. and Willis, E.A. (1936). Stabilized soil roads. *Public Roads, 17*(3), 45–65.

Kirsch, K. and Kirsch, F. (2016). *Ground improvement by deep vibratory methods*. Boca Raton, FL: CRC Press.

Kristiansen, H. and Davies, M. (2004, August). Ground improvement using rapid impact compaction. In *13th World Conference on Earthquake Engineering*, Vancouver, BC, Paper (No. 496).

Kwong, H.K., Lien, B. and Fox, N.S. (2002). Stabilizing landslides using rammed aggregate piers. In *Proceedings of the 5th Malaysian Road Conference*, Kuala Lumpur, Malaysia.

Lambe, T.W. (1958a). The structure of compacted clays. *Journal of the Soil Mechanics and Foundations Division*, 84(2), 1–34.

Lambe, T.W. (1958b). The engineering behavior of compacted clay. *Journal of the Soil Mechanics and Foundations Division*, 84(2), 1–35.

Leonards, G.A., Holtz, R.D. and Cutter, W.A. (1980). Dynamic compaction of granular soils. *Journal of the Geotechnical Engineering Division*, 106(1), 35–44.

Lukas, R.G. (1992). Dynamic compaction engineering considerations. In *Grouting, soil improvement and geosynthetics* (pp. 940–953). Reston, VA: ASCE.

Mackiewicz, S.M. and Camp, III, W.M. (2007). Ground modification: How much improvement? In V.R. Schaefer, G.M. Filz, P.M. Gallagher, A.L. Sehn, and K.J. Wissmann (Eds.) *Soil improvement* (pp. 1–9) Reston, VA: ASCE.

Mayne, P.W., Jones Jr, J.S. and Dumas, J.C. (1984). Ground response to dynamic compaction. *Journal of Geotechnical Engineering*, 110(6), 757–774.

McCabe, B.A., McNeill, J.A. and Black, J.A. (2007). *Ground improvement using the vibro-stone column technique*. Presented Engineers Ireland West Region and the Geotechnical Society of Ireland, NUI Galway, 15th March 2007, Institution of Engineers of Ireland.

McCarthy, D.F. (2002). *Essentials of soil mechanics and foundations basic geotechnics*. Upper Saddle River, NJ: Pearson Education.

Menard, L. and Broise, Y. (1975). Theoretical and practical aspect of dynamic consolidation. *Geotechnique*, 25(1), 3–18.

Mitchell, J.K. (1981). Soil improvement-state of the art report. In *Proceedings of the 11th International Conference on SMFE* (vol. 4, pp. 509–565). Stockholm, Sweden.

Mitchell, J.K. (1986). Ground improvement evaluation by in-situ tests. In Samuel P. Clemence (Ed.) *Use of in situ tests in geotechnical engineering* (pp. 221–236). Reston, VA: ASCE June, 1986, Blacksburg, VA.

Mitchell, J.K., Cooke, H.G. and Schaeffer, J.A. (1998). Design considerations in ground improvement for seismic risk mitigation. In *Geotechnical earthquake engineering and soil dynamics III* (pp. 580–613). ASCE.

Mitchell, J.K. and Huber, T.R. (1985). Performance of a stone column foundation. *Journal of Geotechnical Engineering*, 111(2), 205–223.

Neely, W.J. and Leroy, D.A. (1991). Densification of sand using a variable frequency vibratory probe. In M. Esrig and R. Bachus (Eds.) *Deep foundation improvements: Design, construction, and testing* (pp. 320–332). West Conshohocken, PA: ASTM International.

Olson, R.E. (1963). Effective stress theory of soil compaction. *Journal of the Soil Mechanics and Foundations Division*, 89(2), 27–45.

O'Malley, E.S., Saunders, S.A. and Ecker, J.J. (2004). Slope rehabilitation at the Baltimore-Washington Parkway with rammed aggregate piers. *Transportation Research Record*, 1874(1), 136–146.

Priebe, H.J. (1995). The design of vibro replacement. *Ground Engineering*, 28(10), 31.

Priebe, H.J. (1998). Vibro replacement to prevent earthquake induced liquefaction. *Ground Engineering*, 31(9), 30–33.

Proctor, R. (1933). Fundamental principles of soil compaction. *Engineering News - Record*, 111(13).

Schaefer, V.R., Mitchell, J.K., Berg, R.R., Filz, G.M., and Douglas, S.C. (2012). Ground improvement in the 21st century: a comprehensive web-based information system. In *Geotechnical Engineering State of the Art and Practice: Keynote Lectures from GeoCongress*, 272–293.

Schroeder, W.L. and Byington, M. (1972, April). Experiences with compaction of hydraulic fills. In *Engineering geology and soils engineering symposium, Proceedings of the 6th annual Idaho Department of Highways University of Idaho, Moscow*, Idaho State University, Pocatello.

Seed, H.B. and Chan, C.K. (1959). Structure and strength characteristics of compacted clays. *Journal of the Soil Mechanics and Foundations Division*, 85(5), 87–128.

Serridge, C.J. and Synac, O. (2006, September). Application of the Rapid Impact Compaction (RIC) technique for risk mitigation in problematic soils. In *The 10th IAEG international congress*, Nottingham, United Kingdom (pp. 1–13).

Sexton, B.G., Sivakumar, V. and McCabe, B.A. (2017). Creep improvement factors for vibro-replacement design. *Proceedings of the Institution of Civil Engineers: Ground Improvement, 170*(1), 35–56.

Van Impe, W.F. and Bouazza, A. (1997). Densification of domestic waste fills by dynamic compaction. *Canadian Geotechnical Journal, 33*(6), 879–887.

Watts, K. (2003). *Specifying dynamic compaction.* Building Research Establishment. BRE Report BR458, Garston, BRE Bookshop, UK.

Watts, K.S. and Charles, J.A. (1993). Initial assessment of new rapid ground compactor. In *Engineered fills. Proceedings of the conference, engineered fills '93.* 15–17 September, Newcastle upon Tyne.

Wehr, J. and Sondermann, W. (2012). Deep vibro techniques. In Klaus Kirsch and Alan Bell (Ed.) *Ground improvement* (pp. 28–67). Boca Raton, FL: CRC Press.

Welsh, J.P. (1986). In situ testing for ground modification techniques. In Samuel P. Clemence (Ed.) *Use of in situ tests in geotechnical engineering* (pp. 322–335). Reston, VA: ASCE. June, 1986, Blacksburg, VA.

Yoon, S., Prezzi, M., Siddiki, N.Z. and Kim, B. (2006). Construction of a test embankment using a sand–tire shred mixture as fill material. *Waste Management, 26*(9), 1033–1044.

Zekkos, D. and Flanagan, M. (2011). Case histories-based evaluation of the deep dynamic compaction technique on municipal solid waste sites. In Jie Han and Daniel Alzamora (Eds.) *Geo-Frontiers 2011: Advances in geotechnical engineering* (pp. 529–538). Reston, VA: ASCE.

Zekkos, D., Kabalan, M. and Flanagan, M. (2013). Lessons learned from case histories of dynamic compaction at municipal solid waste sites. *Journal of Geotechnical and Geoenvironmental Engineering, 139*(5), 738–751.

Chapter 5

Consolidation

5.1 INTRODUCTION

Stresses applied to soils from a building or a fill can cause unacceptable settlement under certain site, subsurface, and loading conditions. Unacceptable settlement is more likely when the load is applied to a normally or near normally consolidated soft clayey soil. When the conditions indicate settlement may be unacceptable, traditional deep foundation methods are often considered, e.g. driving piles to a deeper competent layer to support the applied load. Alternatively, ground improvement through pre-consolidation may be a viable option. Pre-consolidation is often accomplished through the application of a surcharge preload. Preloading reduces soil compressibility and increases soil strength (and thus bearing capacity) by consolidating the soft clay prior to the application of the surface load. Since the time for 100% consolidation is theoretically infinite, it is common to use a *surcharge preload* meaning the applied preload is greater than the estimated applied project load. This reduces the time required for the surcharge preload to consolidate the soil enough prior to application of the project load.

Ground improvement by consolidation is defined as decreasing the void ratio while decreasing the water content by increasing the effective stress. Decreasing void ratio translates into deformation. Depending upon the loading conditions, the deformation may be in every direction or vertical only. The vertical movement resulting from consolidation is termed settlement. All stress leads to strain, but, under some conditions, the strain, or settlement, is unacceptable. For example, a building built on soft clay might exhibit structural distress if the total or differential settlement is excessive. *Differential* settlement denotes the difference in the settlement of two locations on a structure. Similarly, a bridge approach embankment might settle excessively compared to the bridge, causing the infamous "bump-at-the-end-of-the-bridge." While any soil can theoretically be consolidated, this method of ground improvement is more likely to be cost effective for saturated soft clayey soil.

An equally important aspect of consolidation is the long-term strength gain that occurs. For instance, a tall embankment built on soft clay might apply shear stresses that would cause a shear failure of the soft clay. However, should the clay consolidate as the load is applied (slowly), the shear strength of the underlying clay will increase, mitigating the chance of failure.

While there are a number of ground improvement techniques suitable for improving the strength and compressibility of soft clays, consolidation can be used to economically reduce post-construction settlements and increase the shear strength of the underlying clayey materials. Consolidation can be induced by the application of a preload or a vacuum. A preload is simply a weight added to a site before construction and soil is the most common preload weight. When the soil is placed, the total stress and the porewater pressure are increased. Over time, the excess porewater pressure dissipates, resulting in an increase in effective stress. Vacuum consolidation (Section 5.6) also applies a load on the ground surface, increasing the effective stress.

One disadvantage of preloading soil to induce consolidation is the time it takes for the clay to consolidate. In fact, theoretically, the time for 100% consolidation is infinite. Fortunately, there is no need to design for 100% consolidation but rather design can target something less, on the order of 95% to 98%. As noted above, *surcharge preloads* are often used to reduce the time before construction. The surcharge increases the preload stress above the stress that will be caused by the design load. For instance, a preload with a surcharge of 110% of the anticipated design load will take less than 100% consolidation to achieve the consolidation expected from the design load.

While a surcharge is commonly used to speed up consolidation during preloading, there are projects when a surcharge alone is insufficient to achieve the project goals in the allotted time. In these cases, it is necessary to speed the rate of consolidation. An expedient way to speed the consolidation is to reduce the time it takes for the porewater to drain which can be accomplished by installing vertical drains. Vertical drains, constructed of either free-draining granular material or geosynthetics, reduce the length of the water drainage path and, importantly, change the preferred drainage flow direction from vertical (toward top and bottom drainage boundaries) to horizontal (toward vertical drains). Since most soils have much lower vertical permeability than horizontal permeability, this can drastically reduce the amount of time it takes for drainage to occur.

Consolidation by preloading, often with a surcharge, with or without vertical drains, can be the optimum ground improvement solution for building on compressible clays. This chapter presents the design and construction details of such systems of ground improvement. A "case" study will be used through a series of example problems to quantitatively illustrate the concepts and calculations.

5.2 CONSOLIDATION FUNDAMENTALS

Chapter 4 described compaction as densification of a soil (decreasing void ratio) at a constant water content. Implicit in this definition is the understanding that during compaction pore air is expelled, resulting in a higher density and, correspondingly, a lower void ratio. Consolidation is the densification at decreasing water content. That is, porewater is expelled, resulting in a higher density and, correspondingly, a lower void ratio. Compaction can occur quite rapidly, as pore air permeability is quite high, whereas consolidation occurs slowly as porewater permeability is relatively low due to water viscosity. Further, there is an equilibrium void ratio associated with any given effective stress for any given clay.

Figure 5.1 illustrates the consolidation (or stress-strain) behavior of soft compressible soil (soil-bentonite backfill from a slurry wall) in one-dimensional compression. In one-dimensional compression (the only case considered in this chapter), the soil is constrained in two dimensions (no strain permitted) and all strain occurs in one dimension. In most cases, one-dimensional consolidation is synonymous with vertical consolidation. On this plot, the ordinate is a measure of vertical deformation. Historically, this was the void ratio, but the vertical strain is more useful in engineering practice and shown in Figure 5.1.

The abscissa of Figure 5.1 is a vertical effective stress plotted to a log scale. Imagine a saturated clay obtained by undisturbed sampling from the subsurface environment and brought into the laboratory for one-dimensional consolidation testing. In the subsurface environment, the original vertical effective stress, σ'_{vo}, can be estimated as described in Chapter 2. For soils where the original vertical effective stress, σ'_{vo} is equal to the maximum past pressure, σ'_p, the sample is said to be *normally consolidated*. For soils where σ'_{vo} is less

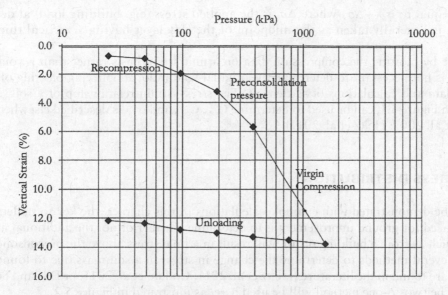

Figure 5.1 Consolidation of a soil bentonite backfill (unit strain basis)

than σ'_P, the sample is said to be overconsolidated. The degree to which a soil is overconsolidated is quantified by the overconsolidation ratio (OCR), defined by equation 5.1. For a normally consolidated soil, the OCR is unity.

$$OCR = \frac{\sigma'_p}{\sigma'_{vo}} \tag{5.1}$$

Loading a soil at stress levels below σ'_p is termed recompression (see Figure 5.1). As the applied stress approaches the preconsolidation pressure (shown approximately in Figure 5.1), the load-settlement relationship changes slope and again becomes linear at stresses greater than σ'_p. This portion of the loading curve beyond the preconsolidation pressure is called virgin compression. The change in inflection occurs at the preconsolidation pressure.

With this conceptual background, and skipping the derivation, which can be found elsewhere (e.g. Holtz et al. 2011), the settlement of an overconsolidated clay due to an applied load that results in a final stress in the normally consolidated range can be computed by equation 5.2.

$$S_{\text{ult}} = C_{r\epsilon} H_0 \log \frac{\sigma'_p}{\sigma'_{v0}} + C_{c\epsilon} H_0 \log \frac{\sigma'_f}{\sigma'_p} \tag{5.2}$$

where
S_{ult} is the ultimate settlement of the layer which occurs at a time equal to infinity
$C_{r\epsilon}$ is the modified recompression index of the layer
$C_{c\epsilon}$ is the modified compression index of the layer
H_0 is the initial height of the layer
σ'_p is the effective preconsolidation stress of the layer
σ'_{v0} is the initial effective vertical stress in the layer and
σ'_f is the final effective vertical stress in the layer

σ'_f is equal to $\sigma'_{v0} + \Delta\sigma_z$ where $\Delta\sigma_z$ is the applied stress (e.g. building load) at depth z. Depth z is generally taken as the midpoint of the clay layer having a vertical thickness H_0.

Notice by plotting the compression data on a unit strain basis rather than a void ratio basis, it is not necessary to determine the original void ratio, simplifying plotting of laboratory data and calculations of settlement. The stress–strain relationship of a soil, e.g. as shown in Figure 5.1, can be used to determine C_{re}, C_{ce}, and σ'_p as described elsewhere (e.g. Fratta et al. 2007; Holtz et al. 2011).

5.3 STRESS DISTRIBUTION

As will be demonstrated with example calculations in this chapter, the key to understanding the need for ground improvement is to determine whether or not the additional applied stress, such as from a building or fill, will result in a final stress above the preconsolidation stress. Several methods to determine the change in stress in a soil mass due to foundation loading are found in the literature (Holtz et al. 2011, Coduto et al. 2011, or Das and Sobhan 2013). The two-to-one method will be used here, as illustrated in Figure 5.2.

Referring to Figure 5.2, the applied foundation stress, $\Delta\sigma$, can be multiplied by the area of stress application, B times L, to calculate the applied load (a force). This load is divided by the area at the depth of interest, z, namely, $(B + z)$ times $(L + z)$ to arrive at the change in stress at depth z, $\Delta\sigma_z$. In equation form:

$$\Delta\sigma_z = \frac{\sigma_o BL}{(B+z)(L+z)} \tag{5.3}$$

The use of this equation is illustrated in example problem Ex.5.1. Notice in the example problem, the clay layer was treated as a single layer and the stress increase from the applied foundation load was calculated at the midpoint of the clay layer. If desired, the clay layer could have been mathematically divided into multiple sublayers, and the stress increase and settlement calculated for each of the sublayers. The sum of the settlement for each sublayer constitutes the settlement for the entire layer.

5.4 DESIGN APPROACH

The first step in determining the need for a ground improvement system for the consolidation of clay is to determine the expected consolidation settlement under the applied foundation loading. The design approach begins with constructing a design soil profile including the location of the water table, determining the necessary soil properties, determining the applied foundation load, using Equation 5.2 to predict the consolidation settlement under the applied structural loading and, finally, rendering a judgment as to the tolerance of the foundation for the predicted settlement. Information about the tolerance of structures to settlement is found elsewhere (e.g. Murthy 2002; Fang 2013; Das 2015).

Utilizing the design approach just described and employing the fundamentals and equations just presented, example problem Ex.5.1 is presented to calculate the predicted settlement of a loaded, normally consolidated clay layer. The results are then used to decide if ground improvement is needed for the conditions described in the example problem.

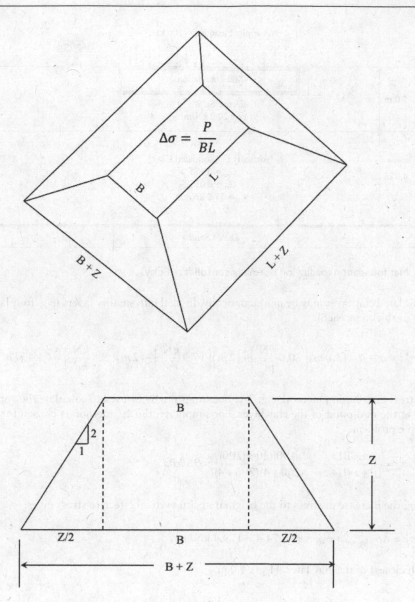

Figure 5.2 Two-to-one stress distribution.

Example problem Ex.5.1: Settlement calculation

A new project is proposed for a large structure supported on a mat foundation. The soil profile, loading conditions, and pertinent properties are shown in Figure Ex.5.1.

Problem: Determine the expected consolidation due to the application of the building load, $\Delta\sigma = 100$ kPa.

Solution:

Equation 5.2, simplified for normally consolidated clay, is:

$$S_{ult} = C_{c\epsilon}H_0 \log \frac{\sigma_f'}{\sigma_p'}$$

First, calculate the original in situ vertical effective stress due to the overburden at the midpoint of the clay layer. For more precision, or thicker layers, the 4.0 m thick normally

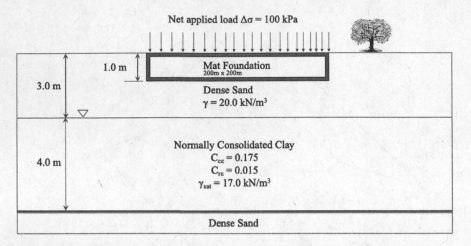

Figure Ex.5.1 Mat foundation loading on normally consolidated clay.

consolidated clay layer may be mathematically divided into smaller layers (e.g. four layers of 1.0 m thickness each).

$$\sigma'_{v0} = \sigma - \mu = (3.0\,m)\left(20.0\frac{kN}{m^3}\right) + (2\,m)\left(17.0\frac{kN}{m^3}\right) - (2\,m)\left(9.8\frac{kN}{m^3}\right) = 74.4\,kPa$$

Any stress distribution method (e.g. 2:1, Boussinesq) may be used to calculate the applied stress at the mid-point of the clay layer. For simplicity, the 2:1 method is chosen for this example problem.

$$\Delta\sigma_z = \frac{\sigma_0\,BL}{(B+z)(L+z)} = \frac{(100)(200)(200)}{(200+4)(200+4)} = 96\,kPa$$

Adding the increase in stress to the original in situ vertical effective stress gives:

$$\sigma'_f = \Delta\sigma_{\text{midpoint}} + \sigma_{v0} = 96 + 74.4 = 170.4 \text{ kN/m}^2$$

The thickness of the clay later, H_o, is 4.0 m.

$$S_{ult} = C_{c\epsilon}H_0\log\frac{\sigma'_f}{\sigma'_p} = (0.175)(4.0)\log\frac{170.4}{74.4} = 0.25\,m$$

For many (most) structures, a total settlement of 0.25 m would be unacceptable, suggesting the need for ground improvement.

5.4.1 Time rate of consolidation

Since consolidation occurs with decreasing water content which requires the porewater to drain, consolidation is time dependent. While the ultimate settlement, S_{ult}, calculated for time equals infinity, may be unacceptable, unacceptable settlement may not occur during the life of the project. If S_{ult} is unacceptable, it is also necessary to calculate S_t, the settlement that occurs at time t.

Upon initial application of additional stress to a saturated clay, it is theoretically assumed that there is an increase in porewater pressure equal to the applied stress, i.e. the porewater is considered incompressible. As the porewater pressure dissipates and the water drains toward drainage boundaries, the applied load is gradually transferred from the porewater to the soil structure increasing the effective stress. At any time prior to 100% consolidation, there is a distribution of porewater a pressures in the soil mass, with greater pore pressure dissipation nearer the drainage boundaries and the maximum residual porewater pressure at the midpoint between the drainage boundaries for a layer drained both top and bottom (doubly drained). At any given time prior to 100% consolidation, there is an average degree of consolidation, U_v, less than 100%. The subscript, v, is used to denote vertical consolidation, which comes from the assumption that the porewater only drains vertically. The settlement at any time, t, can be calculated by first calculating the ultimate settlement (see equations 5.2 and Ex.5.1) and the average degree of consolidation, as follows:

$$S_t = S_{ult} \frac{U_v}{100} \tag{5.4}$$

where S_t is the settlement of the layer at time t, and U_v is the average degree of vertical consolidation, expressed as a percent, within the layer.

In order to determine the average degree of vertical consolidation at any given time, t, Terzaghi (1925), a trained mechanical engineer, naturally adopted solutions to the problem of the time rate of consolidation from existing work on heat transfer. Essentially, time is normalized with respect to the length of the drainage path for the dissipation of excess porewater pressure, and a measure of the rate at which porewater can flow to the drainage boundary is determined, termed the coefficient of consolidation, c_v. The resulting equation is:

$$T_v = \frac{c_v t}{H_{dr}^2} \tag{5.5}$$

where T_v is the time factor for vertical consolidation, c_v is the vertical coefficient of consolidation, t is time, and H_{dr} is the length of the longest drainage path. The relationship between time factor, T_v, and average degree of consolidation is shown in Figure 5.3. Selected values of the average degree of consolidation, U_v, are shown in Table 5.1.

Laboratory methods are available for the determination of c_v (Das 2016). The results in Figure 5.3 and Table 5.1 are used to:

1) determine the time for any given average degree of consolation or
2) determine the average degree of consolidation at any given time.

Since the time required for 100% consolidation is infinity (theoretically), calculations of the time needed for consolidation under a given load are usually performed for a time equal to the life of the project.

Example problem Ex.5.2: Time required for consolidation

Problem: Using the data from Ex.5.1, determine the time required for 95% consolidation. What will be the settlement at that time? The coefficient of consolidation, c_v, for the normally consolidated clay layer is 1.0 m²/yr.

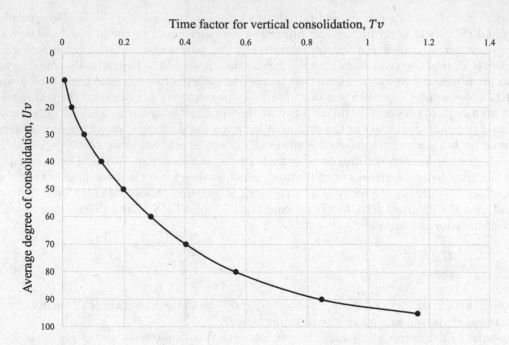

Figure 5.3 Average degree of consolidation as a function of vertical consolidation time factor.

Solution:

Using Equation 5.4, determine the time factor corresponding to 95% of time for 100% (ultimate) consolidation. Substituting into Equation 5.4, $T_v = \dfrac{c_v t}{H_{dr}^2}$, and solving for t, yields a time for 95% consolidation of 4.65 years.

Using Figure 5.3, or Table 5.1, and given the average degree of consolidation, the time factor can be determined. Equation 5.5 can then be used to compute the time for 95% consolidation. Equation 5.4 can be used along with the ultimate settlement calculated in Ex.5.1 to compute the settlement at the time for 95% consolidation.

From Table 5.1, for $U_v = 95\%$, $T_v = 1.163$.

As excess porewater can drain from the clay at both the top and bottom, the clay is said to be doubly drained. Hence, the longest drainage path, H_{dr}, is half of the thickness of

Table 5.1 Time factors for various degrees of consolidation (in percent)

U_v	T_v
10	0.008
20	0.031
30	0.071
40	0.126
50	0.197
60	0.287
70	0.403
80	0.567
90	0.848
95	1.163
100	inf.

the clay layer and here $H_{dr} = 2.0$ m. Substituting into equation 5.4, $T_v = \dfrac{c_v t}{H_{dr}^2}$, and solving for t yields a time for 95% consolidation of 4.65 years.

Using the settlement of 0.25 m computed in Ex.5.1 and an average degree of consolidation of 95%, the <u>settlement is 0.24 m at a time of 4.65 years</u>.

Under most circumstances, a total settlement of 0.24 m in 4.65 years would be unacceptable, opening up the prospect for the application of ground improvement.

5.4.2 Preloading

As shown in the example Ex.5.1, there are times when unacceptable settlement can result from the application of the applied stresses. This is commonly the occasion when the clayey soil is normally consolidated, or near normally consolidated. On these occasions, traditional deep foundation methods are often considered, such as driving piles to a deeper competent layer to support the applied load. Alternatively, ground improvement through the application of a surcharge preload may be a viable option. Clay is a *nonconservative* material; that is clay has "memory" and its behavior depends upon the stress history. As a result of this characteristic, preloading reduces soil compressibility and increases soil strength (and thus bearing capacity) by consolidating the soft clay prior to the application of the planned loading. Consolidation is a time-dependent process and preloading to the stress anticipated in the design would require the preload to stay in place until 100% consolidation is achieved. Since the time for 100% consolidation is theoretically infinite it is common to use a *surcharge preload* meaning the applied preload is greater than the applied building load. Thus instead of needing to achieve 100% consolidation, a surcharge preloading would require something less than 100%, and thus the time required by the surcharge preload to fully consolidate the soil is reduced.

A surcharge preload can be applied by building a fill over the site or by lowering the water table, both of which increase the effective stress, causing consolidation of the underlying soft clay. The details of the application of a surcharge preload are site specific and, even at a specific site, the solution is not unique. There are many combinations of surcharge loading and time to consolidation that may meet the constraints of a project.

The design procedure begins by computing the ultimate settlement of the proposed structure as shown in example Ex.5.1. The next steps depend upon which variable will be controlled (time available or height of fill). For example, if the time available is constrained, the time factor for vertical consolidation, T_v, can be computed using the time available. With T_v known, the average degree of consolidation, U_v, can be computed (see example problem Ex.5.2). Knowing both U_v and S_{ult}, the ultimate settlement of the preload can be calculated as follows from a reconfigured version of Equation 5.4 presented here as Equation 5.6:

$$S_{ult(pl)} = \frac{S_{ult}}{U_v/100}$$

(5.6)

where
   $S_{ult(pl)}$ is the settlement in the layer due to the preload,
   S_{ult} is the settlement in the layer due to the applied building load, and
   U_v is the average degree of vertical consolidation in percent within the layer.

Once the settlement of the preload for a given time is computed, the settlement equation (Equation 5.2) can be used to compute the required applied preload stress, $\Delta\sigma_{pl}$. The final step in the calculations is to compute the height of fill needed to apply sufficient stress to produce the necessary settlement under the preload. The height of fill having a unit weight of γ_{fill} can be calculated as:

$$H_{fill} = \frac{\Delta\sigma_{pl}}{\gamma_{fill}}$$

(5.7)

Alternatively, the designer may select a reasonable fill height for the site and calculate the time this fill will take to achieve the required consolidation. After choosing the fill height, the ultimate settlement of the surcharge preload and applied building load can both be computed (Equation 5.2) followed by the average degree of consolidation (Equation 5.6). Once the necessary average degree of consolidation is computed, the time factor can be determined (Figure 5.2) and the time to achieve the necessary settlement can be computed (Equation 5.5). This time is compared to project constraints to determine if the chosen surcharge preload is feasible.

Example problem Ex.5.3: Surcharge fill height

Problem: For the example problem in Ex.5.1 and Ex.5.2, determine the height of the surcharge preload, of 20 kN/m³ soil, needed to produce a total consolidation settlement of 0.38 m (see Ex.5.1) in 90 days (0.25 years), a project constraint.

Solution: Employing Equation 5.5 and the allotted time of 0.25 years yields:

$$T_v = \frac{c_v t}{H_{dr}^2} = \frac{(1.0)(0.25)}{2^2} = 0.0625$$

Using the computed time factor of 0.0625 and Figure 5.3, or interpolation of Table 5.1, the average degree of consolidation is 28%. Hence, a large enough surcharge preload must be applied to produce 28% of the ultimate settlement of the preload equal to 0.25 m, in 90 days. This concept is expressed in Equation 5.6 as follows:

$$S_{ult(pl)} = \frac{S_{ult}}{U_v/100} = \frac{0.25}{28/100} = 0.893$$

Knowing the ultimate settlement caused by the surcharge preload, the stress needed to produce the settlement is computed. Equation 5.2 can be simplified for normally consolidated clay as:

$$S_{ult} = C_{ce}H_0 \log\frac{\sigma_f'}{\sigma_p'}$$

Solving for σ_f' using the parameters previously provided yields $\sigma_f' = 1404$ kPa.

Finally, convert the needed applied stress to a height of fill using Equation 5.7. Given a unit weight of 20 kN/m³ for the surcharge fill, yields:

$$H_{fill} = \frac{\Delta\sigma_{pl}}{\gamma_{fill}} = \frac{1404}{20} = 70.2 \text{ m}$$

For most projects, a surcharge fill height of 70.2 m is not feasible.

As depicted in Ex.5.3, there are site and project conditions that preclude the use of a surcharge preload without additional steps to speed the rate of consolidation. However, there are many sites where a surcharge preload may be cost-effective. For example, sites with sands interbedded within the clay layer will consolidate much more rapidly than sites with thick, homogeneous, clay layers.

5.5 SPEEDING CONSOLIDATION WITH VERTICAL DRAINS

5.5.1 Introduction

Vertical drains can be used to speed consolidation for site and subsurface conditions that preclude the use of a surcharge preload without additional steps to speed the rate of consolidation. The purpose of vertical drains is to decrease the length of the drainage path, H_{dr}, increasing the rate of consolidation. Since H_{dr} is in the denominator of Equation 5.4, and is squared, any decrease in the length of the drainage path has a compounding effect on the increase in the rate of consolidation. Historically, vertical drains were sand-filled boreholes i.e. *sand drains*. Sand drains have been in use for nearly 100 years with credit for the idea given to Daniel E. Moran of Moran, Proctor, Mueser and Rutledge consulting engineers of New York, New York, USA (Rutledge and Johnson 1958). Moran was granted a patent with the interesting title "Foundations and the Like" for the technique in 1926 (Moran 1926). The first use of the technique was the approach embankment for the San Francisco-Oakland Bay bridge. The laboratory and field studies for this project were published in the 1st International Conference on Soil Mechanics and Foundation Engineering in 1936 (Porter 1936). While sand drains have been successfully employed on countless projects, they require excavation of relatively large quantities of clay and replacement with similarly large quantities of drainage sand and therefore can be quite costly. This cost limitation gave rise to alternative drainage materials including tubes of wood fiber, cardboard *wick drains* (Kjellman 1939) and eventually geosynthetic wick drains, more descriptively called *prefabricated vertical drains* (PVDs).

Figure 5.4 is a two-dimensional schematic of the deployment of vertical sand drains to speed consolidation. Once drains are installed, the longest drainage path for consolidation water is horizontal and equal to one-half of the drain spacing. Hence the time required for consolidation can be significantly reduced.

5.5.2 Consolidation with vertical drains

The drainage path for the consolidation of clays containing vertical drains is primarily horizontal in contrast to consolidation without vertical drains which is primarily vertical. The coefficient of vertical consolidation characterizing the rate of consolidation (c_v in Equation

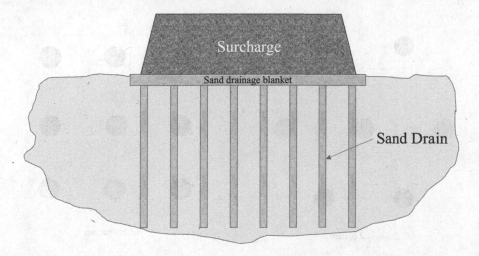

Figure 5.4 Schematic of a sand drain to speed consolidation settlement.

5.4) is most likely much smaller than the coefficient of horizontal consolidation, c_h. This is in part because clayey strata often contain both continuous and discontinuous horizontal sandy and silty layers resulting in a coefficient of hydraulic conductivity that is larger in the horizontal direction than in the vertical direction. In addition, the depositional environment often naturally results in clay particles being horizontally oriented. The substantially shorter drainage distance (exact distance depends upon the vertical drain spacing and layout) coupled with the larger coefficient of consolidation results in substantially shorter consolidation time. In the case of drains, horizontal flow is sometimes termed radial flow.

The two commonly used configurations for the deployment of vertical drains are shown in Figure 5.5.

For the drain layout patterns in Figure 5.5,

$$d_e = 1.06(s) \text{ (for triangular pattern of drain spacing)}$$

$$d_e = 1.13(s) \text{ (for square pattern of drain spacing)}$$

where S is the spacing between the drains.

With the installation of vertical drains, the site soil can drain vertically *and* horizontally. The time rate of vertical consolidation and the governing equations were presented in Section 5.2. Considering the deployment of vertical drains, horizontal drainage can best be analyzed as radial drainage to the vertical drains. The following equations are used for the calculation of the average degree of consolidation for radial drainage (Hausmann 1990; Yeung 1997; Hansbo 1979; Koerner 2012).

$$U_r = 1 - \exp\left(-\frac{8T_r}{\alpha}\right) \tag{5.8}$$

$$\alpha = \frac{n^2}{n^2 - s^2}\ln\left(\frac{n}{s}\right) - \frac{3n^2 - s^2}{4n^2} + \frac{k_h}{k_s}\left(\frac{n^2 - s^2}{n^2}\right)\ln s \tag{5.9}$$

$$T_r = \frac{c_h t}{D_e^2} \tag{5.10}$$

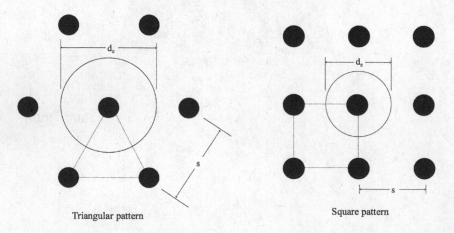

Triangular pattern

Square pattern

Figure 5.5 Plan view patterns for vertical drain layouts.

$$n = \frac{d_e}{d_w} \tag{5.11}$$

$$s = \frac{d_s}{d_w} \tag{5.12}$$

$$d_w = 2(B + t_d) / \pi \tag{5.13}$$

where

U_r is the average degree of consolidation due to radial drainage

T_r is the radial consolidation time factor

c_h is the horizontal coefficient of consolidation

t is the elapsed time since the application of the load

α is a lumped parameter accounting for the geometry of the vertical drainage system as well as the effect of smeared soil around the drain

d_e is the equivalent diameter of soil around each vertical drain

d_w is the diameter of the sand drain or equivalent diameter for a prefabricated vertical drain

d_s is the equivalent diameter of the smeared zone

k_h is the horizontal hydraulic conductivity of the soil

k_s is the hydraulic conductivity of the smeared soil around the drain

n is the vertical drain ratio

B is the width of the wick drain

t_d is the thickness of the wick drain

Note that since prefabricated vertical drains are not circular, the equivalent diameter, d_w, is calculated using the width and thickness of the drain as shown in Equation 5.13.

As can be seen from Equations 5.8 through 5.13, there are a number of variables to be considered in the design of a vertical drain system to speed consolidation under a surcharge preload. Most often, there is a time constraint so the amount of time available for consolidation is fixed in the design process. This fixed time, along with an assumption of a drain spacing, allows for the computation of the time factor in Equation 5.10. If the height of surcharge is constrained by project and site conditions, the average degree of consolidation may be computed using Equation 5.4. At this point in the design process, α can be computed from Equation 5.9. The designer next needs to consider PVD dimensions, the extent of soil smear, and possible PVD spacings to come up with a final design that satisfies equilibrium in Equation 5.8. The following example computation demonstrates one approach to the design process and is a continuation of the case study in example problems Ex.5.1 through Ex.5.3.

Example problem Ex.5.4: Sand drains

Problem: For the data given in example problems Ex.5.1, 5.2, and 5.3, design a sand drain system using 0.25 m diameter sand drains. Assume sand drains are freely drained top and bottom. The maximum fill height specified for the site is 6 m and the unit weight of surcharge preload fill is 20 kN/m³. Assume radial drainage controls so vertical drainage is conservatively neglected. For this case study, $c_h = 10$ m²/yr, $d_s = 0.3$ m, $k_s = 1 \times 10^{-8}$ m/s and $k_h = 1 \times 10^{-7}$ m/s. Using a triangular pattern for vertical drain layout, determine the spacing of the vertical drains needed to produce a total consolidation settlement of 0.25 m (see Ex 5.1) in the 90 days available within project constraints.

Solution:

The first steps are to compute the applied surcharge load followed by a calculation of the ultimate settlement under the applied surcharge preload.

$$\sigma'_f = \Delta\sigma_{\text{preload}} + \sigma'_{v0} = \gamma H_s + \sigma'_{v0} = 20(6) + 74.4 = 194.4 \text{ kN/m}^3$$

$$S_{\text{ult}} = C_{ce}H_0 \log \frac{\sigma'_f}{\sigma'_p} = 0.175(4.0)\log \frac{194.4}{74.4} = 0.29 \text{ m}$$

Knowing the ultimate settlement under the surcharge preload and the ultimate settlement expected under the building load allows for the determination of the average degree of consolidation in percent needed under the preload. Rearranging Equation 5.4 yields:

$$U_{vt} = \frac{S_{\text{ult}}}{S_{\text{ult(pl)}}} \times 100 = \frac{0.25}{0.29} \times 100 = 86\%$$

Next, choose a vertical drain spacing, d. Try 1.5 m in a triangular pattern. The equivalent diameter of soil around each vertical drain can then be computed as:

$$d_e = 1.06d = 1.59 \text{ m}$$

Using Equation 5.10 and allotted time of 0.25 years yields:

$$T_r = \frac{c_h t}{d_e^2} = \frac{(10)(0.25)}{1.59^2} = 0.99$$

To calculate the average degree of radial consolidation (Equation 5.8), first calculate the parameter α (Equation 5.9) which is a function of n and s (Equations 5.11 and 5.12).

$$n = \frac{d_e}{d_w} = \frac{1.59}{0.25} = 6.36 \text{ and } s = \frac{d_s}{d_w} = \frac{0.3}{0.25} = 1.2$$

Solving Equation 5.9 produces $\alpha = 1.006$. Then, solve equation 5.8 to find $U = 0.9996$, meaning the soil would be 99.96% consolidated. This being very close to 100%, suggests the assumption of a drain spacing of 1.5 m is too conservative.

To speed the design process, these variables and equations can be programmed using, for example with MATLAB or Excel, so that the various design choices and their influence on the outcome can be readily investigated. With these calculations, the effect of drain spacing on the average degree of radial consolidation is shown in Figure Ex.5.2.

Example problem Ex.5.4 illustrates the design process with the assumption of exclusively horizontal flow to the vertical drains to speed consolidation. This is often a reasonable, and definitely conservative, assumption, especially for preliminary design calculations. That said, vertical and horizontal consolidations are occurring simultaneously. This effect can be accounted for by using an average degree of consolidation found from:

$$1 - U_{vr} = (1 - U_v)(1 - U_r) \tag{5.14}$$

where
   U_{vr} is the combined vertical and horizontal average degree of consolidation
   U_r is the average degree of radial consolidation
   U_v is the average degree of vertical consolidation

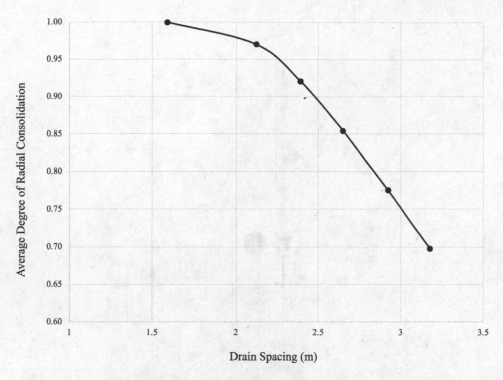

Figure Ex.5.2 Effect of drain spacing on average degree of combined horizontal and vertical consolidation at t = 0.25 years for example problem Ex.5.4.

Using Equation 5.14 and the previously defined equations and processes for U_{vr} and U_r, the average degree of combined vertical and radial consolidation can be computed.

5.6 ADDITIONAL VERTICAL DRAIN CONSIDERATIONS

The foregoing sections provide the fundamentals of ground improvement via consolidation including the use of preloads. Some additional considerations as discussed in this section.

5.6.1 Vertical drain types

The earliest vertical drains were largely sand-filled auger holes made by displacement methods such as a mandrel or jetting. Typical installations were 200–450 mm diameter drains at spacings of 1.5 m to 6 m on center. Typical layout patterns are in Figure 5.5. Notice the columns are sand filled rather than stone filled as they are meant to facilitate drainage and consolidation rather than serve as foundation elements. The applied load could be transferred to stronger stone columns, ameliorating some of the benefits of the columns. Sand drains are a tried and proven method as illustrated by the Kansai International Airport constructed in Osaka Bay where 2.2 million vertical sand drains up to 25 m long were used (Mesri and Funk 2015). Construction is straightforward as shown in Figure 5.6. The mandrel is driven or vibrated into the ground to displace the existing soil and the mandrel casing is withdrawn as the space created is filled with drainage sand. Regardless of whether the sand drain is constructed using augers or displacement methods, smearing of the formation

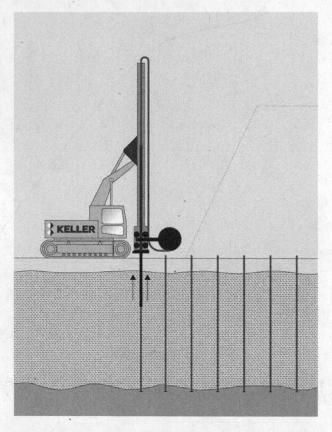

Figure 5.6 Sand drain installation using a mandrel (courtesy Keller North America).

Figure 5.7 PVDs from around the world.

soils is inevitable as is discussed later. Despite the long history of sand drains, PVDs have largely replaced sand drains as a means of speeding consolidation.

PVDs are commonly the method of choice over sand drains for the installation of a vertical drainage system to speed consolidation. PVDs are also known as wick, band, and strip drains. PVDs are a geotextile wrapped plastic or cardboard core that facilitates drainage along the drain axis. The geotextile serves as a filter, allowing water flow while precluding the movement of formation soils. Because sand drains require drilling into the soil followed by disposing of the auger spoil then filling the hole with sand, installation of PVDs can be installed much faster and, normally, at a lower cost. Installation can be quite rapid, about 1 m/s. Typical spacings are 1.5 m on center or less. While Figure 5.7 shows different PVD core designs, the principle remains the same. Generally, PVDs are about 100 mm in width and

Table 5.2 Typical requirements for PVDs (After Hayward Baker 2007)

Specifications for the PVD Assembly		
Discharge capacity	≥ 6 l/min.	ASTM D4716
Tensile strength	≥ 2 kN	ASTM D4595
Specifications for the Geotextile		
Grab tensile strength	≥ 0.6 kN	ASTM D4632
Puncture resistance	≥ 0.2 kN	ASTM D4833
Permittivity	≥ 0.7 s^{-1}	ASTM D4491
Apparent Opening Sizes (AOS)	70 sieve	ASTM D4751

from 3 mm to 6 mm thick. The core as well as the geotextile are flexible and do not impede the downward movement of the formation soil during consolidation, although, with large settlements, the core may kink, impeding drainage.

While project-specific requirements may vary, it is common to specify the discharge capacity and tensile strength of the PVD as well as the grab tensile strength, puncture resistance, permittivity, and apparent opening size (AOS) of the geotextile. Typical values for these parameters are shown in Table 5.2. These geosynthetic parameters are defined and discussed in Chapter 3.

5.6.2 Effect of PVD installation patterns

For a triangular pattern (see Figure 5.5), the radius of a soil cylinder discharging water into the PVD is $0.525(D_s)$ where D_s is the drain spacing. Radial consolidation theory (Section 5.4.1) assumes the drain to have a circular cross section. The PVD does not. As in Equation 5.14, an equivalent radius, r_w, for the PVD as is calculated from:

$$r_w = 2(B+T)/\pi \tag{5.15}$$

where B and T are the width and thickness of the PVD, respectively.

Since PVDs are small in the cross section relative to a sand drain, it is necessary to determine if the PVD can effectively discharge the drainage water. The discharge capacity, q_w, of the PVD can be computed as:

$$q_w = \pi r_w^2 k_w \tag{5.16}$$

where

r_w is the equivalent radius
k_w is the hydraulic conductivity of the PVD

With these data, a discharge factor, D, is calculated that permits assessment of the resistance to flow within the PVD and whether or not that resistance is negligible. The discharge factor (Terzaghi et al. 1996) can be computed thus:

$$D = q_w / k_h l_w^2 \tag{5.17}$$

where

k_h is the horizontal hydraulic conductivity of the consolidating layer
l_w is the maximum drainage length within the PVD (e.g. for single drainage upward l_w would be the length of the PVD)

It is recommended that the discharge factor be greater than five in order for the resistance to flow in the PVD to be negligible (Mesri and Lo 1991).

5.6.3 Effect of soil disturbance (smear)

The effective functioning of a vertical drain requires an undiminished pathway for flow between the soil and the drain. While the vertical drain is installed to speed lateral drainage in the formation, the very act of installing the drain may alter the drainage characteristics of the formation at the drain interface. The soil disturbance reducing the transmissivity at the interface is termed *smear*. It is essential to quantify this effect before the design is done (Basu and Prezzi 2007). The extent to which smear has an impact is also dependent upon the type of vertical drain. Current installation techniques for PVDs may not have a significant smear zone compared with augered installation methods.

There are a number of published methods dealing with the smear zone during design (Indraratna and Redana 1998 and 2000; Rujikiatkamjorn and Indraratna, B. 2009; Chai et al. 1997). In one method, the soil disturbance during the installation of the drain can be considered to be two separate zones. The first smear zone is the soil that is completely remolded and is adjacent to the drain. The second is the transition zone, the disturbed soil between the undisturbed formation soil and the completely remolded smear zone (Bergado et al. 1996). In one of the earlier approaches (Hansbo 1981), smear is accounted for by introducing additional relevant parameters that can be computed and then incorporated into the radial drainage equation. Specifically, to calculate the average degree of consolidation at depth z using equation 5.8, α is replaced by α_s to incorporate the effects of smear. The revised Equation 5.19, accounting for smear is:

$$\alpha_s = \ln\frac{n}{m} + \frac{k_c}{k'_c}\ln m - \frac{3}{4} + \frac{\pi z(2l_w - z)k_c}{q_w} \tag{5.18}$$

where
$n = r_e/r_w$
$m = r_s/r_w$
r_s = diameter of the smeared zone
k_c = horizontal permeability of the undisturbed consolidating soil
k'_c = horizontal permeability of the smeared soil
k_w = axial permeability of the drain
A_w = cross sectional area of the drain
$q_w = k_w A_w$

The difficulty in accounting for the smear zone is not in the use of Equation 5.19 but rather in the selection of the input parameters such as the diameter of the smeared zone and the horizontal permeability of the smeared zone. Short of site-specific data, some guidance is available in the literature (Sathananthan and Indraratna 2006; Sharma and Xiao 2000).

Case studies associated with the use of vertical drains are helpful in understanding the limitations of predicting, a priori, the field performance of the design. Monitoring the preload surcharge settlement allows back calculation of soil parameters during loading. These improved parameters allow an improved prediction of the time for a given settlement.

5.7 VACUUM CONSOLIDATION

Vacuum consolidation or vacuum preloading uses atmospheric pressure to provide the temporary surcharge needed to consolidate soft soils. The idea was first described in 1952

(Kjellman 1952) and is illustrated in Figure 5.8 from the original patent. The concept is straightforward and includes vacuum piping installed beneath an airtight cover. The air beneath the airtight cover is extracted using a vacuum pump resulting in the application of atmospheric pressure to the top of the cover, effectively surcharging the soil. Each of these features is illustrated in Figure 5.8 from the original Kjellman patent. Interestingly, the patent suggests the airtight cover may be of hot asphalt, a layer of ice, sheet rubber, sheet plastic, or impregnated cloth, paper, or cardboard. Limitations in materials for the airtight cover suitable to maintain a vacuum limited the use of the technique.

With developments in geosynthetics, the technique began to show promise and now has been used on a number of projects (Holtz and Wager 1975; Chen and Bao 1983; Bergado et al. 1996; Chu et al. 2008; Indraratna et al. 2005). The technique is practically limited to a vacuum pressure of approximately 80 kPa which is the equivalent load of about 4.3 m of soil. For some projects, the vacuum preloading method can be cheaper and faster than traditional methods using soil cover to preload the underlying soft soils. For example, vacuum preloading may be the method of choice where the placement of a surcharge may not be feasible due to access constraints.

While the original work by Kjellman envisioned sand drains, PVDs (Section 5.6) have been found to be effective. While vacuum consolidation has wide application over a variety of soft and compressible fine-grained soils, it is particularly advantageous for the consolidation of clay slurries that are too soft/weak for a surcharge (Chu et al. 2008).

The details of the deployment of vacuum consolidation systems vary. The most common airtight cover is a geomembrane, often made of PVC. Commonly, a sand layer beneath the geomembrane is used to uniformly distribute the vacuum pressure horizontally over the entire area to be treated. The system can be enhanced by installing PVDs to aid in the vertical distribution of vacuum pressure (Chu et al. 2008).

The nature of the subsurface may limit the applicability of vacuum consolidation systems. For example, if the subsurface contains an interbedded zone of high permeability that extends beyond the area planned for vacuum consolidation, there is a reduction in the effectiveness of the system. Essentially, such subsurface conditions allow "leaks" making it difficult to maintain the vacuum. If appropriate, a slurry trench cutoff wall could be installed around the perimeter to control the vacuum losses. This approach proved successful for vacuum consolidation of the Yaoquing Airport Runway (Tang and Shang 2000).

Oct. 28, 1952   W. KJELLMAN   2,615,307

METHOD OF CONSOLIDATING SOILS

Filed April 29, 1948

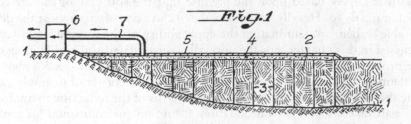

Figure 5.8 Method of consolidating soils (Kjellman, 1952).

Combining the concepts from the previous two paragraphs, the vacuum can be applied directly to the PVD to eliminate the need for a cutoff wall and membrane (Kolff et al. 2004). In this technique, each PVD is a connected plastic pipe which in turn is connected to the vacuum system. With this system, neither the sand layer nor the top geomembrane layer is needed.

The effective depth of vacuum consolidation depends upon the depth to which the vacuum pressure can be applied. The flow of water during vacuum consolidation is controlled by the amount of suction and, in the case of vacuum consolidation employing PVDs, the depth is limited to the length of the installed PVDs.

The success of ground improvement by vacuum consolidation (or surcharge preloading) is normally evaluated using field measurements of porewater pressure and settlement. The degree of consolidation needed is calculated during the design phase in a manner similar to that described for consolidation via preloading. The field measurement of porewater pressure and settlement is used to determine when the needed degree of consolidation is achieved.

5.8 COMBINED VACUUM CONSOLIDATION AND PRELOADING WITH VERTICAL DRAINS

Preloading with vertical drains and vacuum consolidation are effective means to improve the compressibility and strength of soft clay. It follows that these two techniques can be combined in situations where it is desirable to speed consolidation but with consolidation pressures greater than that which would result from either technique alone, or where the combined consolidation stresses are best applied via a combination of both techniques. Combining vacuum consolidation with a reduced-height preload can result in the same rate and amount of consolidation as a large fill without vacuum consolidation. This is attractive when the cost of fill is high (Indraratna et al. 2012).

Modeling of the site performance can be done using either two- or three-dimensional (2D or 3D) finite element methods. For one such case study, it was found that the results from equivalent two- and three-dimensional analyses predicted similar settlements, excess porewater pressures, and lateral movements (Rujikiatkamjorn et al. 2007). Further, the 2D results compared well with the field measurements.

5.9 NATURE'S CONSOLIDATION PRELOADING

The foregoing discussion of consolidation, preloading, and speeding consolidation with vertical drains is largely based upon the premise that the soft clay of the site is normally consolidated or nearly so. Heavily overconsolidated clays can often support the design loads with acceptable settlement, eliminating the applicability of the ground improvement techniques discussed in this chapter. For the normally consolidated soil and loading conditions of example problem Ex.5.1, the calculated settlement was 0.25 m. Ex.5.4 shows the calculated settlement after preloading to be 0.04 m. Both theory and example calculations demonstrate the benefit of soil preconsolidation in terms of the reduction in final settlement.

There are natural causes of preconsolidation. Common environmental (or geologic) processes include glacial loading, erosion of overlying sediments, desiccation, and depression of the water table resulting in changes in effective stress. Even at sites where these, or environmental processes that might cause the clay to become over consolidated are not expected, truly normally consolidated soils are uncommon because of the time effect on the preconsolidation pressure, manifesting itself as secondary compression.

The effect of time on the preconsolidation pressure can be illustrated by first considering the time rate of deformation relationship. Figure 5.9 is a data set of vertical strain plotted as a function of the log of time for a laboratory sample. The three components of settlement (immediate, consolidation, and secondary) are evident. Secondary compression, or creep, is deformation at constant stress. In a typical laboratory consolidation test, a new load is applied every 24 hours and the soil will experience something less than one day of secondary compression. A clay may experience thousands of years of secondary compression in the field.

The secondary compression index on a unit strain basis can be obtained from the slope of the final straight-line portion of the curve. That is:

$$C_{\alpha\varepsilon} = \frac{\Delta\varepsilon}{\Delta\log t} \qquad (5.20)$$

where

$C_{\alpha\varepsilon}$ = secondary compression index
$\Delta\varepsilon$ = change in strain between times t_1 and t_2

Secondary compression settlement is calculated using the secondary compression index as follows:

$$S_s = C_{\alpha\varepsilon} H_0 (\Delta\log t) \qquad (5.21)$$

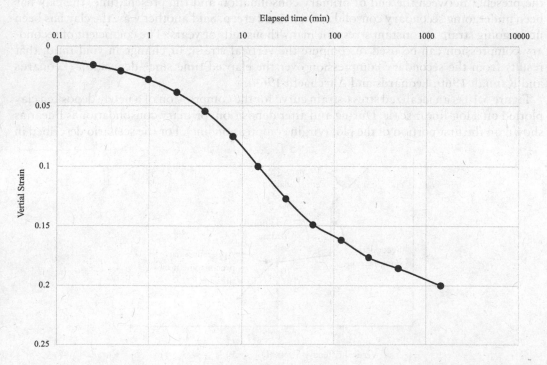

Figure 5.9 Time rate of consolidation on a unit strain basis showing secondary compression (data modified from Holtz and Kovacs, 1981).

Example problem Ex.5.5 illustrates the calculations to compute secondary compression.

Example problem Ex.5.5: Secondary compression

Problem: For the time rate of consolidation curve in Figure 5.10, calculate the deformation in a four-meter thick clay layer due to secondary compression over a period of 10,000 years.

Solution:

First, determine the secondary compression index using Equation 5.20. Figure 5.9 shows the strains are 1.8 and 2.0, for 240 minutes and 1,440 minutes, respectively.

$$C_{\alpha\varepsilon} = \frac{\Delta\varepsilon}{\Delta\log t} = \frac{(0.20-0.15)}{\log 1440 - \log 60} = 0.036$$

Now, using Equation 5.21, calculate the expected settlement over 10,000 years.

$$S_s = C_{\alpha\varepsilon}H_0\left(\Delta\log t\right) = 0.066\left(4.0\right)\log 10000 = \underline{\mathbf{0.58\ m}}$$

Thus, the secondary compression is significant and has an impact upon the apparent preconsolidation pressure.

Where there is an on-going deposition of clays (e.g. the Gulf of Mexico), recently deposited clays may be underconsolidated or normally consolidated. However, for most sites, the clay deposition occurred thousands of years ago. For example, the countless glacio-lacustrine deposits in North America are at least 11,700 years old based upon the dating of the end of the last glacial period. What is the effect of this elapsed time between deposition and the present? Between the end of primary consolidation and the present time, the clay has been undergoing secondary consolidation, termed creep. Said another way, the clay has been undergoing strain at constant stress for many thousands of years. The coefficient of secondary compression can be used to compute the vertical strain, or change in void ratio, that results from the secondary compression over the elapsed time since deposition (Leonards and Ramiah 1960; Leonards and Altschaeffl 1964).

Figure 5.10 is an idealized stress-strain curve for the compression of a newly deposited clay plotted on a log-linear scale. During and after deposition, primary consolidation is linear as shown on the first portion of the plot (virgin compression line). For the scenario described in

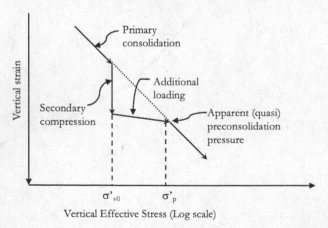

Figure 5.10 Idealized load-deformation curve including a long period of secondary compression.

the preceding paragraph where the clay is undergoing only secondary compression for a long time, the plot shows a straight downward movement of vertical strain at constant effective stress, σ'_{v0}, the present overburden effective stress. If, after a long period of secondary compression, the soil is loaded, the deformation moves along a recompression slope rather than a virgin compression slope. This is labeled "additional loading" in Figure 5.10. Once the load is high enough to intersect the extension of the primary consolidation line, the additional loading results in compression continuing along steeper sloped primary consolidation line, or virgin compression line. The effect of this is that the secondary compression produces a real preconsolidation pressure often termed the apparent preconsolidation pressure or the quasi-preconsolidation pressure. It is a real preconsolidation pressure, just not one formed during traditionally considered geologic processes described above.

Standard laboratory consolidation tests with "undisturbed" samples are often conducted to determine the preconsolidation stress which then can be compared to the calculated in situ effective stress to calculate the OCR (Equation 5.1). Most commonly, laboratory tests are conducted using a load increment ratio (LIR) of one, that is, the applied load increment is equal to the effective stress just prior to the application of the new load increment as follows:

$$LIR = \Delta\sigma / \sigma' = 1.0 \qquad (5.22)$$

This LIR translates into load applications of 24 kPa, 48 kPa, 96 kPa, etc. (0.25 tsf, 0.5 tsf, 1.0 tsf, etc.) for a typical laboratory test. For heavily overconsolidated soils, these LIRs work well and produce curves that are readily analyzable to determine c_v, the coefficient of consolidation. Smaller LIRs are uncommon as they may produce other curve shapes (Leonards and Girault 1961) that are difficult to analyze and may add time and expense to the testing program. Nonetheless, there are times when smaller LIRs are necessary to more accurately define the preconsolidation pressure, particularly with lightly overconsolidated clay.

It is also necessary to note that, as sample disturbance increases, the pressure-strain curve (e.g. Figure 5.1) becomes increasingly rounded and the preconsolidation pressure as

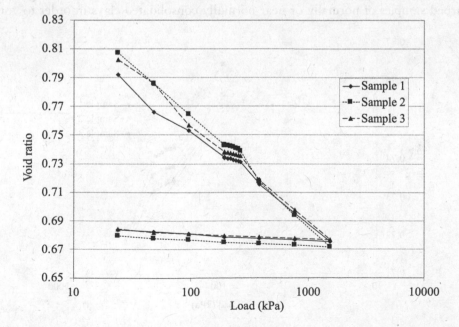

Figure 5.11 Consolidation behavior of soil-bentonite backfill demonstrating the development of a quasi-preconsolidation pressure.

determined with commonly used graphical methods shifts left (i.e. to a lower value). As a result, lightly overconsolidated, standard methods may indicate the soil is normally consolidated. If there is any uncertainty, it is recommended that additional consolidation tests be conducted using a smaller LIR in the vicinity of the expected preconsolidation pressure to better define the shape of the load-pressure curve in this important area of the curve. While it may seem conservative to simply assume the soil is normally consolidated, significant cost consequences from this assumption could result. For example, on a power plant project in Michigan, USA, initial testing indicated the possibility of a normally consolidated clay of lacustrine origin. Pile foundations reaching a depth of over 30 m were under consideration. Subsequent testing of high-quality samples with a small LIR in the range of the preconsolidation pressure indicated an OCR of approximately 1.4. The power plant was subsequently successfully constructed using a shallow foundation.

These phenomena are illustrated in Figure 5.11 using data from a remolded soil-bentonite mixture from a slurry trench cutoff wall backfill (Evans and Ryan 2005). Plots of void ratio versus stress result up to 192 kPa (2 tsf) reveal the typical behavior of a normally consolidated material with a log-linear virgin compression curve. However, for these tests, daily load application was interrupted and the 192 kPa (2 tsf) load was maintained for seven days during which the soil continued to creep. Following seven days at constant stress, additional stress was applied using a small LIR. This region of the void ratio-stress curve is shown in Figure 5.12. A small precompression pressure is visible, the result of the short period of strain at constant stress (creep). Analysis of the virgin compression data (sans the small LIR readings) yields C_c, the compression index, with values of 0.027, 0.030, and 0.033 for the three samples. These values are slightly lower than those reported by Yeo et al. (2005). The rebound portion of the curves is likewise log-linear and with slopes (C_s, the swell index or commonly called C_r the, the recompression index) of 0.0017, 0.0019, and 0.0021 for the three samples. These values are consistent with the sand-bentonite model backfill mixtures tested by Yeo et al. (2005).

To summarize this subsection of this chapter, it is important to secure high-quality, undisturbed samples of normally or near normally consolidated clays in order to conduct

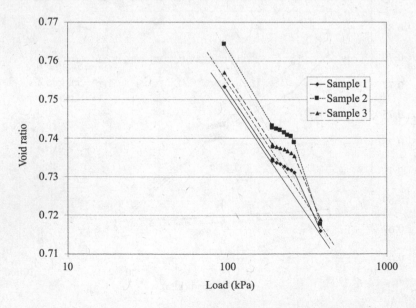

Figure 5.12 Consolidation behavior of soil-bentonite backfill demonstrating the development of a quasi-preconsolidation pressure.

consolidation tests to carefully determine the preconsolidation pressure. Small amounts of sample disturbance can cause significant changes in the calculation of the preconsolidation pressure. In some cases, the use of the standard LIR of one may obscure the actual preconsolidation pressure leading to an incorrect conclusion that the soil being tested is normally consolidated. Increasing sample disturbance leads to a decrease in the measured value of the preconsolidation pressure. Testing should be done as soon as possible after sampling as time in sample storage can likewise lead to sample disturbance and thus a decrease in the measured value of the preconsolidation pressure. Incorrectly concluding the soil is normally consolidated can lead to unnecessarily large foundation costs or unacceptable differential settlements for large structures.

5.10 SUMMARY

Soft, compressible clay can be problematic for many projects. Depending upon the loading conditions, excessive total settlement, differential settlement, and even shear failure can (and does) occur. However, the consolidation and shear strength of clay is one of the oldest and most studied topics in geotechnical engineering. Further, as presented in this chapter, there are a number of ground improvement techniques that are applicable to soft compressible clays. The underlying goal of each is to expedite consolidation which results in a material that is less compressible and stronger. The ground improvement technique of choice will depend upon project requirements and site-specific conditions.

It is especially important to undertake careful sampling and testing to determine the preconsolidation pressures in light of secondary compression that has been ongoing since the clays were originally deposited. On occasion, a more precise quantification of the preconsolidation pressures can reduce the use of more expensive and time-consuming ground improvement techniques.

Soft, compressible clays are often unsuitable for support of buildings, bridges, embankments, and other heavily loaded structures without pretreatment. Ground improvement via consolidation can be applied to these soft clays to increase their strength and reduce their compressibility. These methods include preloading, preloading with a surcharge, vertical drains to speed consolidation, and vacuum consolidation. Often these techniques are used synergistically such as preloading with a surcharge along with vertical drains.

For cases where clay deposits are initially characterized as normally consolidated, it may be possible to demonstrate that these clays are lightly overconsolidated with overconsolidation ratios on the order of 1.4 by using LIRs less than one. In such cases, it may be possible to apply the building loads without experiencing unacceptable settlement, avoiding more expensive and time-consuming ground improvement or deep foundation methods.

5.11 PROBLEMS

5.1 What would be the consolidation settlement in Example problem Ex.5.1 had the clay been substantially overconsolidated $(\sigma'_{v0} + \Delta\sigma_z < \sigma'_p)$? Would additional ground improvement measures be needed to speed consolidation?

5.2 Rework Example problem Ex.5.1 by mathematically dividing the four-meter thick clay layer into four one-meter thick clay layers.

5.3 Using the data shown in Figure 5.1, determine the value of the preconsolidation pressure and the OCR.

5.4 Using the data in Figure 5.12 determine the preconsolidation pressure and the OCR for the soil-bentonite backfill.

5.5 A large building is planned to be built on a mat foundation of 100 m by 100 m as shown in Figure Pr.5.5. The water table is at the top of the clay layer.
   a. Assuming the clay to be normally consolidated, calculate the total consolidation settlement due to consolidation of the clay layer.
   b. How long will it take for 90% consolidation?
   c. Design a wick drain system to speed consolidation such that 90% consolidation occurs in 90 days.
   d. How much settlement would you expect if the clay had an OCR of 1.4 due to 10,000 years of secondary compression?

5.6 A five-meter thick double drained clay layer has a coefficient of consolidation, c_v, of 1.0 m²/yr. How long will it take for 90% of the consolidation to occur after the application of a building load? Design a wick drain system such that 90% of the consolidation is complete in two months. Remember, from a practical standpoint, since buildings are not constructed instantaneously, speeding the consolidation using wick drains allows the building settlement to occur as the building is constructed and minimizes postconstruction consolidation settlement.

5.7 A ten-meter thick clay layer, assumed to be normally consolidated, lies immediately beneath a planned highway embankment to be constructed with a free-draining sandy soil weighing 20kN/m³. The clay has a vertical coefficient of consolidation c_v, of 0.8 m²/yr, a specific gravity of solids of 2.65, a water content of 45%, a vertical coefficient of compression, C_{ce}, of 0.20 and the horizontal coefficient of compression is 10 times the vertical value. The clay layer is underlain by a sand layer and the water table is at the surface. What will be the total settlement along the centerline of the embankment? How long will it take for 90% of the consolidation to occur after the application of the embankment load? Design a surcharge preloading with PVDs or sand drains such that 100% of the expected consolidation settlement will be complete within two months of the final load application.

5.8 Assume the clay in Problem 5.7 is glacio-lacustrine and has been undergoing secondary compression for the last 10,000⁺ years such that careful sampling and testing has revealed the clay has an OCR of 1.4. Recompute the settlement of the clay under the

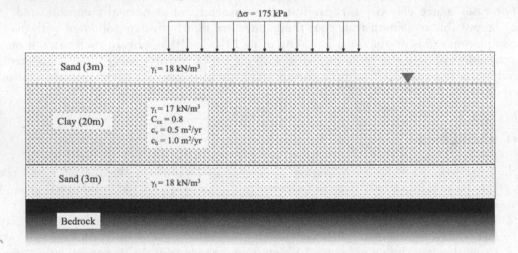

Figure Pr.5.5 Cross-section and soil properties for problem 5.5.

embankment loading. How does this compare to that calculated assuming the clay to be normally consolidated? How does this change the design of the surcharge and vertical drain system?

5.9 PVDs have largely displaced sand drains as the material of choice for vertical drains. Provide three reasons why that is the case.

5.10 Load-deformation curves are typically plotted as e vs log $\sigma v'$ or $\varepsilon v(\%)$ vs log $\sigma v'$. What are two advantages of the latter over the former?

5.11 A five-meter thick clay layer with drainage at both top and bottom of the layer (two-way drainage) reaches 90% consolidation in 4 years. How long would it take if the layer were drained only at the top? Only at the bottom?

5.12 An undisturbed clay sample is tested in a one-dimensional consolidation test in the laboratory and found to reach 50% consolidation in 15 minutes. How long would the same clay take to reach 50% consolidation under field conditions where the clay is 4 m thick with two-way drainage?

5.13 A lacustrine clay was found to have a ratio of c_h/c_v equal to five. Why would the horizontal coefficient of consolidation be five times larger than the vertical coefficient of consolidation?

5.14 How would smear of the clay layer due to construction of vertical drains impact the time rate of consolidation prediction?

5.15 Are the time rates of consolidation values determined in the laboratory under saturated conditions applicable to clay layers above the water table? Explain your answer.

REFERENCES

Basu, D. and Prezzi, M. (2007). Effect of the smear and transition zones around prefabricated vertical drains installed in a triangular pattern on the rate of soil consolidation. *International Journal of Geomechanics*, 7(1), pp. 34–43.

Bergado, D.T., Anderson, L.R., Miura, N. and Balasubramaniam, A.S. (1996). Prefabricated vertical drains PVD. In *Soft ground improvement in lowland and other environments* (pp. 88–185). New York, New York: ASCE.

Chai, J.C., Miura, N. and Sakajo, S. (1997). A theoretical study on smear effect around vertical drain. In *International conference on soil mechanics and foundation engineering* (pp. 1581–1584). Hamburg Germany.

Chen, H. and Bao, X.C. (1983). Analysis of soil consolidation stress under the action of negative pressure. In *Proceedings on 8th European conference on soil mechanics and foundation engineering*, Helsinki, Vol. 2, pp. 591–596.

Chu, J., Yan, S. and Indraranata, B. (2008). Vacuum preloading techniques—recent developments and applications. In *GeoCongress 2008: Geosustainability and geohazard mitigation* (pp. 586–595). New Orleans, LA.

Coduto, D., Yeung, M. and Kitch, W. (2011). *Geotechnical engineering principles and practices*. 2nd ed. Upper Saddle River, NJ: Prentice Hall.

Das, B.M. (2015). *Principles of foundation engineering*. Independence, KY: Cengage Learning.

Das, B.M. (2016). *Soil mechanics laboratory manual*. Oxford, UK: Oxford University Press.

Das, B.M. and Sobhan, K. (2013). *Principles of geotechnical engineering*. Independence, KY: Cengage Learning.

Evans, J.C. and Ryan, C.R. (2005). Time-dependent strength behavior of soil-bentonite slurry wall backfill. In *Waste Containment and remediation: Proceedings of the geo-frontiers 2005 congress*, Geotechnical Special Publication No. 142 (pp. 1–9). Austin, TX.

Fang, H.Y. (2013). *Foundation engineering handbook*. New York, NY: Springer Science+Business Media.

Fratta, D., Aguettant, J. and Roussel-Smith, L. (2007). *Introduction to soil mechanics laboratory testing*. Boca Raton, FL: CRC Press.

Hansbo, S. (1979). Consolidation of clay by bandshaped prefabricated drains. *Ground Engineering*, 12(5), 16–27.

Hansbo, S. (1981). Consolidation of fine-grained soils by prefabricated drains. In *Proceedings of the 10th ICSMF* (Vol. 3, pp. 677–682). Stockholm, Sweden.

Hausmann, M.R. (1990). *Engineering principles of ground modification*. New York, NY: McGraw-Hill.

Hayward Baker. (2007). *Prefabricated vertical drain guide specifications*. Odenton, MD: Hayward Baker.

Holtz, R.D., Kovacs, W.D., and Sheahan, T.C. (2011). *An introduction to geotechnical engineering* (2nd ed.). London, UK: Pearson.

Holtz, R.D. and Wager, O. (1975). Preloading by vacuum: Current prospects. *Transportation Research Record*, 548, 26–29.

Indraratna, B. and Redana, I.W. (1998). Laboratory determination of smear zone due to vertical drain installation. *Journal of Geotechnical and Geoenvironmental Engineering*, 124(2), 180–184.

Indraratna, B. and Redana, I.W. (2000). Numerical modeling of vertical drains with smear and well resistance installed in soft clay. *Canadian Geotechnical Journal*, 37(1), 132–145.

Indraratna, B., Rujikiatkamjorn, C., Balasubramaniam, A.S. and McIntosh, G. (2012). Soft ground improvement via vertical drains and vacuum assisted preloading. *Geotextiles and Geomembranes*, 30, 16–23.

Indraratna, B., Rujikiatkamjorn, C., Balasubramaniam, A.S. and Wijeyakulasuriya, V. (2005). Predictions and observations of soft clay foundations stabilized with geosynthetic drains and vacuum surcharge. In *Elsevier geo-engineering book series* (Vol. 3, pp. 199–229). London, UK: Elsevier.

Kjellman, W. (1939). *Method and means to accelerate the consolidation of clay-ground or other soil*, Application GB632902A.

Kjellman, W. (1951). *Drainage method and device*. U.S. Patent 2,577,252.

Kjellman, W. (1952). *Method of consolidating soils*. U.S. Patent 2,615,307.

Koerner, R.M. (2012). *Wick drains, designing with geosynthetics, Vol. 2*. Bloomington, IN: Xlibris Corporation.

Kolff, A.H.N., Spierenburg, S.E.J. and Mathijssen, F.A.J.M. (2004). BeauDrain: A new consolidation system based on the old concept of vacuum consolidation. In *Proceedings of the 5th international conference on ground improvement techniques*, Kuala Lumpur, Malaysia.

Leonards, G.A. and Altschaeffl, A.G. (1964). Compressibility of clay. *Journal of the Soil Mechanics and Foundations Division*, 90(5), 133–156.

Leonards, G.A. and Girault, P. (1961). A study of the one-dimensional consolidation test. In *Proceedings of 9th ICSMFE*, Paris, 1 (pp. 116–130).

Leonards, G. and Ramiah, B. (1960, January). Time effects in the consolidation of clays. In *Papers on soils 1959 meetings*. Conshohocken, PA: ASTM International.

Mesri, G. and Funk, J.R. (2015). Settlement of the Kansai international airport islands. *Journal of Geotechnical and Geoenvironmental Engineering*, 141(2), 04014102.

Mesri, G. and Lo, D.O.K. (1991). Field performance of prefabricated vertical drains. In *Proceedings of international conference on geotechnical engineering for coastal development-theory to practice, Yokohama, Japan*, vol. 1 (pp. 231–236).

Moran, D.E. (1926). *Foundations and the like*. U.S. Patent 1,598,300.

Murthy, V.N.S. (2002). *Geotechnical engineering: Principles and practices of soil mechanics and foundation engineering*. Boca Raton, FL: CRC Press.

Porter, O.J. (1936). Studies of fill construction over mud flats including a description of experimental construction using vertical sand drains to hasten stabilization. In *Proceedings of the first international conference on soil mechanics and foundation engineering*, Vol. 1 (pp. 229–235), Harvard University, Cambridge, MA.

Rujikiatkamjorn, C. and Indraratna, B. (2007). Analytical solutions and design curves for vacuum-assisted consolidation with both vertical and horizontal drainage. *Canadian Geotechnical Journal*, 44(2), 188–200.

Rujikiatkamjorn, C. and Indraratna, B. (2009). Design procedure for vertical drains considering a linear variation of lateral permeability within the smear zone. *Canadian Geotechnical Journal*, 46(3), 270–280.

Rujikiatkamjorn, C., Indraratna, B. and Chu, J. (2008). 2D and 3D numerical modeling of combined surcharge and vacuum preloading with vertical drains. *International Journal of Geomechanics*, 8(2), 144–156.

Rutledge, P.C. and Johnson, S.J. (1958). Review of uses of vertical sand drains. *Bulletin*, 173, 65–79.

Sathananthan, I. and Indraratna, B. (2006). Laboratory evaluation of smear zone and correlation between permeability and moisture content. *Journal of Geotechnical and Geoenvironmental Engineering*, 132(7), 942–945.

Sharma, J.S. and Xiao, D. (2000). Characterization of a smear zone around vertical drains by large-scale laboratory tests. *Canadian Geotechnical Journal*, 37(6), 1265–1271.

Tang, M. and Shang, J.Q. (2000). Vacuum preloading consolidation of Yaoqiang Airport runway. *Geotechnique*, 50(6), 613–623.

Terzaghi, K. (1925). Principles of soil mechanics, IV—Settlement and consolidation of clay. *Engineering News Record*, 95(3), 874–878.

Terzaghi, K., Peck, R.B. and Mesri, G. (1996). *Soil mechanics in engineering practice*. Hoboken, NJ: John Wiley and Sons.

Yeo, S.S., Shackelford, C.D. and Evans, J.C. (2005). Consolidation and hydraulic conductivity of nine model soil-bentonite backfills. *Journal of Geotechnical and Geoenvironmental Engineering*, 131(10), 1189–1198.

Yeung, A.T. (1997). Design curves for prefabricated vertical drains. *Journal of geotechnical and geoenvironmental engineering*, 123(8), 755–759.

Chapter 6

Soil mixing

6.1 INTRODUCTION

The ground can be improved with soil mixing by introducing additives into the subsurface while mixing (blending) the additives with the in situ soil to create a mixture that has improved physical properties, engineering properties, and/or chemical characteristics. The specific methods used for soil mixing vary as do the additives and, as a result, a variety of names are given to specific techniques to better characterize the method (see Section 6.3). For example, jet grouting (Chapter 11) is a term used for soil mixing where the soil mixing is accomplished using a high-pressure fluid injection (or jet) and the additive is an engineered grout. The broadest use of the term soil mixing refers to methods in which the soil is mechanically mixed with dry or wet additives using augers or other mixing tools.

Specific techniques lend themselves to the improvement of specific soil conditions. Dry soil mixing, for example, is best applied to soft, high moisture soils such as silts, clays, and peats, whereas jet grouting might be more appropriate for loose sandy soils. Soil conditions amenable to soil mixing include soft, loose, and/or contaminated soils that require improvement to meet required properties for future planned construction (Bruce 2001). Some of the points associated with soil mixing include:

1. The majority of projects employ site-specific approaches for soil mixing designs due to a lack of a single, universally applicable design methodology.
2. Soil mixing is an ever-expanding practice that is well received due to its high quality control and assurance, and its ability to mitigate effects of both static and dynamic loading on soft soils (Bruce and Cali 2005).
3. There is a large amount of information and empirical data based on large-scale research and testing that has been performed on soil mixing although much of the data is in the domain of contractors and testing laboratories.
4. While jet grouting (Chapter 11) may be classified as soil mixing, it is important to note that the overall final characteristics of the improved earth not be lumped together with other mixing techniques. The two methods are in fact different and resulting final products cannot be extrapolated between methods.
5. Pre-wetting of soils may be necessary even when it would appear that there is enough moisture in the soil, based upon calculations and previous experience as some of the water may not be available.
6. Along with soil type, total water content, amount, and type of additives, the blade rotation number (total number of rotations passing through a given unit of depth), or any quantification of mixing energy applied, has been proven to be a prime contributor to the final strength and properties of the mixed soil. These factors need to be given consideration during design.

7. Based on all the issues raised in points one through six above, it is important that soil mixing be performed by experienced personnel and companies.

This chapter presents a variety of soil mixing means and methods and categorizes the techniques into discrete groupings for the purposes of presentation. Each technique will be introduced followed by a discussion of construction methods and equipment. A general overview, mostly relevant to all construction types, will follow covering reagent additives, design methodology, and quality control. Example calculations are presented. The chapter concludes with a problems section and a reference list for pursuing a given technique in more detail.

In much of soil mixing literature and practice, the terms deep mix method, deep soil mixing, and soil mixing are used interchangeably. The reader can assume the same interchangeability of terminology in this chapter.

6.2 HISTORY OF SOIL MIXING

Soil mixing is one of the newer specialty geotechnical construction techniques, and one that is fast growing. It is becoming a widely used technique throughout the world. The method offers a solution to a variety of ground improvement needs over a large variety of soil types and conditions (Bruce et al. 1998). In situ soil mixing has been utilized in the geotechnical field for over 50 years. The technique is believed to have been first introduced in the United States (US) in the 1950s and was further developed in Europe (most notably Scandinavia) and Japan in the 1960s, 1970s, and 1980s (Bruce et al. 1998). Most consider the Jackson Lake Dam project (Ryan and Jasperse 1989; Ryan and Jasperse 1991), completed in 1987, as the re-introduction of soil mixing into the US market. In this project, the technique was to improve the ground in two separate and distinct ways. First, auger mixing was used to construct six-sided cells upstream and downstream near and beneath the toe of the dam. The minimum strength for the soil mixed composite was 1380 kPa. The design, like modern designs, was based on the fact that in the event of an earthquake and subsequent liquefaction, the liquefied soils would be contained within the soilcrete cells and the rather strong cells would prevent a shear failure in the dam foundation that could result in the collapse of the dam. In addition, auger mixing was used to construct a 610 mm thick cutoff wall along the entire upstream toe of the dam to control under seepage since the original dam was built without any seepage cutoff in the foundation soils. The technique has been refined in the US since its reintroduction (Bruce et al. 1998), particularly as a tool for environmental remediation. As the technique becomes better understood, and an increasing number of companies offer soil mixing construction options, the technique has and will continue to see more widespread use worldwide.

The earliest applications of soil mixing in the US were for the construction of foundation support elements and earth retention structures (Ryan and Walker 1992). Soil mixing for geotechnical ground improvement now includes several technologies and materials all designed to blend soil and additives together into a mostly homogeneous blend commonly referred to as in situ soil-cement (if cement is a component of the mixture), soil-reagent, or sometimes *soilcrete*.

Soil mixing was first used in environmental applications in the late 1980s and early 1990s and is now widely considered a technically sound and cost-effective method for remediating a range of soil contaminants. Soil mixing for the remediation of contaminated sites has seen significant development since the early applications. The technique has become an efficient means of in situ remediation in the US that is accepted by the US EPA and numerous other agencies in a range of applications.

6.3 DEFINITIONS, TYPES, AND CLASSIFICATIONS

Soil mixing may be loosely defined as any technique that is used to mechanically mix soils with or without additives. Commonly, in both geotechnical and environmental applications, soil mixing refers to processes by which reagents are injected and mixed with soil (Day and Ryan 1995). The processes vary widely from in situ to ex situ, dry mixing to wet mixing as shown in Larsson (2005), auger mixing to rotary drum mixing, and single auger to multi-auger. Generally, the purpose remains the same: the efficient creation of a soil-reagent composite with improved properties relative to the in situ soils. Target improvement objectives often include higher strength, lower compressibility, and lower permeability as compared to the unimproved soil. Variability is inevitable as a result of variations in the characteristics of the in situ soil, construction variables (operational parameters-related methods), and variability in properties of the binding product itself (Bruce and Bruce 2003).

Soil mixing for geotechnical ground improvement is primarily performed in situ using augers or other cutting tools (e.g. large chains mounted on trenching machines). The most common reagent used in geotechnical soil mixing is ordinary portland cement (OPC or herein PC). The PC is often added in a water-based suspension referred to as grout or slurry. The most common property improvement objectives include higher strength and/or reduced compressibility to accomplish objectives that might include increased bearing capacity, liquefaction potential reduction, or slope stability improvement.

Soil mixing on environmental remediation projects is often performed in situ using large-diameter single auger soil mixing with wet reagent injection. In this method, a reagent slurry is used as the drilling lubricant and as the final soil additive. The reagent slurry is pumped through a hollow drill stem as a large (0.9 m to 3.7 m diameter) auger is turned in the soil thereby creating treated columns of soil. The columns are installed in an overlapping pattern that ensures 100% coverage of the contaminated target area. Occasionally project or site conditions necessitate the use of dry mixing methods. Soil mixing on environmental projects is typically used to accomplish one of two objectives; stabilization/solidification (S/S) or treatment (herein referred to as in situ treatment or IST) of impacted soils. When S/S is performed in situ, the technique is referred to as in situ solidification/stabilization (ISS). ISS is more common than IST, but IST has seen an increase in use since approximately 2005.

Cutoff walls can also be installed using soil mixing. Using the in situ soils mixed with a slurry is often less expensive than importing materials. Contractors employ a variety of tools for the process of incorporating the slurry into the soil. The fundamental underlying assumption of the technique is that by applying mixing energy and adding slurry to the soil in place, an in situ mixed wall with the desired properties will result. An examination of the underlying assumption reveals two key aspects affecting the completed wall. First, the in situ soils must be amenable to blending with a suitable grain size distribution and capable of producing a final product with the desired characteristics. For example, if the goal is high shear strength, clayey fines in the mixed soil will negatively impact the shear strength, all else being equal. Second, the type and amount of slurry can substantially affect the final product. The slurry may be bentonite-water as described for soil-bentonite (SB) cutoff walls (see Chapter 8) when unconfined compressive strength (UCS) is not required or can include cement, for example, to create a soil-cement mixture. The slurry may also be a cement-bentonite (CB) slurry and/or CB slurry made with granulated ground blast furnace slag (GGBFS) and PC. The terms "deep soil mixing" and "soil mixed walls" (SMW) originally referred to techniques that used a series of overlapping modified auger shafts with a bottom discharge auger for the addition of slurry (grout) capable of blending the soils and slurry into a visually homogeneous mixture. The resulting product was a linear feature, i.e. a wall. Both DSM and SMW were originally proprietary terms used by Geocon and Seiko, respectively, but DSM has come into widespread usage and SMW has fallen out of use. The term DSM is now part of a broader set of means and methods under the common umbrella term deep mix method (Bruce and Bruce 2003). DMM, and other similar acronyms, is a broad term used to refer to a wide variety of techniques and equipment including single augers and other rotary mixing tools. For the purposes of discussions herein, the term "soil mixing" will be used to describe the various construction systems that are used since many terms are used by many contractors and only a few terms refer to any particular proprietary technique.

The construction of a soil mixed cutoff wall can be done with several distinctly different types of construction equipment and tooling. As noted earlier, construction methods aim to blend the in situ soil with a slurry or grout to form a visually homogeneous mixture. To distinguish the different types of soil mixing and to be consistent with the contracting industry's distinction, different acronyms may be used for the different techniques. The principal advantage of an in situ mixed wall relative to other cutoff wall installation techniques is the elimination of the need to bring all of the subsurface soils to the surface. For slurry trenched CB walls (Chapter 8), the material from within the wall length, width, and depth needs to be brought to the surface and then needs to be disposed of or reused elsewhere on the project. Other slurry trench walls, including SB walls, may reuse the in situ materials if the materials are suitable. With in situ mixed walls, there may be some excess due to bulking from the addition of slurry to the mixture, but most of the material is left in the subsurface. This is particularly important when working with contaminated ground which might result in premium disposal costs or worker safety issues. From a sustainability viewpoint, it would seem that less energy is consumed in mixing in situ as compared to excavation and replacement (SB) or excavation and disposal/reuse (CB). The second advantage to the mixed wall technologies is elimination of the continuous trench. In urban areas or on dikes and levees, the length of the open trench largely precludes any beneficial three-dimensional effects on trench stability. Mixed wall technologies do not use trenches full of slurry and are constructed in short panels, so the construction is generally more stable.

There are many different soil mixing techniques, some of which are proprietary to various specialty contractors and organizations (Bruce et al. 1998). Figure 6.1 breaks the soil mixing techniques used today (as of 2021) into broad classifications. The information in this figure flows from applicable depth range to reagent addition type, to mixing method, to stem type, to common acronyms and names (left to right).

The classifications in Figure 6.1 are further defined in the following subsections.

6.3.1 Depth of soil mixing

Applications of soil mixing that are very shallow, typically less than 0.6 m from the ground surface are termed near-surface soil mixing. Most of the applications at this depth are for road construction/repair and agricultural purposes. Some shallow geotechnical (e.g. relieving platforms) and environmental (e.g. soil mixed liner projects) are completed at this depth. This chapter does not explicitly address these very shallow applications.

Applications of soil mixing that are relatively shallow (compared to deep soil mixing) that cannot be completed using near-surface soil mixing equipment are defined herein as shallow soil mixing (SSM) and mixing is performed 0.6 m to 6 m below ground surface. Applications in this depth range include agricultural, geotechnical, and environmental purposes. The acronym SSM is often used in industry to refer to this category.

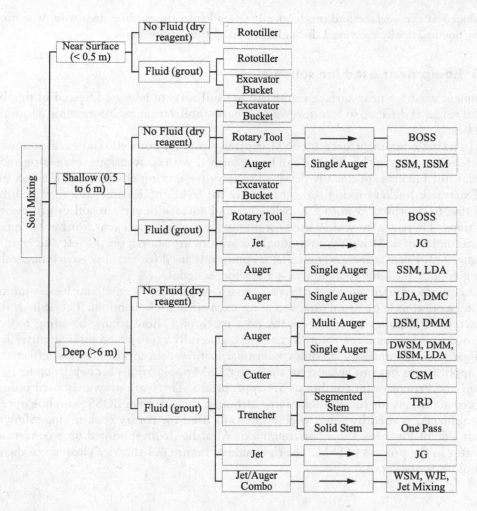

Figure 6.1 Classifications and common acronyms for soil mixing.

DSM refers to applications of soil mixing that are relatively deep (compared to SSM) and are defined herein as soil mixing performed to depths greater than 6 m below ground surface. Applications at this depth are exclusively for geotechnical, environmental, or combination, i.e. geoenvironmental, purposes. As mentioned earlier, the acronyms DSM or DMM are often used to refer to soil mixing in this category.

6.3.2 Methods of mixing reagents

Wet mixing methods are those applications in which the reagent is introduced to the soils in a fluid grout (also termed slurry). In most cases, the reagent is suspended in a water-based slurry and pumped through the mixing tool into the soil while simultaneously mixing the soil and injected slurry. In some cases, the grout slurry is pumped onto the surface of the mixing area and spread into the mixing area using the mixing tool (such as augers and paddles).

Dry mixing methods are those applications of soil mixing in which the reagent is introduced to the soils in a dry form, generally powdered. In this method, the reagent may be introduced at the surface and mechanically spread into the mixing area using the mixing tool or pneumatically conveyed through the end of the mixing tool.

6.3.3 Equipment used for soil mixing

Equipment used for near-surface (very shallow) soil mixing is not addressed in this book and the reader is directed to literature on road base stabilization for information about those processes.

Soil mixing equipment used for SSM is often simply a conventional excavator with a bucket. Soil mixing in this method can be performed with conventional excavator buckets or excavator buckets specifically designed to improve mixing efficiency or the mix product. Excavator bucket mixing is most effective for SSM applications and may be limited to mixing to depths of 3 m to 5 m below ground surface in certain soil or groundwater conditions. There are no widely used acronyms for referring to excavator bucket mixing. Two examples of SSM with conventional excavators are shown on the left and right side of Figure 6.2. Excavator bucket mixing is commonly used for shallow environmental soil mixing applications due to its low cost and short schedule.

The equipment used for SSM can also consist of paddles, stems, and blades mounted on a horizontal axle or drum attached to an excavator stick or boom. The axle or drum is generally hydraulically driven by the host machine. These rotary blending tools are described by some as large rototillers and are generally repurposed mining cutter heads modified to withstand soil mixing conditions. Rotary tool mixing is most effective for SSM applications but can be used for shallower DSM applications for depths up to 10 m if the soil and groundwater conditions are appropriate. There are no widely used industry-accepted acronyms for rotary tool mixing although the acronym BOSS (backhoe operated soil stabilizers) is used by some to describe tools used for rotary tool mixing. Shown on the left side of Figure 6.3 is a horizontal axis rotating drum mounted on a conventional excavator in the process of SSM. The right side of Figure 6.3 shows a close-up of the mixing tool for more detail.

Figure 6.2 Excavator bucket mixing (courtesy of Geo-Solutions, Inc., www.geo-solutions.com).

Figure 6.3 Rotary tool soil mixing (left) and rotary tool (right). (courtesy of Geo-Solutions, Inc., www.geo -solutions.com).

Soil mixing with rotary tools is also a common choice for shallow environmental soil mixing applications due to its cost advantages vs. deeper soil mixing methods (below) and technical advantages vs. conventional bucket mixing (above).

Equipment used for deeper soil mixing includes augers, cutters, trenchers, chains, jet mixing, and combination systems. Each of these equipment types is discussed in the following paragraphs.

Auger mixing equipment for soil mixing consists of horizontally oriented paddles, stems, and blades mounted on one or more vertical axles or stems. The vertical axle(s) or stem(s) are turned by hydraulically or mechanically driven rotary head systems and can be mounted on cranes or excavator bases. Shown on the left side of Figure 6.4 is a multi-axis auger mixing tool showing the mixing paddles extending the length of the auger shaft. The right side of Figure 6.4 shows a close-up of a large diameter single-axis auger. Auger mixing is

Figure 6.4 Multi (left) and single (right) auger soil mixing equipment. (courtesy of Geo-Solutions, Inc., www. geo-solutions.com).

effective for shallow and deep applications in almost any soil or groundwater condition. One widely used acronym for auger mixing is LDA (large diameter auger) and the acronym DSM (as earlier defined) is also commonly used to refer to deep multi-auger mixing systems.

Auger mixing for installation of linear elements is often performed with modified augers installed in a series (Figure 6.4, left side). The augers are advanced into the subsurface, adding slurry as needed to aid the advance. During withdrawal, additional slurry is added, and the augers are moved upward and downward to fully mix the soils with the added slurry. The sequence of penetrations typically used for panel construction methods is shown in Figure 6.5. Here, the first and third panels (or penetrations), termed the primary panels, are in previously unmixed ground. With the first and third panels in place, the second, or overlapping panel, follows. The outermost augers will follow the loosened and mixed ground downward insuring overlap between penetrations. The process continues in a leapfrog pattern, always skipping the space between the most recent penetration and then going back to connect the two initially disconnected panels. The rigidity of the augers, coupled with the propensity of the outermost augers to follow the path of least resistance, provides the control of auger position needed to ensure continuity of the wall.

Auger mixing is also regularly performed using single auger systems (Figure 6.4, right side, and schematically in Figure 6.6), specifically for the installation of soil mixed blocks wherein the columns are overlapped to cover the entire area. This is the most common method of soil mixing for environmental applications and offers many benefits over other, arguably more advanced, techniques, specifically a lower cost.

Auger soil mixing requires a ground free of obstructions (boulders, etc.) and adequate overhead equipment clearance. Boulders, utilities, and old foundations are examples of subsurface and overhead interferences that can preclude soil mixing with augers. Jet grouting,

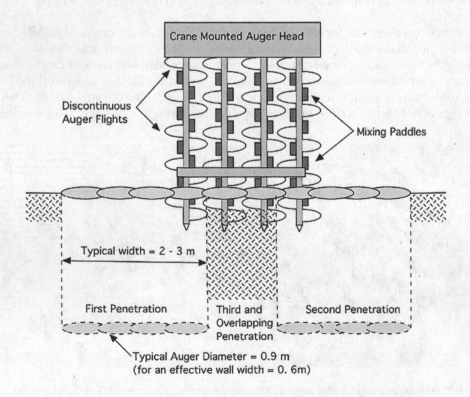

Figure 6.5 Sequence of penetrations for auger mixed cutoff wall (from LaGrega et al. 2010).

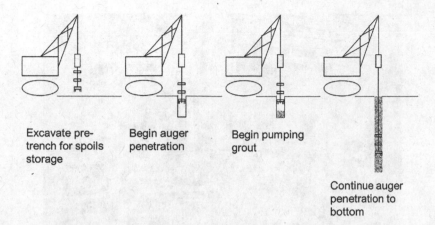

Excavate pre-
trench for spoils
storage

Begin auger
penetration

Begin pumping
grout

Continue auger
penetration to
bottom

Figure 6.6 Wet in situ mixing schematic.

as discussed in more detail in Chapter 11, can be used to supplement auger mixing and other soil mixing techniques in an area of obstructions. Another limitation of auger mixing is the vertical extent of mixing. While the augers and mixing paddles penetrate the full depth of the profile being mixed, the up and down stroke of mixing is generally limited to approximately 1 m to 2 m. As a result, the in situ material at any given depth is blended with material found at that depth plus and minus approximately 2 m. This limited vertical mixing may be significant in highly stratified subsurface environments as the blended soils will be over a relatively narrow range rather than the entire vertical profile.

Advantages of auger mixing include relatively low vibrations, use of the in situ soils to form the soilcrete, little spoil, less contact with contaminated materials, if present, and the ability to use small to medium equipment for the construction.

Equipment used for deeper soil mixing also includes vertically oriented cutter heads or paddles mounted on horizontal axles such as shown in Figure 6.7. Cutter soil mixing (CSM) equipment can be mounted onto hanging lead systems on cranes or to fixed systems on excavator bases as shown in Figure 6.7. The horizontal axles are driven hydraulically by the host machine. CSM is most widely used for DSM performed for geotechnical or geoenvironmental purposes where the soil mixing is being used to install linear elements, e.g. shear walls or cutoff walls, or for the installation of discrete elements, e.g. aerial bearing capacity improvement. Cutter soil mixing is widely referred to using the acronym CSM.

The process for constructing cutoff walls using CSM is similar to that for auger mixing linear elements. The tool is advanced into the ground as the counter-rotating wheels excavate, loosen, and mix the soils with slurry (bentonite-water or CB) that is simultaneously pumped into the ground through nozzles located between the wheels. Once the design depth is reached, the mixing tool is slowly withdrawn while more CB slurry is added. The in situ soils are mixed thoroughly with the slurry and a locally homogeneous mixture results. The result is a soil-mixed panel having a plan area equal to that of the tool and a depth equal to the penetration depth. Vertical soil mixing is essentially limited to the diameter of the CSM wheels (Arnold et al. 2011). CSM, like auger mixing, can be used to create cutoff walls to control groundwater flow and contaminant transport with the primary design consideration being low hydraulic conductivity.

CSM has been used on over 150 projects worldwide (Arnold et al. 2011) and, as an example, was used for a portion of the rehabilitation of the Herbert Hoover Dike (HHD) in Florida, USA, as a means to cut off under seepage and piping. Portions of the subsurface

Figure 6.7 Cutter soil mixing tool.

stratigraphy were ideal for CSM while others were unsuitable. At the HHD, the embankment consists of clean sands, silty sands, clayey sands, with occasional inclusions of cobbles, boulders, gravel, shell, and organic material. Overall, these materials are suited for mixing with CSM. However, the dike is also founded on a peat layer with organic silts, clays, and sands and is up to 4 m thick, underlain by sands and weak rock. CSM mixes the in situ soil with the grout/slurry over a narrow vertical range and thus this peat layer is not suitable for mixing with CSM as organic material can negatively impact the cement reaction. Below the peat layer, there is a decomposed limestone layer which consists of clay, silt, and sand mixtures which is suitable for CSM. Underlying the decomposed limestone is a layer of intact limestone, having unconfined compressive strengths up to 14,000 kPa. (USACE 2008). While advancement rates are slowed in rock such as this, the CSM was able to penetrate the limestone layer. The limestone is underlain by sand and shell layers and sand/shell mixtures readily excavated with CSM. This site illustrates the flexibility of CSM to excavate and mix a wide range of soils including weak rock and the limitation resulting from the limited extent of vertical mixing. Because the peat layer does not provide suitable material for soil mixing, it was necessary to pre-excavate to the bottom of the peat layer and remove the peat from the profile before mixing the entire wall profile with CSM.

Trencher systems, also known as chain tool systems or chain mixing tools, are a class of equipment used for soil mixing that consist of chains that move around a fixed guide. Trench mixing equipment is generally mounted on tracked chassis and may consist of a solid guide (Figure 6.8, left side) or a segmented guide constructed onsite (Figure 6.8, center). A schematic of the segmented guide equipment is shown on the left side of Figure 6.8. Trench mixing is most often used in DSM for geotechnical purposes during the installation of linear elements. This soil mixing method is commonly referred to in the industry as the one-pass system.

Figure 6.8 One pass trenching equipment (left). (courtesy of Geo-Solutions, Inc., www.geo-solutions.com), TRD mixing (center) and TRD schematic (right) (schematic courtesy of Keller, www.Keller-na.com).

Figure 6.9 Photo of TRD soil mixing equipment.

A subset of this one-pass system where the chains rotate vertically around a segmented vertical guide is known as the trench remixing and deep wall method (TRD), e.g. see Evans (2007). A photograph of the cutter bar and teeth of the TRD equipment is shown on the center panel of Figure 6.8 and shown schematically on the right side of Figure 6.8, and the equipment is highlighted in Figure 6.9. The TRD method has been used extensively in Japan (Aoi et al. 1996 and Aoi et al. 1998) with great success and was first introduced in the US in 2006 (Evans 2007). The TRD method has been used for over 30 years in Japan and can be used to install walls to depths of 50 m or more.

Another soil mixing approach using a trencher system, termed herein trencher mixing and also known as the one-pass method (Figure 6.8, left side), has certain characteristics in common with the TRD method. The trencher mixing method of soil mixing is accomplished via track-mounted equipment consisting of a continuous cutting chain equipped with hundreds of cutting teeth that rotate at high speeds on a boom that is inserted into the ground. The boom and chain are near vertical during installation and extend from grade to the desired depth. Trenchers can be equipped with metered additive delivery systems, underground water injection nozzles, pre-mixed slurry injection ports, speed controls for both the mixing chain and track speeds, GPS mapping, and laser guides to control depth and mixing proportions.

The principal advantage of one-pass methods over other in situ mixing techniques (such as vertical-axis mixing tools) is the relatively high degree of uniformity that results. One-pass trenchers fundamentally differ from other commonly used processes such as jet grouting and DSM with vertical-axis mixing tools. With jet grouting (Figure 6.10, discussed in Chapter 11), overlapping columns or panels are constructed by simultaneously cutting and mixing the in situ soil using high-velocity grout, or a combination of air, water, and grout (Croce et al. 2014; Yamanobe et al. 2020). Similarly, auger soil mixing also forms walls by constructing overlapping columns or panels of in situ soil mechanically mixed with injected grout. Thus, all methods have mixing of in situ soils with grout in common. Two key differences between one-pass methods such as the TRD and trenchers are continuity and the zone of mixing. Soil mixing with vertical-axis mixing tools and with jet grouting results in mixing over a limited vertical zone of material at a time. For example, a 2-m thick clay layer located at a depth of 5 m will not be well blended with a 2-m thick sand and gravel layer at a depth of 15 m. Conversely, the trencher systems are a continuous process that promotes vertical homogeneity absent the original soil stratification. Some trencher systems are also applicable in very dense/hard soils and weak rock, without the need for pre-drilling that other methods may require.

In the one-pass methods that use a specialized trencher, the blade and cutting chain are initially rotated from a horizontal position at ground surface to a vertical one at the desired depth. From that point, the equipment moves forward (in the direction of unexcavated trench soils) cutting a continuous trench, on the desired alignment, while mixing the soil in situ with metered amounts of additives (dry mixed or via slurry). In the TRD one-pass method, the cutter bar and chain are inserted vertically and then the equipment moves in the direction of excavation. Because of the speed of rotation of the cutting chain, a relatively

Figure 6.10 Jet Mixing Nozzles (Left) and Exposed Jet Mixed Test Columns (Right). (courtesy of Geo-Solutions, Inc., www.geo-solutions.com).

stiff, heavy mix can be processed. This ensures a uniform mix with no voids or windows. The trenching depth can be varied within a limited range before the withdrawal of the boom and the chain is required to be replaced with a longer or shorter boom and cutting chain. With this trenching process, multi-step installations are condensed into a single-pass operation, hence the one-pass moniker.

The TRD method has been used for nearly 25 years and was also used for a portion of the rehabilitation of the HHD in Florida, USA, as a means to cut off under seepage and piping, as described in Evans and Garbin (2009). As discussed above, this technique was developed and has been widely used in Japan (Aoi et al. 1996; Aoi et al. 1998). The principal advantage of TRD over other in situ mixing techniques is the relatively high degree of uniformity that results from excavating and mixing the entire vertical profile into a single homogeneous mass. One disadvantage is a lack of cold joints which, depending upon temperature and mixing conditions, could leave the wall susceptible to transverse cracking above the water table as the wall shrinks (Evans and Jefferis 2014). At the HHD, the completed wall was cored and photographed and found to be very uniform (see Figure 6.11). Note that the peat layer discussed above was uniformly blended throughout the entire vertical profile and shows up in the photograph as isolated black spots. Deep and shallow grab samples readily met the required strength and permeability requirements. However, due to cracks resulting from hydraulic fracturing during testing, some downhole in situ permeability test results failed to meet the project requirements. Subsequent testing and analysis showed such cracking had no impact on the performance of the completed wall (Cermak et al. 2012).

The original one-pass trenchers were modified versions of trenchers designed for agricultural drain and pipe installations. These tools have been used for soil mixing applications

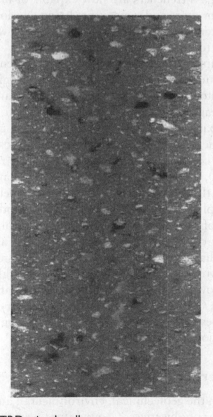

Figure 6.11 Downhole photo of TRD mixed wall.

Figure 6.12 Combination jet/auger mixing (courtesy of Ground Developments Limited, UK, www.grounddevelopments.co.uk).

for decades and the equipment has been improved dramatically through customizations by specialty contractors. One pass trenchers are now capable of installing vertical soil mixed elements to depths of 30 m or more and systems are being developed and used for shallower applications in which the trencher can be used to install very wide linear elements (3+ m) that can also be overlapped to install soil mixed blocks.

Jet mixing, covered in Chapter 11, is the term used for equipment and methods that are used for soil mixing utilizing a fluid stream. The equipment for jet mixing typically consists of a specialized drill bit and monitor mounted on a vertical drill stem and a pump capable of pumping large volumes of fluid reagent at very high pressure. The left panel of Figure 6.10 shows the business end of a jet mixing system with the jets above the ground surface to aid in the visualization of the role of the jets. Jet mixing is effective at almost any depth but is generally limited to erodible soils, e.g. cohesionless materials. To show the results of jet mixing, refer to the right panel of Figure 6.10, showing jet-mixed columns that have been revealed by excavating the unmixed soil around the columns. Jet mixing is most widely referred to as jet grouting.

DSM can also be accomplished with equipment and methods that combine techniques including the combination of a jet fluid stream and augers. Equipment for combination jet and auger mixing typically consists of a specialized drill bit and monitor mounted on a vertical drill stem that also has horizontally oriented paddles, stems, or blades, and a pump capable of pumping large volumes of fluid reagent at very high pressure. This combined system is effective for shallow and deep applications and suitable for a wide range of soil types (Quickfall et al. 2014; O'Brien 2019). There is no industry consensus for naming this technique, but it is sometimes referred to as turbojet soil mixing. An example of equipment used in this application is shown in the photo in Figure 6.12.

6.3.4 Treatment patterns

Soil mixing is widely used in the geotechnical, environmental, and geoenvironmental industries to accomplish a number of objectives. Ground improvement objectives may include:

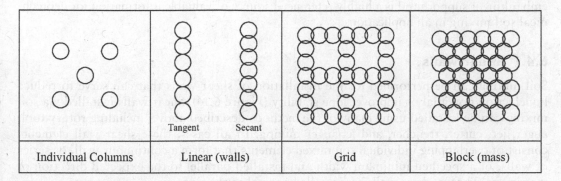

Tangent   Secant

| Individual Columns | Linear (walls) | Grid | Block (mass) |

Figure 6.13 Basic deep mixing treatment patterns.

1. increasing shear strength to improve the bearing capacity of the material,
2. reducing compressibility to improve the material to support additional construction activities such as placement of a final cover, and
3. reducing permeability either linearly to construct a subsurface cutoff wall or in mass to reduce leaching and contaminant transport.

Figure 6.13 shows conceptual patterns or layouts, of soil mixing completed with augers. Similar layouts are available for other soil mixing techniques. Notice columns can be overlapping secant columns or adjacent tangent columns.

The layout is selected with recognition of the construction method, equipment, and application. That is, the layout design reflects the project objectives such as those previously listed. The resulting subsurface properties should reflect the project needs such as structural integrity, low permeability, or other design objectives.

6.4 APPLICATIONS

There are many applications for which soil mixing can be used, including hydraulic cutoffs for seepage control in waterfront or marine applications, structural walls, tunneling, ground treatment/improvements for foundation design or excavation support, liquefaction mitigation, and environmental remediation. The following subsections define some of the primary applications.

Soil mixing is performed for geotechnical ground improvement through the installation of elements to improve slope stability, reduce liquefaction potential, improve bearing capacity, reduce seepage, reduce internal erosion, provide excavation support, or lengthen seepage flow paths. Soil mixing for geotechnical purposes is almost exclusively accomplished via the addition of a binding agent with the most common binding agent being PC. The most common applications of soil mixing for the accomplishment of geotechnical objectives are for the installation of cutoff walls and ground movement control. Generally, cutoff walls installed via soil mixing are used to accomplish multiple site objectives which may include ground movement control or groundwater flow control by reducing seepage and lengthening flow paths.

Soil mixing for ground movement control is also used to accomplish multiple objectives, including stability improvement (slope, or liquefaction potential reduction) and settlement mitigation (bearing capacity improvement or liquefaction potential reduction). Bruce et al. (2013) describe the use of soil mixing for specific geotechnical applications, primarily for embankment support and is a highly referenced source of valuable information for geotechnical soil mixing in all applications.

6.4.1 Shear walls

Soil mixing can be performed for the installation of shear walls that can serve to reduce liquefaction potential or improve slope stability (Figure 6.14). Shear walls installed via soil mixing can be installed using many of the methods described above, including rotary tool, auger, jet, cutter, trencher, and jet/auger mixing. In all cases, these shear wall elements consist of overlapping individual soil mixed elements that form a continuous wall, or series of walls, of a specified minimum width and installed parallel to the expected direction of movement. That is, shear walls are installed normal to the slope.

When used for liquefaction potential reduction, the shear walls may be overlapped to form shear boxes. In the design of soil mixed shear walls, the shear walls are typically assumed to take all of the estimated load during the design critical event. Little to no interaction between the shear walls and adjacent soil is accounted for. In a liquefaction mitigation application, the shear boxes are designed to withstand the full load induced by the design earthquake event and are designed to prevent the soils inside the shear boxes from moving laterally as a result of the event. In a slope stability improvement application, including during a seismic design event, the shear walls are installed normal to the slope of concern and are generally designed to move the critical failure plane deeper, improving the factor of safety.

Shear walls are also used directly under levees and flood walls to provide ground improvement. Stability failure modes improved by shear walls include rotational stability and settlement. While shear walls installed beneath a levee and oriented normal to the levee alignment provide improved levee stability, their use requires examination of failure modes within the shear walls including racking, crushing, and extrusion failures (Adams 2011).

An example of shear walls installed via soil mixing occurred over multiple phases between 2013 and 2019 at a Landfill facility as described in Ruffing et al. (2017). The shear walls were installed to stabilize a soft clay layer to ensure stability of the existing landfill during adjacent excavations, unloading, for construction of new cells. The primary design criteria for these projects included a 25% aerial treatment coverage, a maximum shear wall to shear wall spacing of 3.5 times the diameter of the columns, a target composite shear strength of 48 to 72 kPa, and a minimum column to column overlap of 10% of the diameter. Using a UCS to shear strength ratio of 2 and the target aerial coverage of 25%, the strength criteria were converted to a target minimum UCS of 380 kPa to 570 kPa. Over the course of

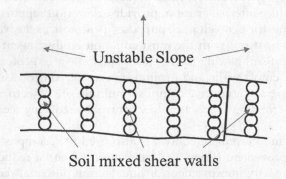

Figure 6.14 Soil mixed shear walls adjacent to the unstable slope.

all phases of the work at this facility, over 700 soil mixed columns were installed for the improvement of more than 25,000 m³.

6.4.2 Aerial bearing capacity improvement

Soil mixing can also be performed to improve the bearing capacity of weak or unsuitable soil layers. Soil mixing in this application is generally accomplished via installation of individual isolated soil mixed elements for DSM applications. The number and layout of individual columns is defined during the design phase. Aerial bearing capacity elements installed via soil mixing can be installed using many of the methods described above, including excavator bucket, rotary tool, auger, jet, cutter, trencher, and jet/auger mixing. Alternatively, treatment can be performed over the entire mass of unsuitable soil for shallow applications using equipment and methods previously described.

Generally, the design of soil mixing in this application is performed by calculating an effective improved area, e.g. 25% or 30%, that achieves the overall composite strength objective. Generally, the untreated soils are assumed to have no strength in this calculation, i.e. the soil mixed elements are designed to take the full load. In many cases, a load transfer platform is needed above the columns to ensure the surface load is spread evenly to the column elements.

Example problem Ex. 6.1:

A project requires ground improvement to produce an average undrained shear strength $(S_{u\,req}) = 70$ kPa. Bench studies have shown that soil mixed material will have a UCS = 700 kPa. For this case study, the soil undrained shear strength is conservatively neglected. Recall from Chapter 2 that the UCS = $2 \times S_u$.

What should be the target aerial improvement (TAI) objective?

$$TAI = (S_{u\,req} / 0.5 UCS)(100)$$

$$TAI = (70 / (0.5)700)(100)$$

$$TAI = (70 / 350)(100) = 20\%$$

Note in the example problem Ex. 6.1 that, in reality, the site soil has some non-zero shear strength. However, when considering stress-strain compatibility, very little of that strength will be mobilized during the movement of the composite system since the soil mixed material is very stiff compared to the site soil. Since the soil mixed material is much stiffer and stronger than the surrounding soils, the soil mixed elements will carry virtually all of the load for small deformations. As the deformation increases, the soil will begin to pick up more of the load, but it is conservative to ignore the soil shear strength unless large deformations are tolerable.

6.4.3 Hydraulic cutoff walls

Soil mixing can also be used to reduce soil permeability (hydraulic conductivity) and create a vertical barrier to reduce the flow of groundwater, lengthen the flow path, reduce internal erosion, and reduce seepage. Not all of the soil mixing methods described above are suitable for the installation of cutoff walls. Of the remaining methods, some are more suited than others. Generally, cutoff walls installed using soil mixing are installed using the auger, cutter, trencher, or jet/auger mixing methods. Jet mixing is also occasionally used to supplement other methods in difficult areas. Soil mixing in this application is accomplished through the installation of a continuous element of a specific minimum width by overlapping individual soil mixed elements. Cutoff wall elements installed via soil mixing can also serve to accomplish strength improvement objectives. One example of this would be when soil mixing is used to install a core wall through a dam or embankment. The goal of all SMW procedures is to produce a nearly homogeneous product having certain desired properties of permeability and strength through the addition of slurry and mixing with the in situ soils. For this reason, the techniques have much in common.

One example, among many, highlights the use of soil mixing for installation of hydraulic cutoff walls. On this project, as described in Walker (1994a,b), flood routing studies indicated the need to raise the existing dam core to the crest of the dam to maintain dam safety. The core extension was accomplished using soil mixing with CB slurry to form an in situ mixed soilcrete wall.

In another application, soil mixing with augers was used to remix, in situ, an SB wall constructed to a depth of 27 m after it was found to contain construction defects. By remixing the original cutoff wall, localized defects were homogenized with the suitable backfill in the trench, thereby removing the defects.

6.4.4 Excavation support walls

Soil mixing can also be used in the geotechnical market for the installation of excavation support walls. These walls can be gravity structures constructed of unreinforced (no tensile reinforcement) soil mixed materials (Ruffing et al. 2012) or soldier pile and lagging walls where the soil mixed material serves as the lagging. Photographs from an example project are shown on the left of Figure 6.15. A simplified schematic of an in situ mixed gravity wall is shown on the right side of Figure 6.15. Depending on the depth of the excavation support, many of the methods described above are suitable for use in this application, including excavator bucket, rotary tool, auger, jet, cutter, trencher, and jet/auger mixing. For shallower applications, excavator buckets, rotary tools, and auger methods are most common. For deeper applications, auger, cutter, and trencher methods are most common. Jet mixing is used to supplement the other methods in logistically challenging areas.

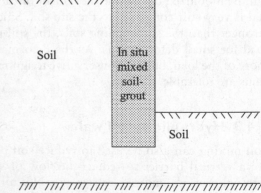

Figure 6.15 In situ mixed gravity wall photos and schematic.

On one project, soil mixing with a four-auger configuration (left side of Figure 6.16) was used to construct a combined groundwater cutoff and excavation support wall for a deep excavation for a highway. The required moment capacity was developed by inserting steel beams in the soil-mixed material immediately after mixing. The result was similar to a conventional soldier pile and lagging wall with the steel beams serving as the soldier piles and the soilcrete serving as the lagging. The top of the wall with the soldier piles is shown on the right side of Figure 6.16.

On another project, soil mixing was used to construct a wall to facilitate the construction of a 1,000-m long highway extension built up to 9 m below grade and below the water table, as described in Bahner and Naguib (1998). The soil-mixed wall was installed to a depth of 12 to 18 m and was required to allow a maximum groundwater infiltration rate of 4.3 l/min per meter of the wall with lateral wall movements not to exceed 25 mm. The glacial geology offered a variable profile with layers of fill, sand, silt, and clay, all underlain by a very stiff glacial till. Jet grouting (Chapter 11) was used to complete the soil mixing around utilities. Using a four-shaft (each 0.9 m in diameter) drilling rig similar to that shown in Figure 6.16, left side, the wall was formed by mixing the in situ soils with CB slurry. Before the resulting soil mixed material set, steel soldier beams were set into place with a small vibratory hammer on 1.4 m spacing. The soldier piles were restrained with internal bracing and with tiebacks. The CB slurry for this project was prepared in a batch plant schematically shown in Figure 6.17.

Using field mixed samples cast into molds and cured for 14 days, the hydraulic conductivities were in the range of 1×10^{-6} and 9×10^{-8} cm/s, better than the target of 1×10^{-5} cm/s. The target strength of 500 kPa was exceeded by a factor of 2 to 4 after only three days of curing. Such performance is not uncommon for in situ mixed barriers as contractors and designers recognize the intrinsic variation in field mixing and are normally, and properly, conservative in the development and selection of mix designs. Field measures of groundwater and deflections (using inclinometers) found the project met both groundwater control and inclinometer (deflection) criteria.

Additional details of the performance of SMWs under other design and geological conditions are available in the literature, e.g. O'Rourke and O'Donnell (1997), Taki and Yang

Figure 6.16 Auger mixed wall showing augers (left) and steel beams inserted for moment capacity.

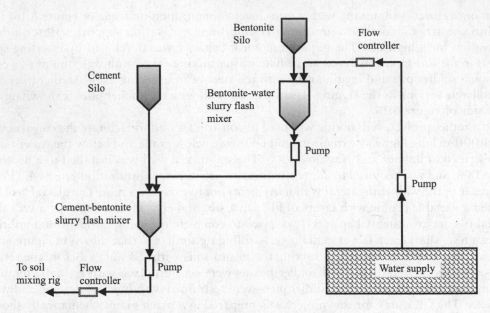

Figure 6.17 Soil mixing batch plant flow diagram for a CB slurry/grout.

(1991), and Yang (2003). Methods of analysis to determine the structural load and resistance resembles that for soldier pile and lagging design, which are also available in the literature (Rutherford et al. 2007).

6.4.5 Environmental soil mixing

Soil mixing is also performed for environmental ground improvement. The improvement may result from solidifying the waste material, stabilization of the waste material to minimize contaminant transport, and from contaminant mass destruction via processes including chemical oxidation or reduction. Contaminant transport may be reduced by soil mixing as a result of one or more mechanisms including microencapsulation, macroencapsulation, absorbtion, adsorption, precipitation, and/or detoxification (LaGrega et al. 2010). As an example, one of the more common applications of soil mixing for environmental remediation is the use of in situ stabilization/solidification for controlling the movement of difficult contaminant matrices, e.g. nonaqueous phase liquids (NAPLs) like those found at manufactured gas works sites (Jayaram et al. 2002). On many environmental projects, there are also secondary geotechnical improvement objectives. For instance, most environmental soil mixing projects include a minimum strength improvement objective that is needed for efficient future site use or for the placement of a final cover over the site.

The term solidification generally refers to the process of mixing additives (also known as reagents) with the waste to result in a solidified block of material. Solidification, in and of itself, increases the material strength and reduces compressibility and generally reduces or limits the mobility of contaminants in the subsurface through encapsulation by incorporation of the contaminant(s) into monolithic blocks. The term stabilization generally refers to the process of mixing additives to reduce the hazardous nature and/or mobility of the contaminants. Stabilization generally relies on chemical reactions to make contaminants less mobile or less harmful by reducing toxicity. Mobility can be assessed by leaching tests. Leachable reduction may be due to binding the contaminants to less mobile particles as a result of mechanisms such as precipitation and adsorption or by reducing contaminant mass via detoxification due to oxidation or reduction.

The terms stabilization and solidification are often used synonymously and simultaneously along with the preface in situ to describe processes of treating or encapsulating contaminants in the subsurface to reduce impacts on the surrounding soil and groundwater. By definition, ISS can be used to describe nearly every type of soil mixing and can even be broadened to describe most geotechnical applications, although the term is generally reserved for environmental soil mixing. With ISS, the processes of stabilization and solidification both contribute to the end result, reduced contaminant impact on the subsurface. Generally, no effort is made to evaluate or assess the relative contribution of each. The most common reagent used in ISS is PC, but other common reagents include ground granulated blast furnace slag, activated carbon, bentonite, and lime. The most common construction methods used for ISS are bucket mixing, rotary mixing, and auger mixing. Bucket and rotary mixing are reserved for shallower applications and auger mixing is used for the deeper applications. Jet mixing is sometimes used to supplement another method in logistically challenging areas of a site where one of the other techniques would not be convenient or feasible.

The term in situ chemical oxidation (ISCO) is used frequently within the environmental remediation industry to describe processes in which the reaction with oxygen to a contaminant molecule is used to render the contaminant inert or less harmful. This term describes the use of oxidants for the in-place treatment of a contaminant mass. Common reagents used in soil mixing applications of ISCO include potassium permanganate, hydrogen peroxide, sodium persulfate, and ferrous sulfate. Auger mixing is generally the preferred construction technique for ISCO via soil mixing, but bucket or rotary mixing can be used as well if the oxidant addition can be controlled and if it is safe to do so. ISCO performed via soil mixing is a subset of IST.

The term in situ chemical reduction (ISCR) is frequently used within the environmental remediation industry to describe the removal of oxygen from a contaminant molecule to render the contaminant inert or less harmful. This term describes the use of reducing agents for the in-place reduction of a contaminant mass. The most common reagent used in soil mixing applications of ISCR is zero-valent iron (ZVI). Calcium polysulfide is also used. Auger mixing is generally the preferred construction technique for ISCR via soil mixing, but recent projects have shown the cost-effectiveness of bucket or rotary mixing for shallow applications. ISCR performed via soil mixing is a subset of IST.

In environmental applications, soil mixing is a means of delivering various reagents to the subsurface in a way that ensures exposure of the reagents to the contaminated soil or groundwater almost independent of lithology. Soil mixing in this application is only limited by the natural limitations of the equipment used (ever improving) and the creativity of the designer or contractor charged with developing the reagent mixture (Ruffing et al. 2021).

A detailed overview of stabilization and solidification of contaminated soils is presented in Bates and Hills (2015) which covers the topic from an international perspective with contributions from a range of authors spanning regulators, designers, and contractors. Bates (2010a,b) covers performance specifications for stabilization and solidification and provides an overview of the use of S/S for US Superfund sites.

6.4.6 Geoenvironmental soil mixing

Soil mixing for combined geotechnical and environment purposes is termed geoenvironmental ground improvement. For the purposes of this discussion, this category of soil mixing refers to soil mixing used on an environmental site for strictly geotechnical improvement or on a geotechnical site for accomplishing combined geotechnical/environmental objectives. An example of this type of soil mixing is the installation of a cutoff wall designed for groundwater flow control (contaminant movement control is not the primary objective but may be accomplished via advective flow control) or improvement of contaminated sludge or sediment for access and capping. Another example is the installation of an excavation support wall through contaminated soils wherein the soil mixing also solidifies the contaminants, reducing long-term environmental impacts.

Example problem Ex. 6.2: Shear strength of soil-mixed sludge

A site requires soil mixing to improve the shear strength of a weak sludge-like material for future access and cap placement. For preliminary planning purposes, the client would like to understand what the strength of the soil mixed material would need to be for safe equipment access and for placement of a 1 m soil cap. Assume: the bearing pressure of the largest piece of equipment is 100 kPa and the soil for the cap weighs 18 kN/m³.

Solution:

For preliminary planning purposes, it should be acceptable to assume that the shear strength of the soil mixed material will need to be twice the bearing pressure of the equipment or soil load, whichever is greater. Based on the above assumptions, the 100 kPa equipment load will control as the soil load would be smaller at $(18 \text{ kN/m}^3)(1 \text{ m}) = 54$ kPa. In this case, the UCS of the soil mixed material should be approximately 400 kPa. In providing this preliminary response it is important to relay to the client that this is based on a rudimentary analysis with many simplifying assumptions and no consideration for long-term or dynamic load impacts. In practice, it is not uncommon to use simplified analyses like this to determine a preliminary result which can then be checked through a more rigorous analysis within the context of the larger, more complicated problem.

6.5 DESIGN CONSIDERATIONS

In general, soil mixing project design progresses through the following steps:

1. Determine project needs.
2. Translate project needs into target design parameters.
3. Select addition rates for reagent(s) needed to achieve target design parameters.
4. Develop construction objectives to achieve objectives in (2) and (3).
5. Evaluate construction objectives for overall compliance with the needs identified in Step (1).

Because of the specialty nature of soil mixing and the relative inexperience of the majority of general design practitioners and contractors, soil mixing design often necessitates the involvement of a specialty designer and/or construction team in the design process or a design-build project approach. If the project is approached in a design-bid-build manner, a construction team may be engaged to assist with target parameter selection in step 2 or will almost certainly be involved by steps 3 or 4. The following sections break down the design process used for soil mixing projects into greater detail, identifying specific considerations.

6.5.1 Determine project needs

The project needs determination for soil mixing projects follows the same principles and procedures as any subsurface construction project but varies depending on whether the site objectives fall strictly into the areas of geotechnical, environmental, or geoenvironmental concerns or if there are cross-disciplinary objectives. Chapter 2 provides approaches and fundamentals for ground improvement design which should be used in the early stages of soil mixing projects as well. Specific items that should be considered at this stage of a soil mixing project may include:

1. What type of soil mixing can/should be used to accomplish the overall site objective(s)? Given the specialized nature of soil mixing and the subtleties associated with the various construction methods, it is important to consider the expected construction approach early in the project for the development of a design that harmonizes with the anticipated construction approach. Conversely, it is also important to maintain flexibility in the design for construction team optimization at the bid stage to prevent unnecessary construction costs associated with an unreasonably strict design.
2. Are there other methods that are or could be more cost-effective?
3. What aspects of the site are going to control the construction cost and are there ways to reduce cost or schedule through modified design? Attention must be paid to the presence of challenging layers (e.g. organics), site logistical constraints (space), and underground/overhead obstructions or utilities.

6.5.2 Select target design parameters

Once the project objectives are clearly defined, and one or more soil mixing construction techniques that can accomplish the objectives have been identified, the objectives must be converted into measurable parameters. The most commonly evaluated parameters for geo-technical soil mixing projects are strength and permeability, although the elastic modulus and/or Poission's ratio may be needed for higher-level designs (not covered here). For environmental projects, the common parameters are strength, permeability, and leachability. In both applications, other parameters may include freeze/thaw or wet/dry durability and seismic performance (shear wave velocity, dynamic strength performance). Table 6.1 shows

Table 6.1 Summary of typical properties of soil mixed with cement by wet and dry methods

Property	Wet Methods	Dry Methods
Unconfined compressive strength (UCS, MPa)	0.3–10	0.2–5
Permeability (cm/s)	10^{-6}–10^{-7}	Approximately equal to soil permeability
Elastic modulus (E_{50})	350–1,000 times UCS – lab samples 150–500 times UCS – field samples	100–600 times UCS
Shear strength	40–50% of UCS ratio may decrease as UCS increases	
28-day UCS (vs. 7-day UCS)	50–100% greater depends on soil type and w/c ratio	
60-day UCS (vs. 28-day UCS)	10–50% greater Long-term strength gain (years) can be significant, e.g. 300%, Grouts with high w/c ratios exhibit less strength gain beyond 28 days.	
Tensile strength	10–20% UCS	
Strain at failure	1–3% Strain at failure decreases with increasing UCS	1–2%

various geotechnical parameters and representative achievable ranges for soil-reagent blends created via soil mixing with dry and wet binder addition (some information pulled from Bruce et al. 2001 and Bruce and Bruce 2003).

Bruce et al. 2013 define commonly used acronyms for key features of geotechnical soil mixing project design and a step-by-step process for soil mixed system design for embankment support. This book refers to soil mixing properties according to common fundamental material property names (e.g. see Chapter 2) or commonly used industry terminology.

6.5.2.1 Strength

Most, if not all, soil mixing projects include a target strength. Most often the target strength parameter is a minimum UCS, with some projects also including a maximum UCS. Occasionally, the strength requirement is in terms of undrained shear strength. However, representing strength in terms of shear strength is not recommended because this will require the construction team to make an assumption about the relationship between shear strength and UCS (see discussion of Mohr's circles in Chapter 2). By definition, the undrained shear strength is 50% of the peak UCS so it would seem that specifying undrained shear strength or UCS is interchangeable. While, by definition, the undrained shear strength is 50% of the UCS, 30% to 40% are sometimes used in practice to account for uncertainty in shear strength measurement. The decision to use a shear strength to UCS ratio that is less than 50% hinges upon how the problem is approached, primarily whether an allowable stress or a load resistance factored design procedure is being used. Using a UCS to shear strength ratio less than 50% will result in conservatism in the calculated target UCS, whereas the UCS calculated from the actual relationship has no additional safety factor.

Sometimes soil mixing project designs include a maximum UCS. This is not recommended and should only be included when there is ample bench-scale data to suggest that a "window" between the minimum and maximum target UCS values can be consistently achieved.

As described in Ruffing et al. 2017 and Ruffing et al. 2019, among other sources, it is important to recognize that soil mixing inherently results in a material with highly variable properties, with especially large variability in strength. A soil mixed product's strength variability results from variability in the soils being mixed, variability in the distribution of reagents being mixed, and variability in the energy of mixing during the construction process. The soil mixed product strength should be expected to vary more than the base soils themselves due to the cross mixing of horizontal and vertical soil strata and the potential interfaces between soil mixed materials containing different or intermixed base soils. The strength of the product of DSM has been found to have a coefficient of variation of 0.3 to 0.8 (Filz and Navin 2010) with an average of about 0.55 (Navin and Filz 2006). This statistical variability is influenced by the inherent variability in the base materials (see above) and the inherent variability of the sampling and testing procedures. Design strengths for soil mixing should be selected with this variability in mind. The resulting specifications should incorporate flexibility for dealing with this variability during construction.

The design strengths for soil mixing performed on environmental sites are generally selected to bring the mixed soil strength up to or slightly above a UCS that could be considered equivalent to the soils prior to treatment. Most in the industry consider a minimum strength of 200 to 350 kPa suitable for accomplishing that objective. For environmental projects where subsequent grading and construction is required, it would seem reasonable to consider a maximum UCS such that future site use is not negatively impacted by an overly strong stabilized material. However, hitting a window is challenging and this may result in challenges during construction. As a result, absolute maximum UCS requirements are not recommended. The preferred approach is to consider future site use and build areas of clean fill within the soil mixed monolith. This is preferable because this approach (1) prevents damage to the soil mixed monolith associated with future site activities, (2) prevents future workers from coming into contact with contaminants that are locked in the soil mixed materials, and (3) should result in a lower cost of future construction because of easier working conditions.

Example problem Ex.6.3: Strength statistics

If strength tests on a soil mixed material show a coefficient of variation of 0.5 and the standard deviation is 100 kPa, what is the mean strength? What are the median, minimum, and maximum strength?

Solution:

The coefficient of variation is the ratio of the standard deviation to the mean so the mean strength is 200 kPa. The information provided tells nothing of the median, minimum, or maximum.

In developing strength objectives and methods for measuring strength, it is important to consider scale. For instance, strengths measured on wet grab samples are often ignored because of a belief that the strengths of these samples are uncharacteristically high. However, strengths of wet grab samples have actually been shown to be less than those of cores. Since wet grab samples are much easier and less expensive to obtain, this method of evaluation should be considered. Further, the soil and rock inclusions that are inevitably part of small wet grab samples (see Figure 6.18) actually should make the samples representative of macro-scale soil mixed columns with unrealistically large (> 0.6 m) inclusions that are not allowed anyway. For this reason, the strength of a wet grab sample may actually underrepresent the macro strength of the soil mixed material.

Factors affecting strength development for a soil mixed material are identified in Table 6.2 (modified after Terashi 1997). The final strength of mixed soils is directly dependent on what binder is chosen, the type of mixer used, the soil being mixed, the mixing duration, other factors affecting curing of the binder, and characteristics of the native soil prior to improvements.

One of the variables affecting strength is pH. A lower pH in the native soil porewater decreases strength increase from PC by about 50% (Babasaki et al. 1996).

In general, strength increases with time for all binding reagents and native soils with few exceptions. This increase is variably linked to the quantity and type of binding agents used to stabilize and treat the subsurface (Ahnberg 2006). Soil mixing with dry binders decreases water content, increases density, causing permeability to initially decrease or increase in some instances due to the effect of the specific binder. However, over time (independent of curing time), for cementitious binders, the permeability, discussed further in a subsequent section, of the treated soils will be less than the initial native permeability (Ahnberg 2006).

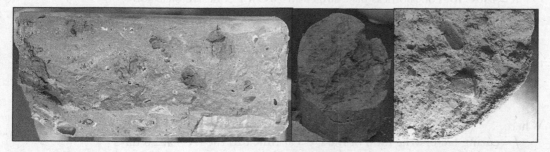

Figure 6.18 Cross-section photos of wet grab samples of soil mixed materials showing inclusions (Ruffing et al. 2017, Ruffing et al. 2021).

Table 6.2 Factors affecting the strength of soil mixed composites

Category	Details
Reagent/Grout	Type of reagent
	Reagent dosage
	Mixing water properties
	Additives (e.g. dispersants)
Base soil	Physical properties
	Mineralogical properties
	Organic content
	Porewater chemistry
	Moisture content
Mixing	Degree of mixing
	Timing of mixing and remixing
Curing	Temperature
	Curing time
	Humidity
	Wet/dry and/or freeze/thaw cycles

There are varying degrees of improvement of soil mixed strength based on differences in the base soils and the reagents (Ahnberg and Johansson 2005). Further, as observed in other techniques, e.g. self-hardening slurry trenching materials, the length of time to achieve the "final" mix properties depends on the amount and type of reagents used. For instance, mixes containing slag tend to improve considerably after 28 days of curing whereas the improvement of mixes containing only ordinary PC improve marginally after 28 days. In the referenced study, the relationship for all shows an increase in strength with time. This is extremely variable based on the type of native soil, and type and quantity of binder used. The overall importance of the results of that study is: taking a time factor (typically due to curing) into account for estimating the strength of a stabilized soil is crucial for design. Mitchell (1976) offers the following relationship for estimating the UCS as a function of time for soil mixing with PC:

$$UCS(t) = UCS(t_o) + K \log \frac{t}{t_o} \qquad (6.1)$$

where

$UCS(t)$ = unconfined compressive strength at t
$UCS(t_o)$ = unconfined compressive strength at t_o
$K = 480C$ for cohesionless base soils and $70C$ for cohesive base soils
C = cement content, % by mass

Bruce et al. (2013) offer an alternate equation for calculating strength as a function of time through calculation of a curing factor, f_c:

$$f_c = 0.187 \ln(t) = 0.375 \qquad (6.2)$$

Where

f_c = curing factor = the ratio of the UCS at time t to the UCS at 28 days
t = curing time (days)

Other relationships for strength vs. time exist and it is best to develop site-specific curves in a bench-scale or pilot study for use during the full-scale construction.

The tensile (flexural) strength of soil mixed materials is a small fraction of the compressive strength, estimated at a ratio of 5% to 30% of the compressive strength with the lower ratio applying to higher strength mixes and the highest ratio applying only to very low strength mixes.

6.5.2.2 Hydraulic conductivity

In the realm of soil mixing, the terms hydraulic conductivity and permeability are often used interchangeably. For the purposes of this text, hydraulic conductivity will be used and is defined in Chapter 2. The value of hydraulic conductivity, k, is determined in a permeability test. Commonly, potable water is used as a permeant. In some cases, the samples are permeated with site groundwater to determine k that is reflective of the site groundwater properties including the effects of water chemistry due to contaminants and their concentration, salts, minerals, and pH. Since soil mixed materials have inherent property variability, the expected variability in hydraulic conductivity should be considered at all stages of design and construction. Experience with conventional hydraulic conductivity testing has shown that hydraulic conductivity can be expected to vary by plus or minus one-half an order of magnitude between specimens that would otherwise be considered identical. That is to say, duplicate specimens cast from the same bulk sample set using similar casting methods, stored in a similar environment, and tested in a similar manner could yield hydraulic conductivities that differ by as much as one order of magnitude.

Hydraulic conductivity objectives for geotechnical projects vary. For shear walls or aerial bearing capacity improvement, hydraulic conductivity is often not specified. For cutoff walls or excavation support, specified hydraulic conductivity values in the 1×10^{-6} cm/s to 1×10^{-7} cm/s range are common. Most environmental applications of soil mixing include a specified maximum hydraulic conductivity of 1×10^{-6} cm/s. Specifying a target maximum hydraulic conductivity value lower than 1×10^{-7} cm/s is not recommended for several reasons. Such low values are difficult to consistently achieve in the field. Similarly, laboratory tests seeking to measure hydraulic conductivity values less than 1×10^{-7} cm/s are difficult to perform reliably and consistently. Finally, lower hydraulic conductivities may not improve the product performance because at values of 1×10^{-7} cm/s or lower the dominant contaminant transport mechanism is diffusion, not advection. Thus, lowering the value of hydraulic conductivity below 1×10^{-7} cm/s does not change the contaminant transport within the stabilized mass in any meaningful way.

In general, the hydraulic conductivity of mixed and treated soil decreases over time. The decrease in hydraulic conductivity with time is particularly evident when GGBFS is used in the mixture. Occasionally, the hydraulic conductivity may increase with time, depending on the amount and type of binder used. Brandl (1999) shows the influence that the amount and type of binder have, with respect to curing time, on the hydraulic conductivity of clays mixed with various amounts of cement and lime. That study showed that, as the amount of lime increases or the amount of cement increases, and curing time increases, the hydraulic conductivity of the soil mixed material decreases.

The ratio of the hydraulic conductivity of the soil-reagent blends to the unstabilized soils is directly related to water content before and after stabilization, and the UCS of the soil. Note that hydraulic conductivity measured in the laboratory should be considered a lower limit due to large-scale variations which cannot be accounted for in laboratory testing (Ahnberg and Johansson 2005).

Hydraulic conductivity evaluation of soil mixing can result in large variability in test results depending upon sampling and testing methods. Laboratory tests of hydraulic conductivity on bulk samples often produce lower values than those on core samples due to difficulties in obtaining a truly undisturbed sample (core). Likewise, in situ hydraulic conductivity test results may be variable due to damage of the borehole wall during drilling, hydraulic fracturing, and difficulty in maintaining an adequate seal with the packer (if used). The case study of a soil mixed wall for a seepage cutoff in levee reconstruction provides excellent insight into the difficulties associated with evaluating the as-built performance of a soil mixed material in a cutoff wall (Yang et al. 1993). Careful investigations of hydraulic conductivity at the Herbert Hoover Dike included both laboratory testing and down-hole in situ permeability testing. Down-hole permeability testing was found to induce cracking in some of the tests and, thus, measured values of hydraulic conductivity were much higher than those measured in the laboratory (Cermak et al. 2012).

6.5.2.3 Leachability

Leachability, in the context of soil mixing, refers to the propensity of aqueous phase contaminants to migrate from the stabilized/solidified material into the surrounding soil and/or groundwater. In the US, leachability criteria are specified on approximately less than half of environmental soil mixing projects and rarely specified on geotechnical projects. Leachability is assessed with a variety of standard methods, including the Toxicity Characteristics Leaching Procedure (TCLP) (USEPA 1992), the Synthetic Precipitation Leaching Procedure (SPLP) (USEPA 1994), the "Measurement of the Leachability of Solidified Low-Level Radioactive Wastes" test (ANS 16.1) (ANSI/ANS-16.1-2003), and the more recent and more robust Leaching Environmental Assessment Framework (LEAF, USEPA Methods 1314 to 1316) testing procedures (Garrabrandts 2010; Kosson et al. 2014; USEPA 2017 a,b,c). The LEAF approach is the most robust and complete assessment tool, but it is common for practitioners to use TCLP or SPLP because owners and regulators are familiar with these tests and the tests take considerably less time and are much less expensive. Results from TCLP or SPLP tests end with a limited and often skewed representation of the overall leaching characteristics. Practitioners that use these methods must understand the limitations and evaluate the results with those limitations in mind. This may be difficult to do within a regulatory framework designed for the evaluation of *disposal* wastes not for the design of a soil mixed monolith.

Leachability assessment is commonly performed during the bench-scale studies and then at a reduced frequency or not at all during the fieldwork. If leachability is included as an assessment methodology for the fieldwork, it is often performed as a quality assurance (QA) test, i.e. the costs of the testing and the results are borne by the owner or the owner's representative rather than the contractor, and the contractor is not held responsible for the leachability performance.

6.5.3 Reagent addition rates

After the target objectives have been translated into measurable parameters, the reagent selection process can begin. Reagent selection for soil mixing is generally based on the results of a bench-scale study performed using site soil samples. Although experienced industry practitioners will have some sense of the type and quantity of reagents needed to achieve the target parameters, a bench-scale study will almost always be needed to confirm and refine initial assumptions. The cost of determining and optimizing the reagent addition rate is small compared to the cost of re-treating soils that fail to meet the target parameters or overtreating due to conservative assumptions.

Bench-scale studies for soil mixing should be completed by an experienced laboratory or construction team. The size, and related cost, of a bench-scale study will depend on the project requirements. For many projects, the study begins with four samples of the site soils. These samples should be representative of the range of properties that will be encountered in the full-scale work. Soil sampling should reflect vertical and horizontal variability. On environmental projects, contaminant concentration and makeup must also be assessed. The samples should be evaluated for natural moisture content, sieve analysis, Atterberg limits, loss on ignition (LOI), soil pH, and single point proctor testing. Contaminated samples should be subjected to baseline analytical testing to determine the initial concentration and type of contaminants. Analytical testing might include an analysis of total contaminant concentrations and/or leachability assessment.

After the index testing, mix development begins using all of the site soil samples subjected to index testing or a smaller group of representative site composites created by combining the soil samples according to the index test results. On most sites, at least two composites are carried forward into the mix development phase: one that is representative of average-case conditions and one that is representative of "worst" case conditions. In this application, the worst-case soil would include high fines content, high clay content, plastic clays, high organic content, very high or low pH, high moisture content, or some combination thereof. Studies conducted using only worst-case conditions will result in unnecessarily conservative reagent mixes. It is possible to consider a zoned treatment approach wherein a different reagent mix can be used in different conditions across the site. For practical purposes, the designer should select one to three horizontal zones and one vertical zone. It is difficult and costly to vary reagent dosage by depth. Furthermore, the reagent combination, i.e. weight of reagent by weight of slurry or grout, should be consistent across the site with only total reagent dosage varied by zone.

Prior to mix development, the designer must consider whether the soil composites are representative of field conditions. First, the sample composites must be well homogenized. This is important to ensure that observations of mix performance yield findings that are due to differences between the mixes, not variations in base material properties. Additionally, the soil composites may need to be conditioned prior to use in the mixes. For example, it is common for samples to dry between sample collection and laboratory use. If the full-scale work is intended to take place in saturated conditions, then the moisture content of the bench-scale composites may need to be conditioned to the field moisture content. Maximum particle size should be considered. The soil composites may include particles that would have an unrealistically large impact on the performance of the relatively small specimens that are used in lab testing. Generally, remove all particles that are greater than or equal to 15% to 25% of the testing specimen diameter. Maximum particle size should be selected using recommendations from applicable ASTM, BS, CEN, or other standards. Finally, in studies for environmental sites, the soil composites may need to be "spiked" with contaminants to ensure that the initial contaminant concentration is consistent with the field concentration. If no specific guidance information is available, the designer will need to use judgment to select the properties for sample conditioning.

Once the site composites have been selected and conditioned, the designer selects the type and quantity of reagents to mix with the site soils. Initial reagent selection is generally based on past experience. In most studies, an initial round of mixture development is used to assess gross material improvement by adding a range of PC doses. Once the designer has a sense of the total reagent addition, related items are evaluated, such as additional reagents in conjunction with, or replacement of PC, optimized reagent dosage, and variable water to reagent ratios. In the absence of a standard method for preparing soil-reagent mixtures, the procedure from Andromalos et al. (2015) is recommended:

1. Sieve the soil sample material to remove unrepresentatively large particles. Generally, a 12.7 mm (0.5 in.) sieve is used in this application.
2. Mix an appropriate volume of slurry using a mixing procedure that mimics field application, e.g. high shear mixer for bentonite vs. low shear paddle mixer for cement, etc. Set aside.
3. Measure appropriate amount of soil.
4. Add slurry from Step 2 to soil from Step 3.
5. Mix slurry with soil by hand, or using table-mounted mixer, until the material is visually homogeneous. Depending on soil and slurry characteristics, this generally requires at least five minutes of continuous effort. Be careful not to use excessive energy to break particles apart that will not break apart in the field.
6. To reduce overall sample preparation, measure slump using a laboratory-sized mini-slump cone (Malusis et al. 2008) or use visual indicators to determine if the soil-reagent mixture is suitably workable. For soil mixing, suitably workable materials generally have a slump greater than 127 mm (5 in.) measured using a slump cone with a height of 305 mm (12 in.). Rule of thumb: if the mixture closes a 12.7 mm (0.5 in.) gap (created by running a tool through the mixed material in the mixing bowl) under its own weight, then it's suitably workable. If the mixture is not suitably workable, add slurry to improve workability. Adding water directly to the mixture is not recommended, as this is not common in the field. The preferable approach is to add slurry or grout or remix the mixture using a higher water to solids ratio slurry or grout.
7. Create individual soil-reagent test specimens by casting soil-reagent mixture in plastic cylinders. Cylinder size should be selected based on the geotechnical laboratory's criteria for each desired test. For example, most laboratories prefer to run permeability tests on 76.2 mm (3 in.) diameter specimens and UCS tests on 50.8mm (2 in.) diameter samples. Steps 8 through 16 are devoted to the casting procedure.
8. Fill 1/3 of the cylinder with the wet soil-reagent mixture.
9. In an effort to remove the air voids, not to compact the specimen, rod the wet mixture in the cylinder 20 to 25 times using a rod with a diameter that is 10% to 15% of the cylinder diameter.
10. Tap the 1/3 full cylinder against a hard surface 20 to 25 times.
11. Fill the cylinder to 2/3 full.
12. Repeat rodding and tapping sequence from Steps 9 and 10.
13. Fill the remaining 1/3.
14. Repeat rodding and tapping sequence from Steps 9 and 10.
15. Screed the surface of the cylinder using a trowel or other sharp edge.
16. Cap the cylinder.
17. Label the cylinder with the sample identification and cast date.
18. Place the recently cast cylinders in a water-filled, sealed, insulated cooler to minimize temperature fluctuations and sample drying and store to cure, undisturbed, prior to testing. Other curing environments may be used if they better represent in situ conditions.

These procedures can be modified by an experienced practitioner to be consistent with the expected full-scale approach. For example, a project that involves jet mixing may require a different bench-scale study approach. Once the mixtures have cured, individual specimens are subjected to laboratory testing to determine strength, permeability, leachability, and other properties.

Example problem Ex.6.4: Cement addition rate

Soil mixed material UCS target = 850 kPa
Total soil unit weight = 18.0 kN/m^3
Gravimetric moisture content, w = 30%
Mix design study strength results (PC by % of dry soil weight):

$$10\% = 800 \text{ kPa}, \quad 15\% = 1400 \text{ kPa}, \quad 20\% = 2100 \text{ kPa}$$

Factor of safety (FS) = 1.5

Solution:
What cement addition rate should be used (kN cement per m^3 of soil)?

$$\text{Dry soil unit weight} = \frac{\text{Total soil unit weight}}{(1+w)}$$

$$\text{Dry soil unit weight} = \frac{18.0}{(1+0.3)}$$

$$\text{Dry soil unit weight} = 13.8 \text{ kN/m}^3$$

Bench Scale Strength Target = (Soil Mixed Material UCS Target) × (FS)

$$= 850 \text{ kPa}(1.5)$$

$$= 1275 \text{ kPa}$$

Assume linear distribution of strength between data points. The addition rate falls between 10% and 15%.

$$\text{Addition rate \%} = 15\% - (1400 \text{ kPa} - 1275 \text{ kPa})/(1400 \text{ kPa} - 800 \text{ kPa})$$

$$= 14.8\%$$

Cement addition rate = Addition Rate % (Dry soil unit weight

$$= 14.8\%(13.8 \text{ kN/m}^3)$$

$$= 2 \text{ kN/m}^3 \text{ (approximately 200 kg/m}^3)$$

6.5.4 Reagent (binder) types and selection

Many different reagents are used for soil mixing. Reagent selection depends on the location of the project site, the soil types, the target objectives, presence and type of contaminants, engineer/contractor preference, and soil mixing construction method. The location of the project site influences reagent selection in various ways, primarily due to the fact that some reagents may not be cost-effective due to the cost of the material in that location. Soil type influences reagent selection because different reagents have varying effectiveness in various soils. For example, slag is commonly added to mixes when sulfate attack is possible, when there is a high degree of contamination, or when organic materials are present. The target objectives play a large role in reagent selection as different reagents are needed to accomplish various soil mixing target objectives. For example, an environmental remediation project with a gross contaminant mass destruction objective might require the use of an oxidizing or reducing agent, whereas a similar project with only a leachability reduction objective might only require a binding agent such as PC. The presence and type of contaminant can influence reagent selection in two key ways: (1) certain reagents perform better than others in the presence of certain contaminants, and (2) contaminants can delay or retard the set time of binding agents which may result in the selection of other reagents or higher addition rates. Table 6.3 (after Andromalos et al. 2012 with information from ITRC 2011; Gardner et al. 1998; Irene 1996; USEPA 2009; U.S Department of Defense 2000; U.K Environmental Agency 2004; Raj et al. 2005; Conner 1990) provides an overview of various reagents used for ISS and IST.

Contractors and engineers engaged in the soil mixing industry draw heavily on their experience in the selection of reagents. This may inherently bias these practitioners towards reagents that they are experienced with, even if better reagent combinations are available.

The expected construction method can influence reagent selection in myriad ways. For wet mixing, PC, bentonite, slag cement, fly-ash, lime, gypsum, sand, and kiln dust are all possible binding agents. In these applications, the grout or slurry may be made from mixing the reagents with water. The type and quality of water will also influence the reagent selection. Many combinations and amounts of each of these binders are used to enhance the properties of the soil in need of improvement. For dry mixing, a combination of cement, lime, or lime and cement are generally used. PC is the most common binder type, but mixes may contain up to 50% quicklime. Where importing lime is too costly, other binders have been used, such as slag, gypsum, and proprietary products. Dry mixing is generally used to treat fairly soft, compressible, or liquefiable materials with high moisture contents (60% to 200%), and high organic contents. In the 21st century, "other binders" have become more prevalent in use in Sweden, including slag in combination with cement used primarily to stabilize organic soils (Ahnberg 2006).

Table 6.3 Example ISS and IST reagents

ISS or IST	Reagent	COCs* Effectively Stabilized or Treated	Underlying Process
ISS	Portland cement	Numerous, MGP** waste, gasoline, and diesel range organics, metals	Binding
	Blast furnace slag	Numerous, MGP waste, gasoline, and diesel range organics	Binding
	Flyash	Metals, organics, and inorganics	Binding
	Cement kiln dust	Metals	Binding
	Activated carbon	Organics, phenolic waste	Adsorption
	Bentonite clay	Organics	Adsorption
	Organophillic clay	Phenolic waste, organics	Adsorption
	Attapulgite clay	Acids waste, metals	Adsorption
	Lime	Inorganics, metals	Binding
IST	Zero valent iron	TCE, arsenic	Reduction
	Potassium permanganate	TCE, acetone, pesticides, VOCs***	Oxidation
	Sodium persulfate	TCE, acetone, pesticides, VOCs	Oxidation
	Ferrous sulfate	TCE, acetone, pesticides, VOCs	Oxidation
	Calcium polysulfide	Chromium	Reduction
	Biological nutrients	Acetone, pesticides	Enhanced bio-degradation
	Hot air	VOCs	Volatilization

*COC – contaminant of concern
**MGP – manufactured gas plant
***VOCs – volatile organic compounds

Table 6.4 shows the relative effectiveness (poor, fair, or good) of various binder reagents in cohesive and organic soils based primarily on European soil mixing experience, predominantly in Sweden (modified after Guide 2010).

Reagent selection and addition rate should be based on experience, stoichiometric calculations (if applicable), a thorough bench-scale study, and, if possible, a field-scale pilot test. If a pilot test is cost-prohibitive, then a test section may be appropriate. A bench-scale study in a laboratory should be performed long before the field application, a field-level pilot test should be performed using full-scale equipment at the project site long before the actual work begins, and a test section should be performed immediately prior to beginning actual work. Bench-scale studies are always warranted for soil mixing projects as the cost to perform such a study is very small compared to the potential reagent cost savings, especially for costly reducing or oxidizing reagents. Pilot tests provide an opportunity for further evaluation of the bench-scale study results in the field conditions but are generally only warranted or feasible on very large projects. Test sections allow for full-scale evaluation of the reagent type and addition rate, though the results may have less of an impact on the actual work because the results are received very close to the beginning of the actual work.

6.5.5 Develop and evaluate construction objectives

After the target parameters have been selected and the reagent dosages needed to achieve the target parameters have been determined, all of this information must be translated into contract documents for construction. Contract documents might include a scope of work (SOW), general memorandum outlining the primary objectives, or a detailed specification. Regardless of the form of this document, it must be written to include flexibility to allow the construction team an opportunity to optimize and to fit the design to their equipment. Looser outlines of the work, like the SOW or general technical memorandum, are well suited for allowing construction team optimization. Specifications are generally more stringent but can be written with built-in flexibility.

Other key documents developed at this stage include plans or drawings showing site features relative to the proposed soil mixing area. Plan and section view drawings should be included so the user can determine the extent of the work in all dimensions. Section view drawings provide an opportunity to establish general baseline soil conditions through the presentation of generalized or specific cross-sections.

At this stage of the project, all target parameters from the previous steps must be converted into quantifiable and measurable objectives that all parties can understand and use

Table 6.4 Binder effectiveness in cohesive and organic soils

	Soil Type	Silt	Clay	Organic Soils	Peat
	Organic Content	0–2%	0–2%	2–30%	50–100%
Binder Type	Cement	good	fair	fair	good
	Cement and gypsum	fair	fair	good	good
	Cement and slag	good	good	good	good
	Lime	poor	good	poor	poor
	Lime and cement	good	good	fair	poor
	Lime and gypsum	good	good	good	poor
	Lime and slag	fair	fair	fair	poor
	Lime and gypsum and slag	good	good	good	poor
	Lime and gypsum and cement	good	good	good	poor

to evaluate the construction work. Ensuring that construction objectives meet this criterion requires experience and knowledge of the construction process. The construction team may be able to help with planning the final details of the work which would then be outlined in a detailed set of plan documents or in a construction team prepared implementation or work plan. This plan would outline all aspects of the construction in detail so that all parties have an opportunity to evaluate the proposed work approach and procedures before the work begins. This plan is typically reviewed and accepted by the designer, owner, and, as applicable, the construction manager.

Once the construction objectives have been outlined, these objectives should be compared with the original project needs identified in Step 1 of the design process to ensure that these original needs are being met, ensuring that none of the original needs were lost as the design progressed. Depending on the style and structure of the plans defining the construction objectives, this final step may also include a review of construction team submittals that are, or will be, included in the contract documents.

6.5.6 Construction

An overview of the various tools and construction methods was provided earlier in this chapter under definitions, types, and classifications. Soil mixing construction can be performed using a variety of different equipment types, each with a unique set of characteristics. Designers and construction teams that are regularly engaged in soil mixing construction understand the inherent characteristics of each approach and must be consulted for the selection of a construction approach at a project site. The key takeaway is that soil mixing construction should only be performed by experienced construction teams according to designs prepared by experienced designers. If less experienced construction teams are selected, then the work performed should be subjected to heavier scrutiny.

Reagent introduction to the soil is generally achieved either through dry addition or in a water-based slurry/grout. Depending on the soil mixing approach, dry reagent addition can be accomplished through direct application of the dry binder on the ground surface or through the mixing shaft via pneumatic transfer. For most soil mixing construction techniques, dry reagent addition at the ground surface results in limited quality control and should be reserved for very shallow applications. Dry reagent addition through the mixing shaft is accomplished by pneumatically moving the dry reagent particles from a storage vessel, through the mixing shaft, and out ports on the mixing tool head. Dry reagent addition is particularly well suited for soil mixing in high moisture content soils. Reagent addition can also be accomplished using a slurry or grout, which is generally termed wet binder addition. In this approach, reagents are typically mixed with water in a water to solids (W:S) weight to weight ratio that typically varies from 0.8 to 2. At the same overall reagent addition rate, as a function of soil weight, lower W:S ratios will generally result in higher strength and lower permeability soil-grout mixtures. However, higher W:S ratios may be needed to achieve an appropriate level of mixing homogeneity.

The quality control (QC) program for a soil mixing project should be developed for the selected construction approach and should contain a combination of process controls with immediate feedback, quality control testing of input materials, laboratory tests on "grab" samples of the mixed material collected in the field, and, when feasible, in situ tests. The process controls will provide a means for immediate feedback that can be used to predict long-term performance and are used to verify that the field procedures are producing a mixture that is consistent with the bench-scale study and/or design.

Process controls are the QC tests and documentation procedures used by the construction team to monitor the soil mixing process in real time. These controls generally include monitoring procedures starting with reagent slurry creation to delivery of the reagent slurry into the soil mass. Specifically, process controls would include the methods used to control and document the volume or weight of each reagent in each unit volume of slurry, the volume of slurry added to each unit volume of soil, and the distribution of slurry in the mixed soil.

Example problem Ex.6.5: Grout density

What is the density of a soil mixing grout made up of 1.5 parts water to 1 part PC, by weight?

Solution:

It is always reasonable to assume:

Density of Water $(\rho_w) = 1,000 \text{ kg/m}^3$
Density of PC $(\rho_{pc}) = 3,100 \text{ kg/m}^3$

To solve the problem, first assume a unit volume of water, in this case, assume a volume of water $V_w = 1.0 \text{ m}^3$

rearranging the equation defining density from Chapter 2, the mass of the water, M_w, can be computed:

$$M_w = V_w * r_w = 1.0 \text{ m}^3 (1000 \text{ kg/m}^3) = 1000 \text{ kg}$$

Since the ratio between water and cement has been specified, the mass of PC, M_{pc}, can be computed:

$$M_{pc} = \frac{M_w}{1.5} = 1000 \text{ kg}/1.5 = 667 \text{ kg}$$

Knowing the mass and density of PC, the volume, V_{pc}, can now be computed as follows:

$$V_{pc} = M_{pc} / \rho_{pc} = 667 / 3100 = 0.22 \text{ m}^3$$

The total mass, (M_t), can be computed as the sum of the masses of the water and PC as follows:

$$M_t = M_{pc} + M_w = 667 + 1000 = 1667 \text{ kg}$$

Similarly, the total volume, V_t, can be computed as the sum of the volumes of water and cement as follows:

$$V_t = V_{pc} + V_w = 0.22 + 1.0 = 1.22 \text{ m}^3$$

Finally, the grout density, ρ_g, can be computed as the total mass divided by the total volume as follows:

$$\rho_g = \frac{M_t}{V_t} = 1667 / 1.22 = \underline{1366 \text{ kg/m}^3}$$

Reagent addition at the batch plant is generally controlled by weight. Commercially available and custom-built batch plants can be programmed to automatically handle multiple dry and liquid reagents, by the weight of each component. The resulting reagent slurry has a known weight of reagent in each unit volume of slurry. Alternatively, other batch plant configurations may include a batch plant operator that manually controls the weight of each component added and documents the weight added per batch. Regardless of the selected batch plant type and configuration, the daily QC documentation at the batch plant should include the weight of each component added to each batch (or series of batches in a semi-continuous mixing approach), the total weight of reagents used, including water, over the course of a day, and the viscosity and density of the mixed slurry. The construction team or engineer can use the information gathered from the batch plant to assess whether or not the correct amount of reagent (by volume) is being added to the treated soil. Density measurements can also be used to monitor the variability of the slurry during batching which are then compared to the theoretical density predicted by an absolute volume calculation (see example problem Ex.6.5).

After it is clear that the batch plant has accurately produced the reagent slurry with a known amount of reagent weight per unit volume of slurry, the construction team can easily calculate the required volume of slurry for each discrete soil mix column or cell. Previously treated sections of the cell or column can be subtracted to reduce redundant reagent addition. In this calculation, the theoretical minimum volume of slurry for each mix cell or column is based on the reagent addition rates determined in the bench-scale study, generally presented in a % of reagent to soil (by weight), converted to a volume of reagent slurry using the known reagent weight per volume of slurry and an assumed in situ soil density. Ideally, the designer provides the construction team with the in situ density for use in this calculation. Calibrated flow meters, capable of reading the flow of fluids with high suspended particle contents, should be used to control the volume of reagent slurry added at each mixing location. Monitoring of reagent slurry volume at the mixing rig is recommended. Available systems for auger mixing allow the drill rig operator to monitor the slurry flow rate, the volume of slurry added over discrete depth intervals, the number of mixing passes, the drill mast inclination, the auger rotation speed, the hydraulic pressure on the rotary head (indirect measure of drilling difficulty), and the current and maximum depth of the bottom of the auger. This information is collected by an onboard computer system that converts the real-time information into a drill "rig report" that summarizes the drilling activity at each location (see example in Figure 6.19). These rig reports, along with operator notes, are used to prepare a portion of the daily QC report. Similar systems are available for many of the other soil mixing techniques outlined above.

Another aspect of soil mixing which has an effect on the properties, especially strength, of a soil-reagent blend is the amount of mixing energy used in the blending process, which is sometimes represented by a calculation called the blade rotation number (BRN). This is the number of rotations a blade on the mixing tool has made throughout a given unit of depth (generally over 1 m).

The direction of injection of binders into the mixed soil can also play a large role in the final properties of the soil-reagent blend created using soil mixing. There are two possibilities: either from top to bottom (penetration) or from bottom to top (withdrawal) during the course of mixing. The penetration injection method results in good mixing of the binder with soft soils; however, installation can be very difficult if a hard stratum is encountered, or great depths of mixing are required. Withdrawal injection typically has fewer problems as the mixer has already achieved the required depth before injection begins. Here, a more uniformly mixed subsurface can be created with fewer problems. In addition, increasing the number of shafts, increasing the rotational speed of shafts, and decreasing the penetration/ withdrawal speed can improve soil mix homogeneity which can result in improved properties, especially an increase in the overall strength of the mixed soil-reagent blend (Porbaha et al. 2001). In all cases, a small amount of fluid, usually the grout or slurry used for the

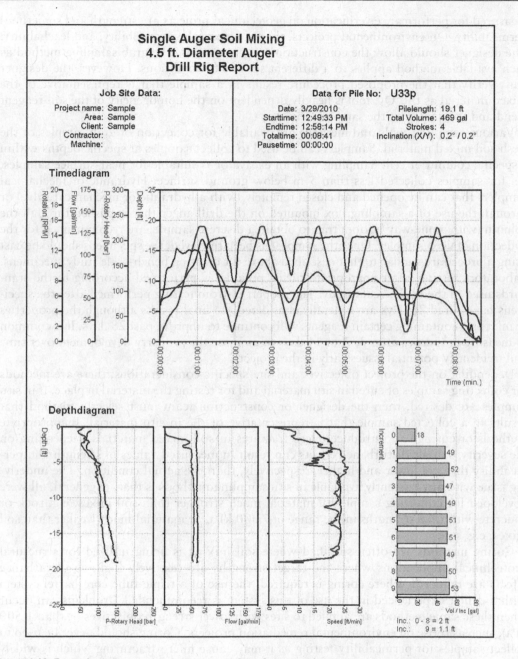

Single Auger Soil Mixing
4.5 ft. Diameter Auger
Drill Rig Report

Job Site Data:

Project name: Sample
Area: Sample
Client: Sample
Contractor:
Machine:

Data for Pile No: U33p

Date: 3/29/2010
Starttime: 12:49:33 PM
Endtime: 12:58:14 PM
Totaltime: 00:08:41
Pausetime: 00:00:00

Pilelength: 19.1 ft
Total Volume: 469 gal
Strokes: 4
Inclination (X/Y): 0.2° / 0.2°

Figure 6.19 Example drilling rig log (courtesy of Geo-Solutions, Inc., www.geo-solutions.com).

soil mixing (though air and water may also be used), must be pumped during penetration to serve as a drilling lubricant.

6.5.7 Sampling

Most soil mixing projects rely on laboratory tests performed on wet grab samples to verify that the final soil mixed product meets or exceeds the established performance criteria. This practice is particularly common in the US (Bruce 2001). The most common properties measured for performance verification on geotechnical projects are strength and sometimes permeability. On environmental projects, they are strength, permeability, and leachability. The designer should allow the construction team to propose the grab sampling method as each available method applies to a different set of site conditions. However, the designer must verify that the proposed procedure results in a sample that is representative of the mixed material as this QC tool is heavily dependent on the homogeneity of the soil-reagent blend and the quality of the sampling tool.

Various sampling tools and methods are available for collection of grab samples of the fresh soil mixed material. Samplers can be used to collect samples at specific depths within a specific column or cell. Sampling with an excavator is suitable for near-surface samples, i.e. for samples collected less than 5 m below ground surface. Hydraulic or mechanical samplers that can be opened and closed remotely (with a hydraulic or mechanical valve) or through the use of a sampling box mounted on the drill auger that is pushed through the column (with a one-way bottom trap to obtain a discrete sample) are recommended for the collection of deep samples. Once the sample has been retrieved, the specimens should be cast using a procedure similar to that described above for creating bench-scale study specimens. Laboratory testing, after appropriate curing periods, is performed according to the standards used in the bench-scale study. The properties from testing performed after the specimens have cured 28 days are generally considered the final results although the properties of mixtures containing certain reagents will continue to improve past 28 days. It is common to measure properties after 3, 7, or 14 days to monitor the property development over time and to identify potential issues early in the project.

Depending on the project objectives and site-specific considerations, there are methods for collecting samples of cured in situ material and for testing the material in place. If in situ samples are desired, then the designer or construction team must select a method that results in a collected sample that is representative of the in situ material. Inappropriate methods or coring can introduce micro-fractures in soil-mixed materials. Depending on the severity, sample disturbance effects can result in measured values of strength and permeability that are lower and higher, respectively, than the actual condition. The underlying issue with any currently available in situ sampling methods is that, in general, all were developed for collecting samples of material much stronger than soil mixes, e.g. rock or concrete with UCS strengths in the range of 7,000 kPa, or material much weaker than soil mixes, e.g. soil.

Coring methods are often specified where soil mixing is being utilized for structural geotechnical applications where the target strengths are relatively high, so disturbance effects are reduced. Where coring is required, the use of a triple-tube core barrel system with a driller experienced in the use of this system is recommended. Problems can occur when these same methods are applied to sites with lower strength objectives (< than 1,500 kPa), such as those on environmental remediation projects. Coring should never be used to collect samples for permeability testing as it may cause micro-fracturing which is widely considered to have too substantial an impact. Despite all of the above, coring is particularly useful for confirming the continuity of the mix and the bedrock contact, if applicable. If the retrieved sample is severely broken, a down-the-hole camera can be used to check the conditions of the in-place material to determine if the coring method contributed to the damaged sample. If coring is specified and feasible, it is common to target 80% to 90% recovery. Strength measurements of cored samples can be 30% to 50% of the strength of laboratory-created samples. If core samples are used for UCS testing, it is best to use undisturbed samples. Here, the target strength should be increased to account for expected variability in the measured results. The acceptance criteria must account for statistical variation of the strength with realistic average and minimum strength requirements. Coring is especially prevalent in Japan. Historically the Japanese have led much of the development of soil-reagent coring techniques (Bruce 2001).

Occasionally, the soil mixed column can be fully exposed and extracted for evaluation. In practice, this is very costly, so large-scale testing of elements is limited. However, full exposure of the top 2 m to 3 m of individual or overlapped test columns is commonly used for qualitative evaluation of mix quality and confirmation of column diameter and overlap, especially during pilot tests or test sections.

Double tube samplers are basically core tubes that are placed in the fluid soil-mixed material. The sampler consists of an inner and outer tube with a lubricant between them. Once the material cures in the ground, the inner tube is removed. If the material is sufficiently intact, these samples can be used in strength and even permeability testing, although many times these types of samples are used solely for visual evaluation.

6.5.8 In situ testing

In lieu of, or in addition to, laboratory sampling on wet grab or in situ samples, the properties of the soil mixed material can be assessed using in situ tests. In situ testing methods used to evaluate soil mixed material properties include any method that may be used for measuring the properties of soil or soft rock, including the standard penetration test (SPT), cone penetration test (CPT), vane shear test (VST), Marchetti dilatometer test, pressuremeter test, and vane pullout test. Essentially, any tool developed for evaluating the properties of soil or rock can be used to measure properties of soil-mixed materials.

Converting the data into useful parameters for direct comparison to the parameters used in the design and translated into the stated objectives is challenging. Many of the equations, tables, or charts used to convert direct measurements from these evaluation tools are empirical or theoretically developed using underlying assumptions that are not appropriate for soil-mixed materials. Some correlations have been developed specifically for use with soil-mixed materials, but, in many cases, the designer or construction team may need to develop a site-specific correlation. This can be problematic since site-specific correlations are based on a limited data set and therefore may be exposed to scrutiny if questions arise.

In situ testing may be performed using geophysical methods to measure shear wave velocity or resistivity. Tests performed for these purposes include the shear wave seismic, borehole resistivity profiling, and low strain sonic integrity/borehole sonar. It is difficult to transfer data from these tests to useful or commonly understood parameters.

6.6 PROBLEMS

6.1 What is the principal advantage of the trench mixing methods, e.g. TRD or one-pass trenchers, versus auger mixing methods for the construction of a vertical cutoff wall?

6.2 What is the difference between jet grouting and soil mixing?

6.3 What is the difference between stabilization and solidification?

6.4 Why do you think PC is a very commonly used reagent? Consider practical, economic, and environmental, and other reasons.

6.5 Organic soils cause problems with PC. Name the problems. Discuss the mechanism (chemical? physical? other?) that causes the problems.

6.6 Discuss the advantages and disadvantages of single and multi-auger blenders.

6.7　What are the advantages and disadvantages of preparing lab-scale test specimens of reagent treated soil before implementing the field program?

6.8　Soil mixing is already specified at a ground improvement project at the site so your client has asked you to use soil mixing to install an excavation support wall. The height of the excavation is 5 m. Determine the approximate width (w) of a soil mixed gravity wall to prevent an overturning failure. Assumptions: native soil unit weight = 18 kN/m3, soil mixed material unit weight = 15 kN/m3, groundwater is below the bottom of the excavation, the soil mixed block extends to 7 m, the friction angle of the native soil is 30 degrees, the shear strength of the soil mixed material is 100 kPa, and the FS for overturning must be > 1.5.

6.9　You are developing a soil mixing design mix for an oxidation/solidification project. The oxidant you are using is sodium persulfate which is catalyzed by basic materials like PC so the additives will be added in a single dose. The contaminant you are targeting requires 1 kg of oxidant for every 1,000 ppm of contaminant and the natural oxidant demand of the soil is 1% (weight to weight). The concentration of the contaminant averages 5,000 ppm. Further, it takes at least 2 kg of cement for every 1 kg of oxidant to create an efficient reaction. Finally, bench scale studies show it takes 5%, 10%, and 15% cement (weight to weight) to get 10, 20, and 30 kPa shear strength. If the post-treatment site requires a UCS of > 10 kPa and enough oxidant to overcome 125% of the calculated demand, what are the oxidant and cement addition rates in kg/m^3 of soil?

6.10　Soil mixing is planned to solidify contaminated sandy soil and groundwater from the ground surface to 3 m below ground. The groundwater table is very near the surface. What method of soil mixing would you recommend and what would be one of your concerns to monitor during implementation?

6.11　A client has asked you to evaluate options to improve the stability of a slope. Based on your preliminary review, soil mixing seems to be a great candidate technology and now you are trying to determine an approximate budget for preliminary comparison to other approaches. High-level calculations indicate that the area requiring improvement is 100 m long × 10 m wide, the critical failure plane in the unsupported slope intersects the mix zone at about 10 m below ground, and the composite strength (UCS) of the improved zone should be 1,000 kPa. A contractor has provided a budget price of $50/m3 plus a mobilization of $100,000. What is a good approximate range for the cost of soil mixing at this site?

6.12　You are tasked with writing a specification for a soil mixed ground improvement project. Based on the designer's modeling, the site requires a soil-mixed material with a shear strength of 50 kPa. In line with industry guidelines, the target strength needs to be presented in terms of UCS and a statistical allowance should be provided (85% of tests above the target strength). The coefficient of variability of the mixed material is expected to be 0.3. What is your recommended target UCS (a), in specifying that target strength, approximately what mean strength are you actually specifying (b), and if it were appropriate to set a minimum strength, what strength would you set? Recall: coefficient of variability is the ratio of the standard deviation to the mean and assume the data is normally distributed.

6.13　Redo example problem Ex.6.1 using imperial units.

6.14　Redo example problem Ex.6.2 using imperial units.

6.15　Redo example problem Ex.6.3 using imperial units.

6.16　Use Equation 6.1 to estimate the strength of a soil mixed material after 28 days of curing with a cohesionless base soil and a 10% cement dosage. Assume the material has no appreciable strength at time zero.

6.17 Use Equation 6.1 to estimate the strength of a soil mixed material after 28 days of curing with a cohesive base soil and a 10% cement dosage. Assume the material has no appreciable strength at time zero.

6.18 Use Equation 6.2 to calculate the curing factor after 7 and 14 days of curing.

REFERENCES

Adams, T.E. (2011). *Stability of levees and floodwalls supported by deep-mixed shear walls: Five case studies in the New Orleans area* (Doctoral dissertation, Virginia Tech).

Åhnberg, H. (2006). *Strength of stabilised soil-A laboratory study on clays and organic soils stabilised with different types of binder.* Lund, Sweden: Lund University.

Åhnberg, H. and Johansson, S.E. (2005). Increase in strength with time in soils stabilised with different types of binder in relation to the type and amount of reaction products. In *Proceedings of deep mixing '05*, Stockholm (pp. 195–202).

Andromalos, K.B., Ruffing, D.G. and Peter, I.F. (2012). In situ remediation and stabilization of contaminated soils and groundwater using soil mixing techniques with various reagents. In *SEFE7: 7th seminar on special foundations engineering and geotechnics*, Sao Paulo, Brazil, June 17–20.

Andromalos, K.B., Ruffing, D.G. and Spillane, V.A. (2015). Construction considerations for ISS bench-scale studies and field-scale monitoring programs. *Journal of Hazardous, Toxic, and Radioactive Waste, 19*(1), C4014001.

ANSI/ANS-16.1. (2003). *Measurement of the leachability of solidified low-level radioactive wastes by a short-term test procedure.* La Grange Park, IL: American Nuclear Society.

Aoi, M., Kinoshita, F., Ashida, S., Kondo, H., Nakajim, Y. and Mizutani, M. (1998). Diaphragm wall continuous excavation method: TRD method. *Korbelco Technology Review, 21,* 44–47.

Aoi, M., Komoto, T. and Ashida, S. (1996). Application of TRD method to waste treatment on the ground. M. Kamon (ed.) *Environmental Geotechnics: Volume 1* (pp. 437–440).

Arnold, M., Beckhaus, K. and Wiedenmann, U. (2011). Cut-off wall construction using cutter soil mixing: A case study. *Geotechnik, 34*(1), 11–21.

Babasaki, R.M., Terashi, T.S., Maekaea, A., Kawamura, M. and Fukazawa, E. (1996). JGS TC report: Factors influencing the strength of improved soil. Grouting and deep mixing. In *Proceedings of IS-Tokyo'96, The 2nd international conference on ground improvement geosystems*, May 14–17, 1996, Tokyo (pp. 913–918).

Bahner, E.W. and Naguib, A.M. (1998). Design and construction of a deep soil mix retaining wall for the Lake Parkway Freeway Extension. In *Soil improvement for big digs* (pp. 41–58). Reston, VA: ASCE.

Bates, E.R. (2010a). Selecting performance specifications for Solidification/Stabilization. In *International solidification/stabilization technology forum*, Sydney, Nova Scotia, June 14–18, 2010, Dalhousie University, Halifax, NS.

Bates, E.R. (2010b). Overview of solidfication/stabilization in the US superfund program. In *International solidification/stabilization technology forum*, Sydney, Nova Scotia, June 14–18, 2010, Dalhousie University, Halifax, NS.

Bates, E. and Hills, C. (2015). Stabilization and solidification of contaminated soil and waste: A manual of practice. *Édit Media H.* September 2015.

Brandl, H. (1999). Mixed-in-place stabilisation of pavement structures with cement and. In *Geotechnical engineering for transportation infrastructure: Theory and practice, planning and design, construction and maintenance: Proceedings of the twelfth European conference on soil mechanics and geotechnical engineering*, Amsterdam, Netherlands, June 7–10, 1999 (p. 1473). Taylor & Francis US.

Bruce, D.A. (2001). *An introduction to the deep mixing methods as used in geotechnical applications, volume III: The verification and properties of treated ground* (No. FHWA-RD-99-167). Washington, DC: Federal Highway Administration.

Bruce, D.A. and Bruce, M.E.C. (2003). The practitioner's guide to deep mixing. In *Grouting and ground treatment* (pp. 474–488). February 2003, New Orleans, LA: GSP 120.

Bruce, D.A., Bruce, M.E.C. and DiMillio, A.F. (1998). Deep mixing method: A global perspective. In *Soil Improvement for Big Digs, Proceedings of Sessions of Geo-Congress 98* (pp. 1–26) December 1998, Boston, MA.

Bruce, D.A. and Cali, P.R. (2005). State of practice report: Session 3 "design of deep mixing applications". In *International conference on deep mixing*, Swedish Geotechnical Institute, Stockholm Sweden. May 2005, volume 2, 685–696.

Bruce, M.E.C., Berg, R.R., Collin, J.G., Filz, G.M., Terashi, M., Yang, D.S. and Geotechnica, S. (2013). *Federal highway administration design manual: Deep mixing for embankment and foundation support* (No. FHWA-HRT-13-046). Washington: DC, United States: Federal Highway Administation. Offices of Research & Development.

Cermak, J., Evans, J. and Tamaro, G.J. (2012). Evaluation of soil-cement-bentonite wall performance-effects of backfill shrinkage. In *Grouting and deep mixing 2012* (pp. 502–511).

Conner, J.R. (1990). *Chemical fixation and solidification of hazardous wastes.* New York: Van Nostrand Reinhold. ISBN 0-442-20511-2.

Croce, P., Flora, A. and Modoni, G. (2014). *Jet grouting: Technology, design and control.* Boca Raton, FL: CRC Press.

Day, S.R. and Ryan, C. (1995). Containment, stabilization and treatment of contaminated soils using in situ soil mixing. In *Geo environ 2000: Characterization, containment, remediation, and performance in environmental geotechnics* (pp. 1349–1365). Reston, VA: ASCE.

Evans, J.C. (2007). The TRD method: Slag-cement materials for in situ mixed vertical barriers. In *Soil improvement* (pp. 1–11) Denver, CO, Feb 2007.

Evans, J.C. and Garbin, E.J. (2009). The TRD method for in situ mixed vertical barriers. In *Advances in ground improvement: Research to practice in the United States and China* (pp. 271–280).

Evans, J.C. and Jefferis, S.A. (2014). Volume Change Characteristics of Cutoff Wall Materials, *Proceedings of the 7th International Congress on Environmental Geotechnics* Melbourne, AU, November (pp. 10–14).

Filz, G.M. and Navin, M.P. (2010). A practical method to account for strength variability of deep-mixed ground. In *GeoFlorida 2010: Advances in analysis, modeling & design* (pp. 2426–2433). Orlando, FL, Feb 2010.

Gardner, F.G., Strong-Gunderson, J., Siegrist, R.L., West, O.R., Cline, S.R. and Baker, J. (1998). Implementation of deep soil mixing at the Kansas City plant. ORNL/TM-13532, Oak Ridge National Laboratory, Grand Junction, CO (United States).

Garrabrants, A.C., Kosson, D.S., Van der Sloot, H.A., Sanchez, F. and Hjelmar, O. (2010). *Background information for the leaching environmental assessment framework (LEAF) test methods.* United States Environmental Protection Agency (US EPA).

Guide, D. (2010). *Soft soil stabilisation: EuroSoilStab: Development of design and construction methods to stabilise soft organic soils.* Taylor and Francis, EP 60, ISBN 10: 1860815995 ISBN 13: 9781860815997

Irene M.C. (1996). Solidification/stabilization of phenolic waste using organic-clay complex. *Technical Paper,* 6 pages. American Society of Civil Engineers (ASCE). Hong Kong University of Science and Technol, Clear Water Bay, Kowloon, Hong Kong.

ITRC (Interstate Technology & Regulatory Council). (2011). *Development of performance specifications for solidification/stabilization.* S/S-1. Washington, DC: Interstate Technology & Regulatory Council, Solidification/Stabilization Team. United States, 2011 www.itrcweb.org.

Jayaram, V., Marks, M.D., Schindler, R.M., Olean, T.J. and Walsh, E. (2002). *In situ soil stabilization of a former MGP site.* Skokie, IL: Portland Cement Association.

Kosson, D.S., Garrabrants, A.C., van der Sloot, H.A. and Brown, K.G. (2014, March). Application of the new US EPA leaching environmental assessment framework (LEAF) to DOE environmental management challenges–14383. In *WM2014 conference* (pp. 2–6), March.

LaGrega, M.D., Buckingham, P.L. and Evans, J.C. (2010). *Hazardous waste management.* McGraw Hill, New York: Waveland Press.

Larsson, S. (2005). State of practice report–execution, monitoring and quality control. *Deep Mixing,* 5, 732–785.

Malusis, M.A., Evans, J.C., McLane, M.H. and Woodward, N.R. (2008). A miniature cone for measuring the slump of soil-bentonite cutoff wall backfill. *Geotechnical Testing Journal, 31*(5), 373–380.

Mitchell, J.K. (1976). The properties of cement-stabilized soils. In *Proceedings of residential workshop on materials and methods for low cost road, rail, and reclamation works*, Leura, Australia, September 6–10. Unisearch Ltd., University of South Wales.

Navin, M.P. and Filz, G.M. (2006). Reliability of deep mixing method columns for embankment support. In *GeoCongress 2006: Geotechnical engineering in the information technology age* (pp. 1–6). Feb 26 to Mar 1, Atlanta, GA.

O'Brien, A. (2019). Some observations on the design and construction of wet soil mixing in the UK. In *Proceedings of the XVII European conference on soil mechanics and geotechnical engineering, geotechnical engineering foundation of the future*. Reykjavik, Iceland, Sept 2019.

O'Rourke, T. D. and O'Donnell, C. J. (1997). Field behavior of excavation stabilized by deep soil mixing. *Journal of Geotechnical and Geoenvironmental Engineering, 123*(6), 516–524.

Porbaha, A., Raybaut, J. L., and Nicholson, P. (2001). State of the art in construction aspects of deep mixing technology. *Proceedings of the Institution of Civil Engineers-Ground Improvement, 5*(3), 123–140.

Quickfall, G., Okada, W. and Morrison, T. (2014). Ground improvement using Turbojet deep soil mixing-case study. *Screening, 2*(3), 4.

Rutherford, C. J., Biscontin, G., Koutsoftas, D., and Briaud, J. L. (2007). Design process of deep soil mixed walls for excavation support. *ISSMGE International Journal of Geoengineering Case Histories, 1*(2), 56–72.

Raj, D.S.S., Rekha, C.A.P., Bindhu, V.H. and Anjaneyulu, Y. (2005). Stabilisation and solidification technologies for the remediation of contaminated soils and sediments: An overview. *Land Contamination & Reclamation, 13*(1), 23–48.

Ruffing, D.G., Andromalos, K.A., Payne, D.R., and Schindler, R.M. (2021). An Overview of US Practice in Environmental Soil Mixing. In *Proceedings of Deep Mixing 2021*, June 1,3,8,10,15,17, 2021, Online Conference.

Ruffing, D., Swackhamer, T. and Panucci, D. (2017). A case study: Soil mixing for soft ground improvement at a landfill. In *31st central pennsylvania geotechnical conference*, Hershey, PA, January 2017.

Ruffing, D.G., Sheleheda, M.J. and Schindler, R.M. (2012). A case study: Unreinforced soil mixing for excavation support and bearing capacity improvement. In *Grouting and deep mixing 2012* (pp. 410–416). New Orleans, LA, Feb 15 to 18, 2012.

Ryan, C. and Jasperse, B.H. (1989, June). Deep soil mixing at the Jackson Lake Dam. In *ASCE geotechnical and construction divisions special conference* (Vol. 5, pp. 25–29). Evanston, IL, June 25 to 29, 1989.

Ryan, C.R. and Jasperse, B.H. (1991). Closure to "Deep soil mixing at Jackson Lake Dam" by Christopher R. Ryan and Brian H. Jasperse (June 25–29, 1989, ASCE Geotech. and Constr. Div. Special Conf., Northwestern Univ., Evanston, IL). *Journal of Geotechnical Engineering, 117*(12), 1978–1979.

Ryan, C.R. and Walker, A. (1992). Soil mixing for soil improvement–Two case studies. In *Proceedings of soil modification conference*, Louisville, KY.

Taki, O. and Yang, D. S., (1991). Soil-cement mixed wall technique. In *Geotechnical engineering congress—1991* (pp. 298–309). ASCE.

Terashi, M. (1997). Theme lecture: Deep mixing method-Brief state of the art. In *Proceedings of 14th ICSMFE* (vol. 4, pp. 2475–2478). Sept 6 to 12, Hamburg, Germany.

US Army. (2008). Corps of Engineers: Specifications, Section 00 31 32, Geotechnical Data Report for Herbert Hoover Dike Reha- bilitation, Reach 1A, Seepage Cutoff Wall, Jacksonville.

US EPA, E.P.A.. (1994). Method 1312: Synthetic precipitation leaching procedure. *Test methods for evaluating solid waste, physical/chemical methods, SW-846*.

UK Environment Agency. (2004). Review of scientific literature on the use of stabilisation/solidification for the treatment of contaminated soil, solid waste and sludges. *Science Report SC980003/ SR2*, United Kingdom, November 2004.

U.S Department of Defense (DoD). (2000). Use of sorbent materials for treating hazardous waste. Environmental Security Technology Certification Program (ESTCP). *Cost and Performance Report (CP-9515)*, United States, March 2000.

U.S. Environmental Protection Agency. (1992). Test methods for evaluating solid waste, physical/chemical methods, 3rd edition. SW-846, Method 3050B. U.S. Government Printing Office, Washington, DC.

U.S. Environmental Protection Agency. (1994). Test method for evaluating solid waste, physical/chemical methods (SW-846), 3rd edition, update 2B. Environmental Protection Agency, National Center for Environmental Publications, Cincinnati, OH.

U.S. Environmental Protection Agency. (2009). Technology performance review: Selecting and using solidification/stabilization treatment for site remediation. National Risk Management Research Laboratory Office of Research and Development, Cincinnati, OH.

U.S. Environmental Protection Agency. (2017a). SW-846 test method 1314: Liquid-solid partitioning as a function of liquid-solid ratio for constituents in solid materials using an up-flow percolation column procedure. Retrieved from https://www.epa.gov/hw-sw846/sw-846-test-method-1314-liquid-solid-partitioning-function-liquid-solid-ratio-constituents.

U.S. Environmental Protection Agency. (2017b). SW-846 test method 1315: Mass transfer rates of constituents in monolithic or compacted granular materials using a semi-dynamic tank leaching procedure. Retrieved from https://www.epa.gov/hw-sw846/sw-846-test-method-1315-mass-transfer-rates-constituents-monolithic-or-compacted-granular.

U.S. Environmental Protection Agency. (2017c). SW-846 test method 1316: Liquid-solid partitioning as a function of liquid-to-solid ratio in solid materials using a parallel batch procedure. Retrieved from https://www.epa.gov/hw-sw846/sw-846-test-method-1316-liquid-solid-partitioning-function-liquid-solid-ratio-solid.

Yamanobe, J., Endo, M. and Komiya, K. (2020). Development of the quick prediction method for the strength of ground improved by jet grouting. *Japanese Geotechnical Society Special Publication*, 8(10), 410–415.

Yang, D.S. (2003). Soil–cement walls for excavation support. In *Earth retention systems 2003: A joint conference presented by ASCE Metropolitan Section of Geotechnical Group, The Deep Foundations Institute, and The International Association of Foundation Drilling*. May 6 to 7, New York, NY.

Yang, D.S., Luscher, U., Kimoto, I. and Takeshima, S. (1993). SMW wall for seepage control in levee reconstruction. *International Conference on Case Histories in Geotechnical Engineering*. 28. June 2, 1993, Rolla, Missouri.

Chapter 7

Grouting

7.1 INTRODUCTION

The term grouting has broad and varied definitions as it is used in a variety of industries and performed using a number of different techniques for many applications. The broadest definition of grouting in the context of ground improvement is the injection of a liquid into voids throughout soil or rock. Figure 7.1 illustrates just one of the methods of grouting, that is, injection of a liquid material, termed grout, into fissures in rock.

Figure 7.2 shows a drilling rig in use to grout rock for a dam project. Notice the inclination of the drilling mast. While many grout holes are drilled vertically as illustrated in Figure 7.1, it is also common to drill inclined holes to intercept vertical fractures.

Grouting is performed to improve the characteristics of the subsurface materials, e.g. to increase shear strength, reduce compressibility and/or reduce permeability. The liquid, termed grout, can be made of many different mixtures of fluids, but the most common grout is cement (ordinary portland cement) mixed in water and often including other additives to improve the grout properties. Grouts may also be foams, organic resins, solutions, and other mixtures (Hausmann 1990). Grouting can be used in many different applications, including penetrating pore spaces in the soil, filling existing fractures in soil or rock, filling voids created by the grouting process itself, or creating a remolded mass of soil and grout. Grouting is one of the oldest methods of ground improvement. As evidenced by the extensive literature available on grouting, many forms of grouting and countless grouting materials have been developed and used with success. Grouting is fundamentally straightforward, but the application, monitoring, and verification of grouting in civil and environmental applications can be quite complex.

The goals of grouting may include increasing shear strength, reducing compressibility, increasing bearing capacity, increasing stiffness, reducing permeability, filling voids or rendering voids inert, improving erosion resistance, and/or decreasing liquefaction risk. The benefits are achieved by hardening of the liquid grout in penetration and fracture grouting (this chapter), hardening of the grout-soil mixture in jet grouting, and/or densification of material surrounding the grouted zone in compaction, compensation, or displacement grouting (Chapter 11). The type of grout and the details of the grout mixture vary depending upon the objectives of the grouting program and the site and subsurface conditions.

Figure 7.1 Schematic of rock grouting (courtesy of Keller; www.Keller-na.com).

Figure 7.2 Photo of rock grouting (courtesy of Christopher Bailey, Gannett Fleming, Inc.).

Various authors and organizations use a range of systems for classifying the practice of grouting. The following groupings of terms are offered to describe the various types and methods of grouting.

Classification by grout type:

1. Suspension (particulate grout) – a liquid mixture typically consisting of finely ground solids (namely, cement) suspended in water that first permeates the pores of granular materials and then hardens (typically through cement hydration) within the penetrated pore spaces. Bentonite is often added to improve grout characteristics.
2. Solution grout – a liquid mixture consisting of materials solubilized in a fluid, normally water. The materials generally consist of various chemical admixtures. The mechanisms of the change of state from liquid to solid differ significantly from the processes controlling suspension grouts.
3. Low mobility grout (LMG) – a stiff, viscous grout usually consisting of cement, fine aggregate (sand), and water. Bentonite and/or flyash may be added to minimize water separation during injection under pressure. This grout, while technically a fluid, typically has the consistency of high slump concrete. See Chapter 11 for more details on LMGs.

Classification by mechanism:

1. Permeation grouting (penetration grouting) – the injection of a liquid mixture into existing pore spaces in the soil or loosely bound rock. Permeation grouting can be performed with suspension or solution grouts although solution grouts are more common due to their ability to flow through and fill smaller voids.
2. Fracture grouting (intrusion grouting) – the injection of a liquid mixture of materials into existing or manufactured fractures, often in rock. Fracture grouting is generally performed with suspension grouts and can include the use of pressures high enough to create or further open fractures, i.e. fracking. Grouts used for fracture grouting vary from low to high viscosity depending on the grouting procedure and fracture size.
3. Compaction grouting (displacement grouting) – the injection of an LMG designed not to penetrate the pore spaces or fractures, but rather to densify the formation materials around the point of grout injection, creating a void that is simultaneously filled with the grout that is allowed to harden in place. Compaction grouting is addressed in Chapter 11: Additional Topics in Ground Improvement.
4. Compensation grouting – a subapplication of any grouting technique (although commonly compaction grouting or fracture grouting) at pressures high enough to remediate surface settlements that might have occurred from natural processes, such as dissolution cavities, or construction activities such as tunneling or open excavations. Compensation grouting is not specifically addressed.
5. Void filling – the application of grouting for the filling of large subsurface voids. These voids are sometimes naturally occurring, e.g. sinkholes or karstic features, or man-made, e.g. pipelines, tunnels, or mines. In either case, the goal of this grouting approach is to fill the void to prevent future collapse that could cause surface deformations that may damage existing or new property.
6. Jet grouting – the application of high-pressure fluid to simultaneously erode and mix soils in situ. The process uses suspension-based grouts (primarily cement and water). Jet grouting is addressed in Chapter 11: Additional techniques in ground improvement.

Further information about each of these terms, including history, introductions to methods and applications, is provided in the remainder of this chapter. The properties of grout are discussed later in the chapter, so the reader first has a better idea of the grouting materials and methods prior to the detailed discussion of grout properties and their measurement.

Example problem Ex.7.1: Grouting thought exercise

Thought exercise: A project requires a cutoff wall from ground surface to 50 m below ground surface. The overburden soil at the site extends to 40 m below grade and is underlain by a fractured shale bedrock. Would you recommend any of the grouting techniques outlined to install the cutoff wall in the overburden or the rock? If so, which ones?

Discussion: Although a permeation grouting program with a solution-based grout may be suitable for use in the overburden, there are likely more cost-effective and verifiable means of installing that portion of the wall, e.g. slurry trenching (Chapter 8), or soil mixing (Chapter 6). Jet grouting may also be an option for the cutoff wall in the overburden and may have advantages because of the depth, but jet grouting would likely not be cost-effective. A grouting program for installing the cutoff wall in the rock would certainly be a good option. A fracture grouting program with a suspension grout would seem reasonable based on the given information.

7.2 HISTORY OF GROUTING

According to Glossop (1968), the use of grouting in modern times can be traced back to at least 1802 when Charles Berigny used grouting to improve a sluice structure in the city of Dieppe, France (Berigny 1832). In this first known application, Berigny used grouts of clay and cement, separately, and pumped the grouts into drilled holes using a piston pump. Early grouting applications were mostly completed with cement-based grouts for dam improvements or to facilitate mining activities.

It is important to note that at the time of the development of grouting techniques, there were many contemporary fundamental breakthroughs in the understanding of the physics of soil, rock, and fluid mechanics that aided in and guided the development of grouting practice. For example, the grouting field was greatly influenced by the publication of Terzaghi's Erdbaumechanik (Terzaghi 1925). Engineers engaged in the practice of grouting were also largely influenced by Maag's (1938) publication, which presented theories about the injection of fluids in granular materials and laid the groundwork for the relationships between injection factors.

The histories of grouting presented below are overviews and separated into the two main categories of grouting addressed in this chapter: suspension and solution grouting. There is significant overlap in the construction methods and theory behind these sub-techniques. Undoubtedly, advances in one category influenced applications in the others, in some cases forming the basis for the subapplication itself, e.g. the extension of conventional grouting techniques to perform compaction via grouting. Most modern grouting techniques incorporate a mix of cement-based and chemical grouting methods, especially since the development of balanced stable grouts which inherently include blends of cementitious reagents, water, and chemical additives. However, there remain site-specific conditions that necessitate the sole application of a suspension-or solution-based technique. These histories are meant to guide the reader through how, and in some cases why, the various grouting techniques developed.

Although the information provided in this book on the history of grouting is limited, it is important to approach the study of grouting with an understanding of its long history. It's particularly essential to recognize that grouting has been developed iteratively and that, even though the process has gotten considerably more scientific over time, the technique still requires a bit of artistry that comes from experience.

7.2.1 History of suspension grouting

There are numerous papers and books that detail the history of grouting. Of particular interest is Stuart Littlejohn's "The Development of Practice in Permeation and Compensation Grouting: A Historical Review (1802–2002) – Part 1 Permeation Grouting" (Littlejohn 2003).

From 1800 to 1850, pozzolanic grouts, pastes, and mortars consisting of cement and/or lime mixed with water were used in England and France to improve foundation conditions on a wide variety of projects. These early applications included alluvial grouting (filling of pore space in gravel foundations), void filling for water control around canals, subsidence prevention for bridge abutments, dam crack sealing, and lock foundation improvement. Toward the end of this period (1845), W. E. Worthen used grouting in the United States to improve the foundation of a flume. Prior to this application for ground improvement, grouting was only used to remediate defective foundations (Nonveiller 2013).

From 1850 to 1900, grouting use was heavily extended to solving seepage control issues, including water ingress into mines and tunnels. There were significant improvements in the injection techniques and batch plants. Developments during this period were happening simultaneously in Europe and the United States. Some important developments during this period include the use of rubber packers (section 7.14.8) for isolating grout delivery points, successive injections, the pre-placement of aggregate materials, and discontinuity exploration. Applications during this period included mine seepage control, void filling around cast iron tunnel linings, rock fissure sealing beneath dams, and cavity infilling in karstic bedrock formations. At this time, most grouting was still based predominantly on cement-based techniques.

From 1900 to 1950, improvements in the field of grouting included:

1. the development of new pumps such as the Joseph Evans pump (Wolverhampton UK),
2. the early uses of silicates (see chemical grouting),
3. the gradual and successive thickening of the grouts,
4. split spacing techniques,
5. simultaneous injection of multiple fluids (e.g. tube-a-manchette (TAM) – section 7.14.2),
6. the development of the Lugeon unit (devised to quantify the water permeability of bedrock and defined as the amount of water injected into a borehole under steady pressure, i.e. loss of water in liters per minute per meter of borehole),
7. improvement of the field measurement of viscosity (e.g. the Marsh funnel),
8. monitoring of uplift,
9. pressure testing,
10. improved understanding of bleed/soil mechanics/cement paste properties,
11. staged grouting, and
12. limitations of cement-based grouts in smaller fissures.

The applications and uses of grouting from this time forward are too broad and substantial to list in this brief history.

From 1950 to 1975, improvements in the field of grouting included:

1. development of closure criteria (pressure and flow limits),
2. higher pressure injection,
3. improvements in laboratory and field investigations,
4. expansive grouts,
5. grouting for waterproofing structures,
6. use of high early strength cements,

7. improved industry standards through state-of-the-art publications,
8. wider use of other additives including bentonite, flyash, and silt to improve physical properties,
9. improved record-keeping, and
10. improved understanding of rheological, physical, and chemical properties of grouts.

From 1975 to the present, improvements in the field of grouting include:

1. improved drilling, pumping, and grout control equipment,
2. multi-pressure Lugeon testing,
3. improved understanding of long-term property behavior (including lessons learned from the forensic evaluation of failures),
4. improved sharing of international knowledge through symposiums and conferences,
5. introduction of real-time computerized monitoring, instrumentation, and control,
6. development of best practices,
7. understanding of the mechanical behavior of grouted soils,
8. introduction of microfine and ultrafine cements,
9. introduction and improvement of dispersants and superplasticizers,
10. the development of the grout intensity number (GIN) a product of the grouting pressure and volume injected, and
11. the development of balanced and stable grouts.

If the history of grouting can be used to predict and influence the future, a key lesson is that modern grouting techniques have been built successively over decades of trial and error and it is essential to incorporate experienced specialists in grouting projects.

7.2.2 History of solution grouting

The following history of solution grouting, herein also known as chemical grouting, relies heavily on the "Chemical Grouting" chapter of ground improvement edited by M. P. Moseley (1993) in which the author, G. S. Littlejohn, acknowledges the paper by Glossop (1961).

Chemical grouting, a logical evolution from cement-based grouting, goes back to at least 1913 at a project in Thorne, Yorkshire, England, where the Belgian engineer Francois used siliconization to grout fine fissures in a porous sandstone. This application was completed using the injection of sodium silicate and aluminum sulphate. The success of this project established chemical grouting as a viable engineering tool. In addition to the injection of the chemical grout, Francois followed the chemical injection with an injection of a neat cement grout. At the time, Francois believed that the chemical grout was acting as a lubricant for the cement grout, but later analyses would conclude that the chemical gel most likely filled the smaller fissures and pores while the neat cement grout effectively filled the larger fractures. Without the chemical gel pretreatment, the cement grouting would have been ineffective because the water would have been driven out of the grout under high pressures.

Early successes of chemical grouting led to further applications, specifically in situations where cement-based grouts were known to be problematic or ineffective, e.g. for grouting of fine-grained soils in which the cement particles were filtered out of the grouts which was hypothesized as early as 1905 by Portier. A patent filed by Lemaire and Dumont in 1909 is the first published reference to a single shot chemical grouting process (Karol 1983). The process would not be perfected until 1925 when the Dutch engineer Joosten perfected a method of staged injection of two fluids, sodium silicate and a brine solution. The invention of the TAM (described in more detail later) in 1933 greatly improved chemical grouting by allowing the injection of grouts with varying properties to be injected through the same borehole in any order.

Many new chemical formulations were developed in the 1930s, 1940s, and 1950s. Additional applications of chemical grouting in the 1950s for dam cutoffs and tunnel support helped establish chemical grouting as a geotechnical tool. In 1963, a state-of-the-art in grouting was prepared and reviewed at the ICE Conference in London (Xanthakos et al. 1994). Further chemical grouting developments in the 1960s, 70s, and 80s included the reduction of grout toxicity (some early grouts were known or later determined to be highly toxic), introduction of grouting materials that react with water, and the introduction and improvement of polyurethane grouts.

Given the higher relative cost, solution grouting is generally applied to unique and challenging situations. As a result, many of the developments in chemical grouting are made in response to those unique conditions. Unfortunately, not all of those developments are widely publicized.

7.3 GROUTING TYPES AND CLASSIFICATIONS

Grout mixtures are composed of many different materials, each added to achieve specific goals. Most grouts start with a mixture of water and cement but may also include (or include in place of the cement), lime, clay (usually bentonite), chemical agents, or some combination of these materials. Depending upon the mixture, grout is classified as either a suspension grout or a solution grout. Grouts with ingredients such as cement and bentonite are suspension grouts (fine particles suspended in water), whereas grouts with ingredients such as silicates (soluble) are solution grouts. The distinction between a suspension and solution grout is evident in the distinction between a suspension and a solution. Each type of grout has advantages and disadvantages.

7.3.1 Suspension grouts

Suspension grouts are mixtures of solids, such as cement and bentonite, in water. Suspension grouts are used in penetration grouting to fill the pores of a soil mass. For this reason, grouts used for penetration grouting are typically very low viscosity suspension grouts, created with materials that have very small particle sizes. Solution grouts are also used for penetration grouting and are addressed in section 7.4. A portion of Chapter 11 addresses low viscosity, suspension grouting with cementitious materials.

7.3.2 Common grout mixtures for suspension grouting

Most grout mixtures start with a combination of cement and water. Other additives are used to control the properties of the grout, including viscosity, density, and filtration. A grout consisting of only cement mixed with water is called a *neat cement* grout. Modern grouting is performed with balanced, stable grouts which are mixtures of cementitious materials and additives used to improve the rheological properties of the grout.

7.3.3 Neat cement grout

The most common neat grouts are mixtures of cementitious materials and water. For permeation and fracture grouting applications in bedrock, neat grouts are generally considered unstable because water can bleed from the mixture and the reagents can be stripped out of the grout through filtration. This results in portions of soil pores or rock fractures being inadequately sealed. Neat grouts are commonly used in other specialty foundation techniques such as the grouting of tiebacks and micro piles, and for soil mixing and jet grouting applications where the grout is mixed with the surrounding soil and bleed/separation is not an issue for the composite soil-cement grout matrix.

7.3.4 Balanced stable grout

Balanced stable grout is so named because of the blend of balanced rheological and physical properties that have been selected for the specific grouting application, typically for permeation and fracture grouting in bedrock. These grout mixtures are stable because the grout is homogeneous and exhibits little to no bleed. Unstable grouts have variabilities in properties (rheology), unknown or poor particle orientation, and high segregation or sedimentation. Unstable grouts are not desirable because their behavior is unpredictable, and, thus, they may not achieve durability objectives. Neat grouts (above) are generally considered unstable grouts. Common additives used to create balanced stable grouts include bentonite, silica fume, flyash, welan gum, dispersants, and plasticizers. Each of these additives results in unique characteristic improvements to the stability of the grout. The commonly identified additives outlined above are presented in Table 7.1 with common dosage rates and characteristic advantages and disadvantages.

Most modern bedrock fracture grouting projects utilize a suite of balanced, stable grouts. Typically, these grouts are labeled A through F with A being the most mobile (lowest viscosity, highest water to solids ratio) and F being the least mobile (highest viscosity, lowest water to solids ratio). The set time of the grouts tends to decrease from the A grouts to the F grouts with A grouts having a set time of greater than 24 hours and F grouts having a set time of 6 to 12 hours. A to C grouts typically have water to solids ratios greater than 1, with A grouts being as high as 1.75 to 1, whereas D to F grouts typically have water to solids ratios less than 1, with F grouts being as low as 0.75 to 1. Most grouts, specifically balanced stable grouts containing bentonite, are thixotropic meaning that the viscosity decreases with increasing shear stresses.

Table 7.1 Commonly used grouting additives (reagents)

Additive	Common Dosage (by weight of cement)	Advantages	Disadvantages
Bentonite	2–8%	Reduces filtration and bleed, enhances stability	Increases cohesion and viscosity
Viscosity modifying polymer (gums)	~0.1%	Reduces filtration and segregation, enhances water repellency	Increases cohesion and viscosity
Superplasticizer	0–2%	Decreases viscosity, allows for lower W: C ratio, increases penetration, enables dispersion of other additives	Delays set time
Other additives:			
Silica Fume	4–8%	Increases penetration, reduces permeability and filtration, enhances durability and water repellency	Increases strength
Dispersant	1–2%	Increases penetration and pumping time	Increases set time

7.3.5 Microfine or ultrafine cement grouting

Microfine cement has a very small maximum particle size, <15 micrometers. The use of this product in grouting applications is relatively new, i.e. within the last 40 years. The product has obvious advantages for grouting in that the small particle size allows for use of suspension grouting methods to be used to fill voids that could not be filled with other types of suspension grouts containing ordinary portland cement. This method started to receive renewed attention in the early 1990s (DePaoli et al. 1992). The use of microfine cements in grouting serves as a technological bridge between solution and suspension methods.

7.4 SOLUTION GROUTS

When the particle sizes of the materials used in penetration (injection) grouting are too large to effectively penetrate the voids, solution "chemical" grouts are used. In practice, solution grouting is very similar to conventional penetration grouting except that the grout mixtures are different. Solution grouting was originally developed for filling small rock fractures or grouting fine sands where the pore or fracture size was too small to be penetrated by the particles in conventional grouts, e.g. cement. Solution grouting can be used effectively to reduce the hydraulic conductivity, increase the strength, and reduce the compressibility of materials in which suspension grouts would be useless. In general, solution grouts are typically several orders of magnitude higher cost compared to cement-based suspension grouts. Given the large quantities of grout typically needed for major civil engineering applications, solution grouts are only used sparingly and generally in small applications. Where suspension grouts are not technically applicable, often the use of a microfine or ultrafine cement-based grouting, or jet grouting, would be used before considering a solution grout.

7.4.1 Types of solution grouts

Various types of solution/chemical grouts are produced commercially. While all chemical grouts have similar basic functions, each grout type has unique characteristics and target applications.

Sodium silicates (waterglass) were the chemical grout of choice during the early 20th century (Mirghasemi et al. 2004). The Joosten two-shot technique involved injecting a sodium silicate solution in one shot followed by an alkali activator in a second shot to form a permanent bond (Bowen 1975). This process was later simplified to a one-shot procedure and the practical use of sodium silicates has continued.

To vary desired properties, products with a range of silica/alkali ratios are commercially available. Silica/alkali ratios typically vary from about 3 to 4 with water in the range of 62% to 72% producing specific gravity values in the range of 1.27–1.40 (Karol 1990).

Depending on the ratio used, it may take up to 4 weeks to reach maximum strength. As a general rule, sodium silicates are considered environmentally friendly and nontoxic (Karol 1990).

Early documented uses of acrylamide-based grouts (acrylics) date back to the 1950s. In many ways, these grouts have better characteristics than silicates, such as a more easily controlled gel time and an easier injection process. However, acrylics are more expensive than silicates. Acrylamide is also known to be toxic and may be linked to cancer. During a two-month grouting project in Romeriksporten, Norway, over 50 workers were observed to have developed symptoms of a deteriorated nervous system which was linked to exposure to the acrylamide (Hagmar et al. 2001). Acrylamide grouts are diluted to ease these issues, but the health, safety and environmental impact of acrylics must be kept in mind at all times (Karol 1990).

Lignochrome grouts are made out of waste produced from wood processing. Specifically, lignochromes are created by removing lignosulfonates from wood scraps and combining them with various chromium compounds, most commonly sodium dichromate. The strength of these grouts can be altered by raising or lowering the pH. Lignochromes are relatively cheap since they are made from wood byproducts, but the chromium compounds used in the grouts are highly toxic (Karol 1990). Because of these toxicity issues, very few lignochromes are still commercially available or used.

Other chemical grout additives include phenolic resins, *aminoplasts*, and polyurethanes.

7.5 PERMEATION (PENETRATION) GROUTING

Permeation grouting is used to fill existing voids, typically pores between soil particles, with a liquid mixture. The liquid mixture in permeation grouting is often a solution grout because the materials used to create solution grouts generally have smaller particle sizes or no particles are present in the grout as the solids are soluble and therefore are in the liquid base of the grout. Hence, these grouts tend to be freer flowing and therefore more easily penetrate the soil pore space. Figure 7.3 shows the penetrability of three grout types for typical soil gradations. The three grout types are labeled with the common names where suspension grouts are also known as cement grouts and solution grouts are also known as chemical grouts.

Example problem Ex.7.2: Choosing a grout type

Thought exercise: A project requires a grouting program in gravelly sand (less than 5% fines, less than 80% sand, less than 20% gravel). Using Figure 7.3, what grout type(s) would you consider for use at this site?

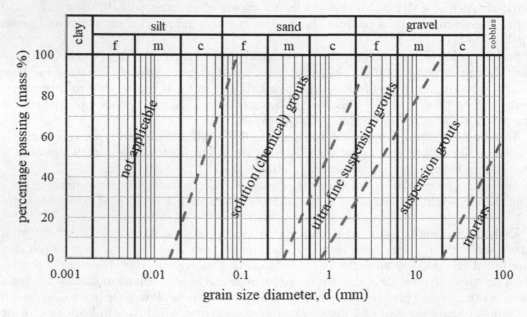

Figure 7.3 Grout material based upon grain size distribution of formation to be grouted (c - coarse, m - medium, f - fine).

Discussion: Without knowing the size of the sand particles (fine, medium, coarse) it seems clear that solution grouting should be considered, but it's possible that a suspension grouting program with an ultra-fine material could work. Both should be considered.

7.6 FRACTURE GROUTING

Fracture grouting is a special application of penetration grouting typically performed using suspension grouts. In this application, grouts are used to fill natural or manmade fractures in rock or soil. Manmade fractures are created by pumping the grout under high pressure to hydraulically fracture the soil or rock mass. Fracture grouting is often performed using successive injections of balanced, stable grouts with increasing viscosity. The early grouts, meant to penetrate the smaller fractures, are followed by thicker grouts meant to fill the larger fractures.

In order to intersect fractures inclined at different angles and located at various depths, fracture grouting is often performed in a split spacing pattern. Split spacing is a method by which grout holes are grouted in an alternating pattern. By using an alternating pattern, grouting is performed at locations between previous grout injections and confinement of the grout is maximized. The initially grouted holes are known as primary holes and subsequent locations between primary grouted holes are termed secondary holes. On some projects, tertiary grout holes are used. Subsequent additional split spacing may be used if necessary. Also, grouting can be performed using inclined grout holes with the inclination chosen to perpendicularly intersect the planes of the rock (the dip) or soil mass being grouted. Fracture grouting is also often performed with the vertical grout interval isolated using packers installed above and below the area being grouted.

7.7 COMPENSATION GROUTING

Compensation grouting is a special application of compaction grouting wherein a thick viscous fluid or semi-solid material is pumped into the ground to densify the soil with the intent of moving the ground surface or moving a structure that may have settled unacceptably. A simple, near-surface example of compensation grouting is *slab jacking* wherein the grout is used to raise a concrete slab to match an adjacent slab or level the slab being jacked.

7.8 VOID GROUTING

Void grouting (void filling) is a special application of grouting wherein grout, generally a rather large volume, is used to fill a subsurface void. Void filling is typically performed to prevent future collapse of the void that could result in surface displacement which damages structures or creates safety issues. Void filling is also used to prevent water from migrating into or out of a void, preventing the migration of soil particles carried by the water. An example of void grouting is the filling of abandoned pipelines or tunnels that would otherwise be impossible or cost prohibitive to remove. These void structures are filled to prevent future collapse that could propagate to the surface and cause issues for future land use or safety concerns.

7.9 GROUT PROPERTIES

Several key properties are monitored and varied for all grouting applications. This section discusses these key properties and their measurement.

7.9.1 Set (gel) time

The time it takes for the grout to noticeably start to change from a liquid to a solid is known as the set time or gel time. Technically there are differences in the set time and gel time, specifically for solution grouts, but, for the purposes of this book, these terms will be considered synonymous. Each grout requires a different amount of time for the chemicals to mix together and form a cohesive unit. Depending on the situation, different gel times may be desired in order to prevent the grout from prematurely hardening, to prevent the grout from seeping into areas it is not desired, or to wash away due to groundwater movement. Depending on the type of grout, the gel time and set time may be very similar, e.g. within 10% of each other (Magill and Berry 2006). Solution grouts usually contain three main components: an activator, an inhibitor, and a catalyst. By altering the mix ratio of these components, the gel time can be adjusted (Karol 1990).

7.9.2 Stability

The stability of a grout is a measure of its ability to remain a uniform mixture. Stable grouts are ones that do not separate either by particles falling out of suspension (suspension grouts) or by precipitation (solution grouts). Stability is an important consideration in grouting because unstable grouts can often lead to incomplete filling of voids or fractures. One measure of the stability of the grout mixture is bleed. *Bleed* is a measure of separation and is typically represented by the volume of water that comes out of the suspension as particles settle from a suspension fluid after fluid movement stops and before it sets. The grain size, shape, and density of suspended particles directly relate to the amount of bleed, with the settlement rate driven by gravity and being directly proportional to the difference in density between the particles and the fluid, and influenced by particle size. Grout additives that change the fluid viscosity and/or particle charge distribution can reduce bleed (one cannot change gravity).

Stability of the grout is partially measured and represented by pressure filtration tests, for example through testing with a filter press. In this test, the grout is forced at a pressure of 690 kPa for 30 minutes through a filter. The quantity of liquid that passes (filtrate loss) is measured. Lower filtrate loss is a measure of the ability of a grout to form a low permeability filter cake at the interface between the grout and the formation.

7.9.3 Viscosity

The viscosity of the grout is the key factor when determining pumpability and penetration into the mass to be grouted. As defined by Newton, viscosity is a measure of a fluid's resistance to deformation due to applied shear stress. For laminar flow, the viscosity can be defined using:

$$\tau = \mu \frac{dv}{dz} \tag{7.1}$$

where

μ = viscosity

τ = shear resistance

$\dfrac{dv}{dz}$ = the velocity gradient

As defined above, μ is the dynamic (absolute) viscosity. The kinematic viscosity can be determined by dividing μ by the density of the fluid, ρ. The units of kinematic viscosity are square meters per second. For turbulent flow, the viscosity must be modified using a dynamic eddy viscosity, denoted η. Given that grouting is purposefully performed at low viscosity, laminar flow conditions are generally assumed in estimating grout penetration into a formation.

The lower the viscosity, the faster and further the grout flows under a given set of conditions. On average, solution grouts have much lower viscosities than traditional suspension grouts, which means they can be pumped into the ground faster and can penetrate further (Bowen 1975).

In the field, the true or absolute viscosity is rarely measured. Rather, indicator field tests like the Marsh Funnel (ASTM D6910) and grout flow cone (ASTM C939) are used as these tests are rapid and can be performed to obtain a general understanding of the grout viscosity, termed the apparent viscosity. These field tests provide a quick measure of viscosity, in units of seconds. By their nature, these sorts of tests are actually measuring a combination of rheological properties. These measurements are inherently performed at a variable shear rate; hence the "apparent" viscosity term (Hausmann 1990).

7.9.4 Permanence

The expected life of the final product is a key consideration for all projects. Due to various forms of degradation, some chemical grouts have a limited service life. The most common form of degradation is caused by freeze/thaw cycles. Freeze/thaw processes can destroy chemical grout structures through the expansion and contraction of residual water in the grout mixture. Chemical degradation can also occur if the groundwater contains materials that react with the grout components. Grout design should consider permanence. The grout should be designed to ensure permanence assuming the site conditions remain unchanged (Karol 1990; Siwula and Krizek 1992; Gemmi et al. 2003).

7.9.5 Toxicity

Because solution grouting involves inserting chemicals into the ground near the water table, grout toxicity must be considered. Any grout must not degrade into, or release, harmful chemicals. Also, while many grouts will not be toxic, once set, the initial handling of the reagents can sometimes lead to construction health, safety, and environmental release issues. In early applications, the toxicity of grouts was often ignored. Now, environmental and health concerns play a key role in grout designs as exposure to some of the chemicals used in chemical grouts has been shown to cause cancer, corrosion, and other irritations (Karol 1990; Chun and Kim 1998; Bonacci et al. 2009). Even the most common reagent, portland cement, can pose toxicity issues in some cases as the cement significantly increases the pH of any water body it encounters, even at relatively low concentrations. Portland cement dust is also a leading cause of silica exposure, a health hazard.

7.10 APPLICATIONS

The many types of grout and methods of grouting can be used to accomplish a variety of objectives. Grouting is widely used for dam and embankment improvements, such as the installation of a grouted cutoff wall in a rock foundation material beneath the dam. Grouting is also widely used for building foundation applications and as a remedial tool in a variety of applications to support other foundation or ground improvement methods.

Applications by grout type include:

1. *Suspension grout uses*. Suspension grouts are used for penetration and fracture grouting, and sometimes in void filling applications (such as the filling of abandoned underground pipes and utilities). These grout mixtures are often used to install permanent cutoff features below dams, including cutoffs in rock and overburden, as a remedial tool for reducing seepage beneath and through dams and embankments, to improve the physical and hydraulic properties of materials surrounding tunnels (during construction and for permanent protection), and for increasing the strength of rock masses.
2. *LMG Uses*. These grout mixtures are prepared in a very stiff (25 mm to 75 mm slump) consistency for compaction grouting applications to densify loose, unconsolidated soils and non-engineered fills. In a higher slump consistency (100 mm to 300 mm slump) low mobility grouts can be used for the underground filling of large open voids such as abandoned mines and limestone solution cavities.
3. *Solution grout uses*. The most common use for solution (chemical) grouts is to prevent seepage from occurring in a given area. Because solution grouts are liquid solutions, not suspensions like cement grouts, they can be pumped more easily into very small voids to seal water flow pathways (Kazemian and Huat 2009). Solution grouts can also be used in place of or in conjunction with cement grouts to create a grout curtain to cut off the groundwater flow under dams or levees (Mirghasemi et al. 2004). The exclusive use of solution grouts is not as popular as exclusively used cement-based grouts, due to higher cost. Solution grout types are sometimes the only option in applications for materials with small pores or fractures. Solution grouting can also be used to add cohesion to cohesionless soils (sands) to increase the shear strength of the soil. Since shear strength depends on the soil type and in situ conditions, it is very difficult to associate any particular shear strength with a particular chemical grout (Karol 1990). Finally, solution grouts can be pumped in successive stages to accomplish specific objectives, e.g. low viscosity penetration followed by the addition of a catalyst to speed gelling.

Applications by mechanism include:

1. *Permeation/penetration grouting uses*. Permeation grouting is used to increase the shear strength of a soil mass for improved liquefaction resistance or improved bearing capacity, to reduce seepage through a soil mass for the containment of groundwater (contaminated or otherwise), to improve the stability of a soil mass ahead of a tunnel boring machine, to reduce or eliminate soil particle migration through pipeline bedding materials, to improve erosion resistance of granular materials, and as a remedial tool for completion of cutoff walls installed using other methods.
2. *Fracture grouting uses*. Fracture grouting can be used to fill existing fractures in rock or to fill fractures created by the grouting process (hydrofracturing or "fracking"). The most common use of fracture grouting is for the creation of permanent cutoffs below dams or embankments to reduce underseepage and improve dam performance/permanence. Fracture grouting may also be performed to fill fractures in rock below existing

or new construction to prevent settlement or to reduce water infiltration through rock during dewatering for an excavation.

3. *Void grouting uses.* This application of grouting is used to fill sinkholes, abandon underground structures (tunnels, pipes, buildings) that cannot be accessed or removed cost-effectively, or fill large underground caverns such as those that may exist in a karstic bedrock. Specific example uses of void filling include the filling of long, deep natural gas pipelines that would otherwise be too costly to remove, filling of abandoned coal mine shafts, or filling of unused power plant cooling tunnels located at an inaccessible depth.

7.11 DESIGN CONSIDERATIONS

Design for grouting projects requires an in-depth understanding of the grout, the geologic conditions of the grouting project site, and the interaction between the grout and the geologic features.

7.11.1 Understanding grout physics and preliminary planning

Grouts range from viscous to free-flowing. The grouting practitioner must understand the physics controlling the flow along the mobility spectrum. In order to do so, the grouting practitioner should understand, for instance, the difference between types of fluids or semi-plastics, or the difference between Newtonian and non-Newtonian fluids, in order to classify the subject grout accordingly. A short discussion is included here. Newtonian fluids have a linear relationship between shear stress and velocity gradient; non-Newtonian fluids have a nonlinear relationship between the shear stress and the velocity gradient. A mechanical model to describe fluids is known as a rheological model. If a fluid material has initial shear stress that must be overcome, τ_o, and a linear increase in μ as the velocity gradient increases, the substance is referred to as a Bingham body. Technically speaking, a Bingham body is not actually a fluid, but rather a viscous semi-solid. Nonetheless, the term *Bingham fluid* is used widely in grouting (Hausmann 1990). The behavior of a Bingham body can be defined by the following equation:

$$\tau = \tau_o + \mu \frac{dv}{dz}$$
(7.2)

where

τ = shear stress
τ_o = initial shear stress
μ = absolute viscosity
$\dfrac{dv}{dz}$ = velocity gradient

The initial shear (yield) stress may also be referred to as the rigidity, cohesion, or flow limit (Hausmann 1990). Flow behavior that follows the Bingham fluid rheological model (Eq 7.2) is also referred to as *ideal plastic* behavior. Fluids, or more correctly semi-solids, that have increasing viscosity with agitation are known as thixotropic. Suspensions of bentonite clay in water are thixotropic. Thixotropy is a key component of stable grouts as the bentonite helps to prevent settling and bleeding prior to the cement hardening. Relatedly, bentonite suspensions (or any clay suspension) are non-Newtonian fluids vs. water which is a Newtonian fluid. For grout design purposes, it is common to assume that a cement and clay grout behaves as a Bingham body.

Once the grout rheology is understood, grout flow can be modeled similarly to modeling flow in pipes. The practitioner must be careful to use the appropriate calculations for either a Newtonian or non-Newtonian (Bingham body) fluid. These calculations will always be limited, as flow in soil or rock is significantly more complicated than flow in pipes. Yet, the calculations are useful in understanding the controlling mechanics. For instance, these calculations will help draw conclusions about grout intake, which is proportional to the applied pressure and inversely proportional to grout viscosity, and grout flow, which is heavily influenced by void size (based on grain or fracture size). Equation 7.3, for calculating the flow rate of a Newtonian fluid in a pipe, highlights the strong relationship between flow and void size:

$$Q = \frac{\gamma i \pi r^4}{8\mu} \tag{7.3}$$

where

Q = flow rate
γ = unit weight of grout
i = hydraulic gradient ($\Delta h/\Delta x$)
r = radius of void or "tube" (between grains, or fracture opening)
μ = absolute viscosity of grout

Equation 7.3 could be useful for estimating the flow of a solution grout in soil or rock. Suspension grout flow is considerably more complicated due to the rheological properties of the grout. Equation 7.4, from Hausman (1990), can be used to calculate flow of a Bingham substance (grout) in a "pipe" assuming laminar flow conditions and rough boundaries (shear stresses develop along the walls of the "pipe"):

$$Q = \frac{\gamma i \pi r^4}{8\mu}\left[1 - \frac{4r_c}{3r} + \frac{\left(r_c/r\right)^4}{3}\right] \tag{7.4}$$

where

Q = flow rate
γ = unit weight of grout
i = hydraulic gradient ($\Delta h/\Delta x$)
r = radius of void or "tube" (between grains, or fracture opening)
μ = absolute viscosity of grout
r_c = radius of the fluids' "rigid core" = $\dfrac{2\tau_o}{i\gamma}$

If the rigid core, r_c, is larger than the radius of the pipe, no flow will occur. Relatedly, the minimum gradient, i_{min}, can be determined from Equation 7.5:

$$i_{min} = \frac{2\tau_o}{r\gamma} \tag{7.5}$$

This minimum gradient corresponds to the head difference needed to just overcome the initial yield stress at the boundaries of the pipe so flow is initiated.

In all of the above discussions, it is noteworthy that the velocity of the grout in the void is not uniform: the velocity in the center of the pipe is highest and velocity along the boundaries is lowest. Newtonian fluids (water or most solution grouts) will have a parabolic velocity profile whereas the non-Newtonian fluids, the Bingham substances (suspension grouts), have a uniform maximum velocity in the center of the pipe corresponding to the diameter of the rigid core with parabolic curves extending to the pipe boundaries where the minimum velocity occurs. In the case where all flow conditions are equal and the viscosity of the grouts are equal, a Newtonian grout will travel more quickly through a pipe than a Bingham grout. All of this relates to grouting in that the velocity of the grout will decrease as the grout moves away from the injection point and grout movement will stop when the grout sets or the gradient drops below the minimum needed to maintain flow.

More information about the flow of the grout can be determined using equations for spherical flow through porous media, after Hausmann (1990). For instance, the net pressure, p_e, to maintain a flow, Q, for a Newtonian fluid flowing from sphere of radius R_o under laminar flow conditions is:

$$p_e = \frac{Q\gamma\mu}{4\pi R_o k\mu_w} \tag{7.6}$$

where

Q = flow rate
γ = unit weight of grout
k = permeability of soil to water
μ = viscosity of grout
μ_w = viscosity of water

In grouting performed following the spherical grout flow model in Equation 7.6, the grout will travel a distance dr in time dt and so the grout taken in time (t) is (by integration):

$$Q\,dt = 4\pi r^2 n\,dr \tag{7.7}$$

Where

n = porosity of the soil

Looked at another way, the t needed to travel a distance R from an initial spherical cavity with radius R_o is defined by Equation 7.8:

$$t = \frac{4\pi n}{3Q}\left(R^3 - R_o^3\right) \tag{7.8}$$

Note that a Newtonian fluid will continue to travel outward through the porous media as long as there is a net driving force (excess pressure) or until the grout gels (viscosity increases). The calculations above for spherical flow are helpful for determining the time needed to create a "bulb" of grout in a porous media with a known radius in a given time. In reality, the bulb will likely not have a uniform diameter as the porous media, soil, will not have uniform properties in all directions. However, the calculation is useful in determining the approximate radius of influence for grouting performed from a drill tip.

A more accurate representation of grout flow from a borehole can be modeled using equations developed for radial flow from a cylindrical cavity, after Hausmann (1990). The equivalent equations to Equations 7.6 and 7.8 are, respectively:

$$p_e = \frac{Q\gamma\mu}{2\pi mk\mu_w}\ln\frac{R}{R_o} \tag{7.9}$$

$$t = \frac{\pi mn}{Q}\left(R^2 - R_o^2\right) \tag{7.10}$$

where
   p_e = excess pressure needed to maintain Q at a distance R
   R_o = radius of borehole
   m = thickness of grouted layer

A few key conclusions can be drawn from Equations 7.9 and 7.10. First, the flow rate determines how long it takes to grout a set distance away from the injection point. Second, a higher pressure or a lower viscosity grout can be used to increase the flow rate (reduce grouting time). Third, when designing the grout, the setting time must be greater than the amount of time it takes for the grout to reach the required distance from the injection point. Finally, larger boreholes (R_o) require higher pressures to reach the required influence distance (R). Recall, 7.9 and 7.10, as with 7.6, 7.7, and 7.8 are all equations for determining flow rate, influence distance, and time for grouting performed with a Newtonian fluid in laminar flow. Thus, these equations will not accurately model conventional suspension grouting in soil because the grout in that case is non-Newtonian. For a better understanding of grouting with a visco-plastic substance (conventional suspension grout), see Hausmann (1990), Lombardi (1985), or other publications devoted solely to grouting.

Example problem Ex.7.3: Grouting calculations

Grouting is planned for use in a 4 m thick sand layer with a porosity of 40%. The contractor is proposing to use a borehole with a diameter of 0.1 m which will have an open screened interval of 4 m (corresponding to the top and bottom of the target layer) and the contractor plans to pump the grout at 0.5 m³/min. Approximately how long will it take for the grout to reach 3 m from the borehole (Part 1)? Approximately how far will the grout travel after 30 mins of pumping (Part 2)? Use Equation 7.10 (recall, this is based on laminar flow through porous media from a cylindrical cavity).

Solution:

PART 1

Equation 7.10 defines the relationship needed to solve this problem:

$$t = \frac{\pi mn}{Q}\left(R^2 - R_o^2\right)$$

Inputting the information in Equation 7.10:

$$t = \frac{\pi(4)(0.4)}{0.5}\left(3^2 - 0.1^2\right)$$

$$\underline{t = 90 \text{ mins}}$$

PART 2

Solving for R yields:

$$R = \sqrt{t\,\frac{Q}{\pi mn} + R_o^2}$$

Inputting the information into this reorganized version of 7.10:

$$R = \sqrt{30\,\frac{0.5}{\pi\,(4)(0.4)} + 0.1^2}$$

$$\underline{R = 1.7\,\text{m}}$$

Once the flow properties and mechanics are understood, the grout must also be evaluated for short- and long-term toxicity and long-term permanence. Generally, the grout, or grouts, that will be used are developed in a bench-scale (laboratory) study performed with samples of the materials expected at the project site. These studies will confirm (or not) that any expectations from past experience or research hold true for locally available materials and will help to establish target properties for the full-scale application. If time and budget allow, a full-scale pilot test, or test section, should be performed prior to the full-scale construction work. This allows an opportunity to test the selected grout or grouts in the actual ground conditions, helping confirm expectations of both the grout performance and the geologic conditions. Success in a grouting bench-scale study, or even during the pilot study, does not guarantee success in the field.

The observational method (Peck 1969) is paramount in grouting construction. Grouting requires constant vigilance and flexibility to adjust to observations during construction including changes in subsurface conditions. Planned monitoring strategies and procedures for field modifications based on observations should be established, or at least considered, in the design phase.

Well-designed grouting programs provide a feedback loop. Construction observations (which are ostensibly detailed subsurface investigations) are incorporated into the conceptual understanding of the subsurface. The enhanced understanding of the subsurface is then used to verify or modify the approach as the grouting program progresses. Most grouting is performed in a split spacing pattern preferably with multiple rows. Figure 7.4 shows a plan view schematic of a two-row grout curtain including primary, secondary, and tertiary grout holes.

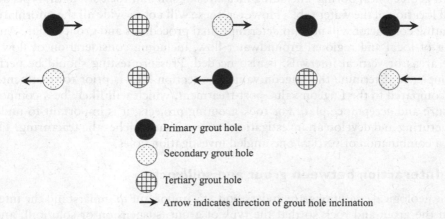

● Primary grout hole

◌ Secondary grout hole

⊞ Tertiary grout hole

→ Arrow indicates direction of grout hole inclination

Figure 7.4 Typical grout hole layout (plan view) for a rock grout curtain.

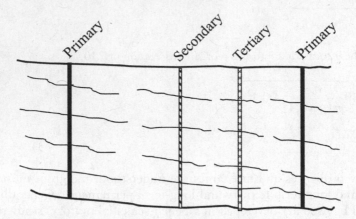

Figure 7.5 Section view of primary, secondary, and tertiary grout holes for grout curtain in rock.

Figure 7.5 shows primary, secondary, and tertiary holes in section view. Observations during the primary grouting phase should be used to inform the grouting procedures for the secondary phase. The tertiary phase, if needed, can be used for confirmation. Note that angled grout holes are often used to maximize intersection of fractures based on geological conditions.

7.11.2 Geological conditions and site investigations

A detailed understanding of the geologic conditions at the project site, and how those conditions could change across the site and with depth, is essential for the successful design of a grouting program. As with geotechnical design and construction in general, it is as important to understand what you don't know as it is to understand what you do know. Important things to understand for rock grouting include fracture size, spacing, and orientation, water location and movement, existing sedimentation in fractures, and the chemical makeup of the rock and groundwater. Important things to understand for soil grouting include porosity, permeability, and particle size of the soil, and the water location, movement, and chemistry.

Site investigations for grouting projects should be designed to collect information for geotechnical design purposes and also for use in developing the construction approach. Standard geotechnical borings including measures of soil and rock strength, type, and density, and location of the water table. However, these will not provide all the information that the grouting contractor will need to determine final procedures and grout design. An understanding of local and regional groundwater flow, including consideration of flow within discrete areas or vertical intervals, is also needed. Pressure testing should be performed, for example, to determine the Lugeon value (see section 7.13.1) prior to treatment, which can be compared to the Lugeon value post-treatment, which will likely be a component of the closure and acceptance plan. For rock grouting projects, it is important to understand rock fracturing and develop an investigation strategy that matches the fracturing. This may include a combination of vertical and angled investigation holes.

7.11.3 Interaction between grout and soil/rock

Once the geological conditions are understood, it's important to understand the interaction between the grout and rock so that the type of grout (suspension or solution), and grout makeup can be selected. At this stage, it is important to understand physical interactions in the short term, e.g. pore and fracture size related to grout viscosity, and the long term, e.g. particle settlement prior to set in large void structures which could leave gaps, or grout shrinkage that may result in voids along the interface of the grout and soil or rock. In addition to physical interactions, it's important to understand chemical interactions, e.g. dissolution of soil or rock during grouting, or long-term permanence issues associated with incompatibilities between the soil and rock and grout. Details associated with the various interactions deserve site-specific consideration.

7.11.4 Grout mix design

Independent of the type of grout (solution or suspension), the grout or grouts must be customized for each project site. In fact, the first step in grout design is to select the grout type based on an understanding of the project objectives and geological conditions. For example, if the primary objective is to fill large voids in rock, it's clear early on that a suspension grout is the right choice. Alternatively, if the objective is to grout soil pores, the project will likely require a solution grout or at least a specialty suspension grout with microfine cement. Bench-scale laboratory studies are an important part of grout design.

The grout design, in addition to the equipment and grouting procedures, can and should also be refined or modified from the results of a full-scale pilot test or test section performed at the site. Since many grouting projects utilize a successive installation approach, there are typically multiple grout designs for each project site with the grout used at any point being selected based on the needs at that time. For instance, thinner (lower viscosity) grouts may be used first to fill small voids or fractures followed by successively thicker grouts used to close off remaining voids and achieve closure criteria (flow and/or pressure). In many cases, the grouting program can be designed to allow successive grouting from within the same boreholes. However, other sites may utilize different borehole patterns to accomplish the successive installation of thicker grouts, by, for example, applying thinner grouts in the outer rows, or primary holes, and thicker grouts in the inner rows, or during installation of secondary/tertiary holes.

7.12 CONSTRUCTION

7.12.1 Pre-grouting

Pre-grouting is defined as a program of grouting done in advance of either a different construction process or as a first stage in a larger grouting program. For example, pre-grouting is sometimes used as a pre-treatment for another ground improvement technique such as grouting subsurface profiles containing open features (cobbles, boulders) prior to installing a cutoff wall. Pre-grouting is typically performed with relatively low viscosity grouts that are allowed to flow under low pressure into small and distant voids. Pre-grouting termination criteria is typically a volume per vertical increment, i.e. a set volume of grout is pumped over a discrete vertical interval. Since the pre-grouting step is always a precursor to either additional grouting or another construction technique, pre-grouting is never intended to achieve final closure objectives.

The procedures used for pre-grouting are similar or identical to the procedures used for successive grouting, i.e. a hole is drilled into the ground and grout is pumped into the hole and allowed to flow into the soil or rock either over the entire length of the hole or a discrete vertical interval. Pre-grouting may be performed using pressure to drive the grout into the voids or may be performed with little or no pressure on the grout.

7.12.2 Suspension and solution grouting

Suspension grouting is performed by pumping grouts of various viscosities into soil or rock through a borehole with or without applied pressure. All grouting construction begins with the preparation of grout. Although grout preparation will vary by project, as the grout consistency varies by project, there are similarities and generalities. Reagents for grouting can be used in a dry or liquid form. Most of the major grout components such as portland cement and bentonite are in a dry powder form. Lesser volume components, typically property modification additives, are in a liquid form. Dry reagents are stored on site in silos to prevent moisture contamination. Silos used for grout mixture ingredients often contain metered augers to accurately move the reagent from the silo into the batch plant.

Grouting batch plants come in many shapes and sizes, including fully enclosed systems built within steel shipping containers and systems of components combined onsite. Batch plants typically consist of a measuring tank, a mixing tank, a holding tank, and a pump. Some plants combine the measuring, mixing, and holding tanks, or some combination thereof, depending on output needs and space constraints. Generally, the reagents are pumped or metered into the mixing tank and tracked by the weight of the reagent added. Depending on the grout makeup, there may be an order to the addition of reagents, or all reagents may be added at once. In other cases, certain components may be pre-mixed, as with the mixing of bentonite and water to form a slurry ahead of the addition of the portland cement. Understanding the need and reasons for the order of reagent addition is a key skill of an experienced grouting practitioner. Once the reagents are added to the mixing tank, the components are blended in a high shear mixing environment to ensure adequate hydration, to improve suspension, or to allow time for reagents to dissolve in a solution grout. In other cases, the reagents are purposely not sheared to allow for hydration or dissolution to occur in the ground. The subtleties associated with grouting require mix and procedure customization for each project. Once the reagents have been mixed together properly, the mixture is either pumped to a holding tank for future use or pumped to the borehole.

Prior to, or concurrent with, the grout batching, the borehole is drilled. There are various methods of borehole creation ranging from simple uncased auger excavations to complicated casing methods with various annuluses between casings. Drilling and casing selection is based on the grouting program objectives. In some cases, grouting is performed through the drill tooling while the tooling is in the ground and subsequently withdrawn. More commonly, the drilling is used to install a casing that is left in the ground for subsequent grouting. Here, the casing is installed ahead of time, the grouting is performed, and the casing is either backfilled or removed. Depending on the casing type, it can be very expensive and is often reused.

After the grout has been mixed and the casing or borehole advanced, the grout is pumped into the borehole and allowed to enter the soil or rock through the borehole walls or holes in the casing. There are many different ways the grout is forced to exit the hole ranging from full depth grouting, wherein the grout is forced through holes in the casing that extend from the ground surface (or near) all the way to the bottom of the hole, to grouting performed through the bottom of the casing/borehole only. Grouting can be performed as the tool is advanced (downstage grouting), or withdrawn (upstage grouting), or in a staged approach wherein multiple grouts are pumped into the same hole at different times.

> **Example problem Ex.7.4: Grouting pressures**
>
> Imagine you are grouting in a 50 m thick sand layer. The soil has a unit weight of 18 kN/m³ and the grout has a unit weight of 14 kN/m³. Grouting is being performed at 25 m below ground surface. The pressure gauge at the surface reads 100 kPa. What is the net pressure

at the grouting tip if (a) groundwater is at the ground surface and (b) groundwater is 10 m below the ground surface? Ignore friction losses in the grouting rods. (assume water unit weight is 9.8 kN/m³)

Solution:

a. The total pressure in the grout is the pressure being applied by the pumping system as measured by the gauge at the ground surface plus the elevation head. The total pressure in the grout is therefore 100 kPa + (14 kN/m³)(25 m) = 450 kPA. The net pressure is the total pressure minus the water pressure pushing back against the grout. The water pressure is 25 m (9.8 kN/m³) = 245 kPa. So, the net pressure is 450 kPa – 245 kPa = 205 kPa.

b. The total pressure in the grout is the pressure being applied by the pumping system as measured by the gauge at the ground surface plus the elevation head. The total pressure in the grout is therefore 100 kPa + (14 kN/m³)(25 m) = 450 kPA. The net pressure is the total pressure minus the water pressure pushing back against the grout. The water pressure is 15 m (9.8 kN/m³) = 147 kPa. So, the net pressure is 450 kPa – 147 kPa = 303 kPa.

Depending on how the grouting is performed, vertical sections of the borehole or casing can be isolated using bladders filled with air or water, called packers. Packers can be installed above and below a target grouting zone and moved up or down the holes as the project dictates. In other cases, there may be multiple casings with slots at select vertical increments to allow for successive grouting from within the same hole without other forms of isolation. The TAM grouting method can be used for successive grouting installations from within the same hole. In this method, the manchette tube, made from PVC or metal, has rubber sleeves located in the pipe at specific intervals. The Manchette tube is installed in the holes in the subsurface and grout is pumped to a packer pre-installed in the tube. The packers are used to isolate the zones within the tube to force the grout through the pre-drilled holes in the tube and into the target grout zone.

As the grout is pumped into the hole, pressure and flow are monitored, to determine when the grouting objectives have been completed. Once the objectives have been completed, the hole may be grouted shut, backfilled, or washed out in preparation for a successive stage depending on the project needs.

No separate section for solution grouting construction is provided as the methods are very similar to suspension grouting. As may be evident from a review of the above information, all grouting projects require slightly different and nuanced construction approaches depending on the specific project needs.

7.12.3 Drill rigs

Drill rigs for grouting come in various sizes and configurations designed to meet specific needs, including headroom limitations, depth objectives, and drilling and lift rate control. The type and configuration of the drill also vary according to what is being drilled, such as rock, soil, and concrete. Drills for concrete are designed to core through the concrete. In grouting applications, concrete coring is typically started near or at the ground surface, for instance when extending grout holes beneath a dam spillway. This portion of the drilling may be performed with coring tooling on the grouting drill or an entirely different drill. Figure 7.6 shows a concrete core barrel used to drill through a concrete surface layer.

Other drills are designed for drilling through the overburden to install a casing that extends from the ground surface through the overburden and into or to rock. Overburden drilling is usually performed using a rotary percussion drill that allows the sinking of a

Figure 7.6 Core barrel for drilling through rock for grouting (courtesy of Geo-Solutions, Inc., www.geo
-solutions.com).

casing as the bit is advanced. The casing is coupled to the drill rod and is sunk through the overburden as the bit is advanced, the casing rotating and reciprocating with the drill rod in the process. The end of the casing includes a casing bit that helps to advance the casing through the soil. Beyond the end of the casing is a rock bit that acts as a "pilot" bit or guide for the casing and drill rod. In this type of drilling, the casing may be referred to as the "outer pipe" and the drill rod may be referred to as the "inner pipe."

Drilling is often performed with a flushing medium that may be air, water, or a slurry (commonly bentonite and water). The flushing medium helps to lubricate the drill bits and also flushes cuttings from within the borehole to the surface. The flushing material is pumped through the inner pipe out through the drill bit and is allowed to discharge from the borehole at the "mouth" (surface). The loosened drill cuttings are carried by the flushing medium up through the gap between the inner and outer pipe. This system can be referred to as a wet rotary duplex system. There are many configurations of this approach to accomplish drilling in different subsurface conditions. Figure 7.7 is an example drill rig used for grouting.

The overburden drilling eccentric system (ODEX) is a system for simultaneously drilling and advancing casing through overburden. In this system, an eccentric bit extends below and beyond (outside) the casing exterior so that drilling of the soil can be performed to allow casing advancement. The special component of the ODEX system is a retractable bit that can extend beyond the casing and can be folded in and out to allow the bit to either be smaller, as wide, or wider, than the casing so the same bit can be used to drill ahead of the casing and can be retracted through the casing. Figure 7.8 is a schematic of the ODEX bit.

Rock drilling is performed with similar equipment except that the tooling is developed for drilling in rock instead of soil and, depending on the rock makeup, there may not be a need for casing. Wet rotary drilling can be used to drill rock. This technique uses a rotating drill bit to cut through the rock and is often performed with some sort of drilling lubricant which also serves as the flushing medium. Here, the drill bit is slightly wider than the drill rod, so the flushing medium and drill cuttings are allowed to flow to the surface in the annular space between the drill rod and the outside of the hole in the rock.

Figure 7.7 Rig for drilling through overburden for grouting (courtesy of Geo-Solutions, Inc., www.geo-solutions.com).

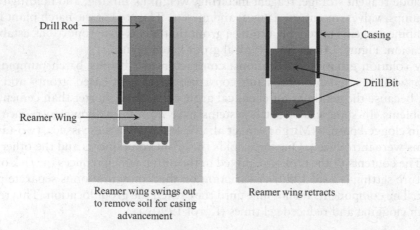

Reamer wing swings out
to remove soil for casing
advancement

Reamer wing retracts

Figure 7.8 Schematic of ODEX Drilling Bit.

For rock that is too hard to drill with a rotary technique, a down-the-hole (DTH) hammer can be used to break up the rock ahead of the drill rod. These hammers attach to the bottom of the drill string, with the hammer being driven pneumatically or hydraulically.

In some applications, casing can be driven or pushed into the ground through percussion or hammering of the pipe. In these applications, the end of the casing can be fashioned into the drive point or a disposable drive point or drive tip may be installed at the head of the casing. Here, the casing is generally advanced using a hammer or percussion drive (see Figure 7.9).

Figure 7.9 Hammer-driven pipe with disposable drive point.

7.12.4 Mixing (batch) plants

Mixing or batching plants for grouting come in all shapes and sizes, ranging from small units capable of fitting on a standard pallet, to self-contained units in sea boxes, to very large setups assembled from individual components at the job site. In general, all batch plants include reagent storage, reagent metering, weighing, mixing, and holding tanks, and various pumps, valves, monitors, gauges, and meters. The goal of the batch plant is to accurately combine reagents to reliably produce grout that meets the proportions established for the application. Figure 7.10 shows a typical grout batch plant.

In early solution grouting applications, contractors tried using batch pumping systems similar to what is used for producing conventional cement-based grouts and concrete. However, because the gel times for chemical grouts are much shorter than cement, this can cause problems. In some cases, batch systems have even been shown to reduce gel times resulting in clogged pumps (Mirghasemi et al. 2004). To solve these issues, two-tank pumping systems were introduced. One tank holds the grout components and the other holds the catalyst. The contents of the tanks are mixed in the pump which reduces the risk of clogging or premature setting (Karol 1990). A variation on this configuration is separate pumps for each tank. The components do not mix until they reach the grout location. This removes all issues with clogging and reduced gel times (Karol 1990).

7.12.5 Pumping systems

Pumping systems are key to successful grouting. Without the pressures supplied by the pumping system, it would not be possible to get the grout deep underground or into small pores and fractures. Solution grouts require particularly close attention to the selected pumping system.

There are many different types of pumps used in grouting, though a detailed description of pumping systems is beyond the scope of this book. Pump selection depends on the type of grout, the depth of grouting, the purpose of the grouting, and the design of the grouting system. Pumps are selected to meet pressure and flow rate objectives with the selected grout

Figure 7.10 Grout batch plant (courtesy of Geo-Solutions, Inc., www.geo-solutions.com).

which will have its own unique viscosity and other properties affecting flow rate and pressure. High-flow, low-pressure applications may be completed using simple piston pumps whereas low-flow, high-pressure applications may be completed using progressive cavity expansion pumps.

7.12.6 Packers

Packers are bladders that can be expanded and contracted in the hole from controls located at the surface. In grouting, packers are used to isolate vertical increments of the borehole for targeted grout application. Figure 7.11 is a photo of a grouting packer and Figure 7.12 is a schematic showing a packer installed in a formation for grouting of a specific zone of the formation.

7.13 QUALITY CONTROL

7.13.1 Flow measurements

Quality control (QC) for grouting projects is designed to ensure the grout has the required properties, is injected at the right locations and at the right pressures, and ultimately accomplishes the overall grouting objectives. If flow control is the primary objective, the grouting is typically assessed for completion by measuring the Lugeon value. The Lugeon test is a measure of the amount of water that can be injected in a vertical interval of a borehole under a steady pressure. The Lugeon value is defined as the loss of water (litres per minute) per meter of borehole at a pressure difference (over-pressure) of 1 MPa. The pressure difference

Figure 7.11 Grouting Packer (courtesy of Christopher Bailey, Gannett Fleming, Inc.).

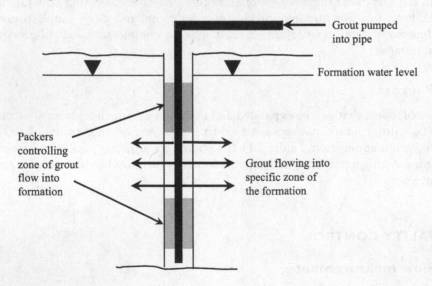

Figure 7.12 Schematic of packer use in grouting.

is a measure of the net difference between the pressure in the grout, a function of gauge pressure measured at the pump, grout density, and elevation difference, and the pressure in the porewater, a function of elevation difference and water density. An over-pressure of 1 MPa means that the pressure in the grout is 1 MPa greater than the pressure in the surrounding formation. The Lugeon is a common unit of measure in grouting. Those practicing grouting must intuitively understand what the unit means. Equation 7.11 shows the calculation of the Lugeon unit:

$$LU = \frac{q}{L}\left(\frac{P_o}{P}\right)\left[1\frac{l}{\min(m')}\right] \tag{7.11}$$

where

LU = lugeon unit
q = flow rate (liters per min)
L = length of borehole (m)
P_o = reference pressure of 1 MPa
P = test pressure (MPa)

7.13.2 Monitoring

Monitoring programs for grouting projects are built around a series of tests, gauges, and meters selected to understand the properties of the grout, the grout flow, and the driving forces behind the grout flow.

Pressure gauges are installed around the grouting system to understand the pressure in the grout at various points in the process, particularly between the pump and the borehole. As the grout penetrates and fills the voids, the pressure in the grout builds and is measured by a pressure gauge. Closure criteria for grouting programs often includes a maximum pressure.

Flow meters are installed around the grouting system to understand the flow of the grout at various points in the mixing process and as the grout moves from the mixing plant into the ground. As the grout flows through the flow meter into the borehole, the cumulative volume is tracked. Closure criteria for grouting programs often include a maximum flow (volume) per vertical increment.

There are several proprietary commercial grouting programs that were developed by engineers or contractors. One of the first such systems was created by a joint effort of a geotechnical design firm and a grouting contractor (Bruce 2015). These systems can be used to automate the monitoring and documentation of grouting programs. Since these systems vary by manufacturer, or are proprietary systems developed by practitioners, these systems can only be understood through direct exposure or through review of more detailed manuals.

7.13.3 Automated Monitoring Equipment

Complete grout monitoring systems can be used to monitor important grouting parameters in real time. The parameters generally include pressure, flow, and injected volume. Instrumentation companies customize measurements as well as recording and display features. A photograph of a grout monitoring system is presented as Figure 7.13.

A number of comprehensive proprietary grout monitoring systems have been developed by engineers and contractors in the grouting industry (Bruce et al. 2008). The goal of simple and complex monitoring systems is the same: collection and organization of real-time data during construction to inform decisions. These systems compile information from the individual tests, monitors, and gauges into a data management system that is often hosted online and can be viewed by all party stakeholders including the owner, engineer, and contractor. The data management system can then be used to remotely monitor the active progress of the grouting program or review past work to inform decisions about current and future work. Since many grouting programs are designed as successive/staged systems with each stage informed by the previous one, good data that is easy to review is critical for

Figure 7.13 Automated grout monitoring system (courtesy of Geo-Solutions, Inc., www.geo-solutions.com).

successful grouting. Unfortunately, these systems are expensive to develop and implement so the most robust and complete systems are generally applicable only to large grouting projects. Monitoring on smaller grouting projects is performed using more rudimentary systems.

7.14 VOID GROUTING, A SPECIAL APPLICATION

Void grouting models pressure or fracture grouting except that the void being filled is very large in comparison to voids in soil or rock. Common applications of void grouting include pipeline abandonment, mine filling/abandonment, and sinkhole remediation. The procedures and equipment used for void filling are similar to other grouting applications, with differences in the grout makeup, monitoring, and setup of void filling projects due to the inherent differences of filling large voids vs. small voids. Key components of void filling include ensuring the void is actually filled (including access locations and monitoring locations), venting port locations relative to the pumping location, and long-term stability of the grout. Although the theory behind void filling is simple, and many construction firms are capable of placing the grout, it is important to engage designers and construction companies that understand the subtle aspects of these applications to ensure a successful project in the short and long term.

Example problem Ex.7.5: Grouting design

A project requires void grouting to close a series of abandoned mine shafts. All of the mine shafts are 3 m by 3 m squares (cross-section). The bottom of mine shaft A is located at EL 100, the bottom of mine shaft B is located at EL 105, and the bottom of mine shaft C is located at EL 95. Mine shafts A, B, and C are 10, 15, and 20 m long, respectively. There are equally sized (cross section dimensions) vertical shafts connecting the horizontal shafts. What is the neat volume of grout (a) that it will take to fill the mine shafts?

Where would you begin pumping and where would you put monitoring points and vent shafts for the grouting program (b)? Describe your reasoning for part b.

Solution:

a. The volume of the shafts themselves are A = (3)(3)(10) = 90 m³, B = (3)(3)(15) = 135 m³, C = (3)(3)(20) = 180 m³. The vertical shaft connecting A to B is EL 105 – EL 100 – shaft height = 2 m long so the volume is (2)(3)(3) = 18 m³. The vertical shaft connecting B to C is EL 105 – EL 95 – shaft height = 7 m long so the volume is (7)(3)(3) = 63 m³. The total volume of the shafts is 90 + 135 + 180 + 18 + 63 = **486 m³**.

b. A self-leveling grout would be pumped into the lowest area first so the pumping port should be drilled into shaft C. Venting/monitoring ports could be placed in all three shafts, but would definitely be needed in shafts A and B. Depending on the orientation of the shafts, a second and/or third pumping port may be needed in shafts A and B. Typically, all of the ports are the same so the venting and monitoring points in shafts A and B could be used for grouting, too.

7.15 PROBLEMS

7.1 Why aren't all grout holes vertical?

7.2 What is a Lugeon? Describe in your own words.

7.3 What is bleed and why is it important to understand and monitor it?

7.4 What sort of long-term items should be considered in the selection of a grout?

7.5 What information from a conventional drilling and sampling program would help you when preparing a grouting program at a site?

7.6 Describe the difference between permeation grouting and fracture grouting. When would you consider each?

7.7 Grouting is planned to create a cutoff wall in a gravel layer located beneath a clay layer. The planned cutoff wall should have a minimum width of 1 m and the grouting program is planned to have primary/secondary holes on a 5 m center to center primary spacing. The top of the gravel layer is located 5 m below ground surface and the layer is 5 m thick. What volume (a) and pressure (b) closure criteria would you recommend for the grouting program? Note: at this site it is important not to cause ground heave. Ignore friction losses in the grouting rods.

7.8 You are developing a grouting program to grout vertically oriented fractures in a 25 m thick rock layer that starts 100 m below the ground surface. The drill mast will be oriented at 30° from vertical. How much drill rod do you need to reach 5 m below the deepest point? How far away horizontally from the drill rig is the actual grouting taking place at the midpoint of the rock layer? How far away from the working area at the surface would you recommend monitoring structures?

7.9 You are grouting in a 100 m thick sand layer. The soil has a unit weight of 20 kN/m³ and the grout has a unit weight of 12 kN/m³. Grouting is being performed at 50 m below ground surface. The pressure gauge at the surface reads 100 kPa. What is the net pressure at the grouting tip if (a) groundwater is at the ground surface and (b) groundwater is 25 m below the ground surface? Ignore friction losses in the grouting rods and assume the water unit weight is 9.8 kN/m³.

7.10 You have been asked to design a grouting program for groundwater control in the Lockport Dolomite, a limestone rock with a well-developed fracture flow system in the Niagara Falls, NY, USA, area. The rock is encountered at a depth of 3 m below the ground surface to a depth of 30 m below the ground surface. The water table is essentially at the surface in a very dense glacial till overburden. Identify the category of grout you believe would work the best (penetration, displacement, compaction, or jet) and explain the basis of your decision.

7.11 Assuming the equations for radial flow from a cylindrical cavity apply, what is the excess pressure needed to maintain a flow rate of 1 m^3/min at a distance of 2 m from the borehole if the grouted layer thickness is 5 m, the borehole is 0.25 m, the soil permeability (to water) is 1×10^{-6} cm/s, and the grout viscosity is 1.5 times the viscosity of water?

7.12 Calculate the excess pressure using the applicable information in problem 7.12, but assuming flow from a spherical cavity (point injection).

7.13 Calculate the time required for the grout to get to the distance of 2 m in problem 7.12, assuming a soil porosity of 40% and radial flow from a cylindrical cavity.

7.14 Given the data in problems 7.12 and 7.14, calculate the time required for the grout to get to the distance of 2 m, assuming a soil porosity of 35% and spherical flow from a spherical cavity.

7.15 Using the information provided in problem 7.12, calculate a new distance from the borehole that the grout will reach in 30 minutes for a flow rate of 2 m^3/min, all other things being equal and assuming radial flow from a cylindrical cavity.

REFERENCES

ASTM C 939–10. (2010). *Standard test method for flow of grout for preplaced-aggregate concrete (flow cone method)*. West Conshohocken, PA: American Society of Testing and Materials.

ASTM D6910/D6910M – 19. (2019). *Standard test method for Marsh funnel viscosity of construction slurries*. West Conshohocken, PA: American Society of Testing and Materials.

Bérigny, C. (1832). *Mémoire sur un procédé d'injection propre à prévenir ou arrêter les filtrations sous les fondations des ouvrages hydrauliques*. Paris: Chez Carilian-Goeury.

Bonacci, O., Gottstein, S. and Roje-Bonacci, T. (2009). Negative impacts of grouting on the underground karst environment. *Ecohydrology: Ecosystems, Land and Water Process Interactions, Ecohydrogeomorphology*, 2(4), 492–502.

Bowen, R. (1975). *Grouting in engineering practice*. New York: John Wiley & Sons.

Bruce, D.A. (2015). Remedial cutoff walls for dams: Great Leaps and Wolf Creek. In *IFCEE 2015*, San Antonio, TX, March 17–21.

Bruce, D.A., Dreese, T.L. and Heenan, D.M. (2008, April). Concrete walls and grout curtains in the twenty-first century: The concept of composite cut-offs for seepage control. In *USSD 2008 Conference*, Portland, OR, April.

Chun, B.S. and Kim, J.C. (1998). The evaluation of toxic effect of grouting materials by Fish Poison Test. *Journal of the Korean Society of Civil Engineers*, 18(3_4), 531.

De Paoli, B., Bosco, B., Granata, R. and Bruce, D.A. (1992). Fundamental observations on cement based grouts (1): Traditional materials. In *Proceedings of grouting, soil improvement and geosynthetics*, New Orleans, LA (pp. 25–28).

Gemmi, B., Morelli, G. and Bares, F.A. (2003). Geophysical investigations to assess the outcome of soil modification work: Measuring percentile variations of soil resistivity to assess the successful modification of foundation soil by jet grouting. In *Grouting and ground treatment* (pp. 1490–1506). New Orleans, LA, February 10 to 12, 2003.

Glossop, R. (1961). The invention and development of injection processes part II: 1850–1960. *Geotechnique*, 11(4), 255–279.

Glossop, R. (1968). The rise of geotechnology and its influence on engineering practice. *Géotechnique*, 18(2), 107–150.

Hagmar, L., Tornqvist, M. and Nordander, C. (2001). Health effects of occupational exposure to acrylamide using hemoglobin adducts as biomarkers of internal dose. *Scandinavian Journal of Work, Environment and Health*, 27(4), 219–226.

Hausmann, M.R. (1990). *Engineering principles of ground modification*. New York: McGraw-Hill Publishing Company.

Karol, R.H. (1983). *Chemical grouting*. New York: Basel, Inc.

Karol, R.H. (1990). *Chemical grouting* (2nd edition, revised and expanded). New York: Marcel Dekker, Inc.

Kazemian, S. and Huat, B.B. (2009). Assessment and comparison of grouting and injection methods in geotechnical engineering. *European Journal of Scientific Research*, 27(2), 234–247.

Littlejohn, S. (2003). The development of practice in permeation and compensation grouting: A historical review (1802–2002): Part 1 permeation grouting. In *Grouting and ground treatment* (pp. 50–99). New Orleans, LA, February 10 to 12, 2003.

Lombardi, G. (1985). The role of cohesion in cement grouting of rock. In *Commission Internationale des Grands Barrages, 15eme Congres des Grands Barrages*, Lausanne (pp. Q.58–R.13).

Maag, E. (1938). Ueber die Verfestigung und Dichtung des Baugrundes (Injektionen). *Course on soil mech., Zurich Tech. School.*

Magill, D. and Berry, R. (2006). Comparison of chemical grout properties, which grout can be used where and why. *Avanti International and Rembco Geotechnical Contractors.*

Mirghasemi, A., Heidarzadeh, M., Etemadzadeh, M. and Pakzad, M. (2004). Results and experiences obtained from chemical grout testing in part of conglomerate foundation of Karkheh Dam – Iran. In *New developments in dam engineering* (pp. 627–634). Najing, China, October 18 to 20, 2004

Moseley, M.P. (1993). *Ground improvement*. Boca Raton, FL: Chapman and Hall.

Nonveiller, Ervin. (2013). *Grouting theory and practice*. Amsterdam: Elsevier Science Publishers B.V.

Peck, R.B. (1969). Advantages and limitations of the observational method in applied soil mechanics. *Geotechnique*, 19(2), 171–187.

Siwula, J.M. and Krizek, R.J. (1992). Permanence of grouted sands exposed to various water chemistries. *Geotechnical Special Publication*, 2(30), 1403–1419.

Terzaghi, K. (1925). Principles of soil mechanics. *Engineering News - Record*, 95(19–27), 19–32.

Xanthakos, P.P., Abramson, L.W. and Bruce, D.A. (1994). *Ground control and improvement*. New York, NY: John Wiley & Sons.

Chapter 8

Slurry trench cutoff walls

8.1 INTRODUCTION AND OVERVIEW

Slurry trenching is the act of creating a slot in the ground (trench) using a slurry to maintain trench stability. The slurry in the trench is then either left to harden in place or replaced with other materials. The terminology for slurry walls varies and, at times, the same terms are used for different materials and/or methods. For example, *slurry wall* is used to denote both soil-bentonite cutoff walls and structural diaphragm walls, the first of which has no unconfined compressive strength and the second of which is made of a concrete backfill with a strength that may exceed 20,000 kPa. This chapter will identify and define terms with particular care to highlight multiple definitions or uses for the same term. Topics include construction methods, materials and mix designs, analysis, and field verification.

Slurry trenching is both widely used and used in a wide variety of applications. Some examples include:

1. Flow beneath a flood control dike in Birdsboro, Pennsylvania, was reduced by a soil-bentonite (SB) slurry trench cutoff wall installed prior to dike construction (Ruffing and Evans 2010).
2. A site development project near Heathrow Airport, London, used a cutoff wall composed of bentonite, ordinary portland cement (PC), and fly ash to isolate landfill gas and contaminants from previously buried waste materials from a portion of the site planned for residential construction (RWE Power International 2011).
3. Levees along the Sacramento River were improved by using cutoff walls composed of PC, soil, and bentonite to both strengthen and reduce flow through and beneath the levees (Owaidat et al. 1999).
4. Excavation for the Westminster Station, London, employed a slurry trench, later filled with concrete and reinforcing steel, to minimize movement of surrounding buildings and utilities resulting from excavations and to give both short- and long-term support of the excavation (Glass and Stones 2001).
5. Cement-bentonite (CB) shear walls were installed to improve the stability of the Tuttle Creek Dam under earthquake loading (Bellew et al. 2012).

These examples of subsurface walls installed using slurry trenching illustrate the wide variety of projects for which slurry trenching is used. They illustrate the wide range of materials used in the trench depending upon the desired final properties of the wall. The examples also show the two major ground improvement goals for cutoff walls: reducing hydraulic conductivity, increasing shear strength, or both.

Slurry trenching is a term that defines the means of construction. That is, the trench is excavated under a head of slurry and the role of the slurry is to maintain trench stability. Vertical cutoff wall, or simply cutoff wall, is a term that defines the finished product. The cutoff wall may result from hardening of the slurry used to excavate the trench or from materials placed in the trench to replace the slurry.

8.1.1 Functions of slurry trench cutoff walls

Cutoff walls are used to control horizontal flow of groundwater by reducing permeability in the subsurface. A special case of groundwater flow control is that of a contaminated land site where controlling groundwater flow also reduces advective contaminant transport. Groundwater cutoff walls may be temporary, as in those to control groundwater flow to excavations, or permanent such as those used in dams and levees.

By constructing a wall with substantial shear strength, vertical cutoff walls can also be used for earth retention. In many cases, vertical cutoff walls serve both functions, that is, reducing groundwater flow and providing a high strength vertical wall for earth retention such as in the Westminster Station project cited above.

8.1.2 History of slurry trench cutoff walls

Slurry trenching began in the middle of the 20th century. It is reported that the US Army Corps of Engineers conducted field trials of slurry trench cutoff walls to control seepage under and through levees along the Mississippi River (Jefferis 1997). At about the same time, it is reported that the US Navy employed slurry trenching in California (di Cervia 1992). It is also reported that a slurry trench was excavated using a trenching machine in California to construct a barrier to saltwater intrusion (Nash 1974). All of these mid-century slurry trenches in the Unted States were used to construct soil-bentonite groundwater barriers. The 1970s saw slurry trenching become more common. The increased frequency of use was driven by the need for vertical barriers in environmental containment projects. Developments in Europe occurred in a similar time frame. The first European patents on slurry wall construction were in Italy in the late 1940s (di Cervia 1992). Interestingly, slurry trenches developed in Europe were largely made of cement-bentonite slurries that hardened in place. A paradigm shift in slurry trenching came in 1974 with the research directed by Stephan Jefferis demonstrating substantial property improvement in slurries using ground granulated blast furnace slag (Jefferis 1981). The first cement-bentonite wall with slag was constructed in 1975 (Jefferis 1997) followed by the first environmental application in 1983 (Jefferis 1993). A further review of the slurry wall developments and a comparison of techniques are given in Evans and Dawson (1999).

Vertical cutoff walls are frequently denoted by their method of construction. Hence, the term slurry wall identifies techniques that employ slurry to maintain trench stability during cutoff wall construction. Here, slurry walls and slurry trenches include SB, CB, and diaphragm (or structural) slurry walls. In all slurry wall techniques, slurry is used to maintain trench stability while subsurface materials are excavated from the trench. In contrast, in situ mixed walls are those where slurry is added to the subsurface and mixed in place with the ground. In situ mixed walls include those constructed with single and multiple augers, cutter bar/chain devices, and other down trench cutting and mixing tools. Slurry trenching is explored in this chapter including detailed design, construction, and performance monitoring while in situ mixing is covered in Chapter 6, Soil mixing.

8.1.3 Slurry trench cutoff walls as a ground improvement technique

Vertical cutoff walls constructed using the slurry trench method of construction provide two principal areas of ground improvement: reduction in hydraulic conductivity and/or increase in shear strength. Groundwater cutoff walls are designed and constructed to have a relatively low hydraulic conductivity. As a result, horizontal groundwater flow is substantially reduced (usually by orders of magnitude), making vertical cutoff walls useful in reducing seepage into construction excavations, reducing seepage through and beneath levees, and reducing advective contaminant transport from zones of subsurface contamination. Vertical cutoff walls to improve the strength of the ground can vary from modest strength improvement, such as CB walls, to substantial strength improvement, such as diaphragm walls. Diaphragm walls are those slurry trenches where the slurry is displaced with concrete and steel resulting in a wall with significantly higher strength than the surrounding ground. Unconfined compressive strengths are zero for SB cutoff walls (without cement) up to 20 MPa (or more) for diaphragm walls. This strong, vertical wall can then be used as part of the excavation support system or even as a component of the final structure forming the basement wall. Perhaps the most well-known case study of a diaphragm wall is that of the World Trade Center's bathtub in New York (Tamaro 2002).

In this chapter, SB walls are covered in more detail than other techniques. Note that there is considerable overlap between uses, design, construction, and quality control testing between SB and other techniques. Thus, many of the topics explained in this section, are applicable to the techniques discussed later in this chapter. In addition, the SB technique receives the highest level of focus herein as that is generally the most cost-effective and common installation technique for non-structural vertical barrier walls in the United States. CB walls are more common in the UK (Evans and Dawson 1999) and are also discussed in considerable detail in this chapter.

8.2 SB SLURRY TRENCH CUTOFF WALLS

An SB slurry trench cutoff wall is constructed by excavating a vertical trench while simultaneously filling the trench with slurry. The slurry, an engineered fluid, is usually comprised of bentonite and water and serves to support the sidewalls of the trench, i.e. maintain trench stability. The use of slurry for this purpose is a natural extension of the use of slurries designed to keep boreholes open during drilling and is simply a construction technique to prevent the trench from collapsing by forming a very low permeability filter cake along the sidewalls of the trench, against which the water (slurry) pressure acts, stabilizing the trench. Figure 8.1 shows a backhoe excavating a slurry trench while bentonite-water slurry is being pumped in from a remote mixing plant. Notice the filter cake above the slurry level and along the sidewall of the trench.

Once an adequate portion of the trench has been excavated, backfill is placed in the trench in such a manner that the backfill displaces the slurry without substantial mixing of the two. The backfill forms the completed SB cutoff wall. The backfill comprises a mixture of soil (either from the excavation or from a borrow area) and slurry (from the trench or from the slurry mixer). Figure 8.2 illustrates the consistency of the backfill, similar to that of a high slump concrete, and the flow of backfill into the trench. The excavation and backfill placement operation continue until construction is complete. Figure 8.3 represents the entire process.

Figure 8.1 Excavation of a slurry trench cutoff wall.

Figure 8.2 Backfill mixing and placement for an SB cutoff wall.

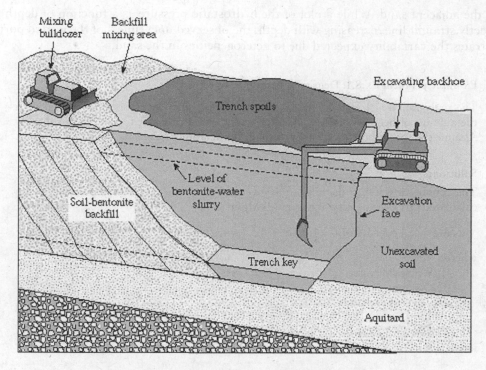

Figure 8.3 Schematic of SB excavation and backfill (from LaGrega et al. 2010).

8.2.1 Excavation stability

Slurry-supported excavations are stable because of the properties of the slurry. For SB walls, this slurry is a mixture of bentonite and water. Bentonite is a high swelling, naturally occurring smectitic clay (for more information about bentonite, see Mitchell and Soga 2005). When mixed with water at a ratio of approximately 5% bentonite and 95% water by mass, the resulting liquid demonstrates viscosity, density, and filtrate loss properties desirable for stable slurry trenching. The pronounced benefit is illustrated in a laboratory demonstration shown in Figure 8.4. The right-hand side of the photo illustrates a vertical trench in sand excavated using the slurry to maintain trench stability. In contrast, the left-hand side of the photograph illustrates the sloughing of the sand to its angle of repose without the use of slurry. The triangular nature of the wetted portion of the sand that shows up darker than the dry sand also illustrates the general hydrostatic trend expected from the slurry filtering

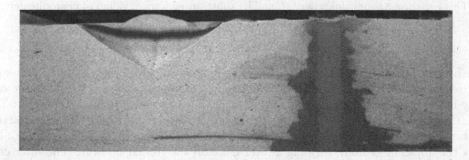

Figure 8.4 Model slurry wall excavation in sand.

into the adjacent sand. While a plot of the hydrostatic pressure as a function of depth is a perfectly straight line increasing with depth, the observed unevenness of the wetted portion illustrates the variability expected due to heterogeneities in the sand.

Example problem Ex.8.1: Density of bentonite-water slurry

The analysis of trench stability reveals a slurry density of 1.05 g/cm^3 is required. Calculate the amount of bentonite in Mg added per cubic meter of water to achieve this density. Assume the specific gravity (density of solids) of bentonite is 2.77.

Solution:

The volume of water in the mixture is given and the mass of the water can be readily calculated given the density of water, 1.0 Mg/m^3.

$$M_w = V_w / \rho_w = 1.0 / 1.0 = 1.0 \text{ Mg}$$

where

M_w = mass of water
V_w = volume of water
ρ_w = density of water

The desired total density is the cumulative mass of all ingredients divided by the cumulative volume of all ingredients including both the water and bentonite:

$$\rho_T = M_T / V_T = \left(M_b + M_w \right) / \left(V_b + V_w \right) = 1.05$$

and

$$V_b = M_b / \rho_b$$

where

M_b = mass of the bentonite
V_b = volume of the bentonite
M_T = total mass of the slurry
V_T = total volume

Substituting the volume of the bentonite written in terms of the mass and density of the bentonite results in an equation with one unknown i.e. the mass of the bentonite, as follows:

$$\rho_T = M_T / V_T = \left(M_b + M_w \right) / \left[\left(M_b / \rho_b \right) + V_w \right]$$

This equation has one unknown, M_b and by substituting known values of bentonite density, water mass, and volume we find:

$$\underline{M_b = 0.0805 \text{ Mg} = 80.5 \text{ kg}}$$

The hydrostatic pressure of the bentonite-water slurry in equilibrium with the active earth pressures from the adjacent ground provides trench stability. The slurry level in the trench ideally is maintained at an elevation higher than the adjacent groundwater level and, as a result, there is a hydraulic head difference that causes the slurry to exfiltrate from the trench into the surrounding ground. In so doing, the bentonite particles in the slurry are filtered out by the adjacent ground, forming a filter cake on the trench sidewall. The filter cake is a relatively thin layer (a few mm) of hydrated bentonite of very low hydraulic conductivity

Figure 8.5 Bentonite filter cake in a slurry trench.

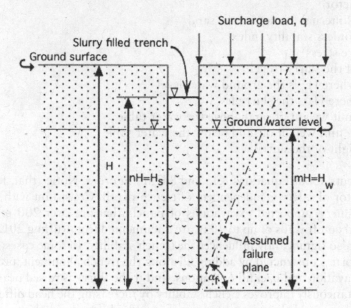

Figure 8.6 Schematic of slurry trench stability (after Filz et al. 2004).

($< 10^{-10}$ cm/s). A photo of the filter cake in a slurry trench is presented in Figure 8.5. The hydraulic head in the trench and the filter cake on the trench walls are key to trench stability.

The analysis of trench stability can vary from a straightforward equilibrium analysis to a very sophisticated analysis (e.g. finite element) that could include three-dimensional effects or the impact of a weak soil layer. The factor of safety for trench stability can be expressed in terms of force (or stress) equilibrium as a ratio between resisting forces and driving forces. There are a number of formulations of this basic approach. For cohesionless ground and adequate filter cake formation, the approach presented by Filz et al. (2004) is commonly used. This approach allows for a uniform surcharge adjacent to the trench, variable slurry levels, slurry unit weights, and soil strength. The equilibrium formulation is illustrated in Figure 8.6. In the field, dimensions are normally measured downward, including depth of the trench, depth of groundwater beneath the surface, and depth of slurry in the trench. For the mathematical formulation, these parameters are identified as heights all measured from the same datum, that is, the bottom of the trench. Hence, Figure 8.6 identifies the height of the trench, height of the groundwater, and the height of the slurry in the trench as well as the angle of the assumed failure surface.

The equations from Filz et al. (2004) result from a comparison of the driving forces (an active wedge sliding into the trench) and the resisting forces (the slurry fluid pressure in the trench). The result are equations for the factor of safety defined in terms of the soil strength as presented in Eqs. 8.1a and b.

$$F = \frac{2\sqrt{B}}{B-1} \tan\phi \tag{8.1a}$$

$$B = \frac{2q + H[(1-m^2)\gamma_m + m^2\gamma']}{H(n^2\gamma_s - m^2\gamma_w)} \tag{8.1b}$$

where:
F = safety factor
ϕ = angle of internal friction of the sand
B = dimensionless stability index
q = surcharge stress
H = depth of the slurry trench
m = H_w/H where H_w is the ground water level
n = H_s/H where H_s is the slurry level
γ_m = moist unit weight of sand above the water table
γ' = effective unit weight of sand below the water table
γ_s = unit weight of the slurry

A few comments about Equations 8.1a and 8.b are in order. Notice that, for no surcharge loading, the factor of safety is independent of trench depth. Consistant with the theory, safe installation of slurry supported excavations of panels that are over 200 m deep and long slurry-supported cutoff walls of up to 40 m are documented (e.g. Ruffing 2012). The depth of the water table also significantly influences the trench stability. In some cases, construction of a working platform, constructed by adding fill along the trench alignment, provides a suitable base for the excavation and backfill operations, and if the slurry is raised nearer the platform elevation, simultaneously improves trench stability by increasing the head differential between the top of the slurry and the groundwater surface. Fluid density in the trench has a major influence on trench stability. Computations using the as-mixed slurry density (see example Ex.8.1) may show a safety factor of approximately 1.0 in situations where the water table is near the surface. However, the process of excavating under the slurry invariably results in the entrainment of formation soils in the slurry and a corresponding increase in slurry unit weight. Example Ex.8.2 demonstrates the influence of slurry density on trench stability.

In developing the solution above, the angle of the failure plane, α_f, must be assumed. The critical failure plane, which offers the least resistance to failure, is given by:

$$\alpha_f = 45° + \left(\frac{\phi}{2}\right) \tag{8.2}$$

measured from the horizontal. Equation 8.2 is useful for approximating the lateral extent of possible surface deformations during slurry wall construction. The same geometric calculation can be used to calculate the horizontal distance beyond which surcharge loadings will have a negligible influence on the trench stability. In the absence of site-specific data or a more sophisticated calculation, it is common practice to keep surcharges a distance of the trench depth horizontally away from the slurry trench during construction.

Example problem Ex.8.2: Slurry trench stability

A 20 m deep SB slurry wall is planned for a silty, sandy site (ϕ = 30°) where the ground-water table is 2 m below the ground surface. The total density of the sand is 1.8 g/cm³ and can be assumed to be the same above and below the water table. The project specifications required the contractor to maintain the slurry level within 0.60 m of the top of the trench with an initial slurry density of 1.05 g/cm³. The slurry density is expected to rise to 1.35 g/cm³ during excavation. Compare the safety factor for trench stability for a slurry density varying from 1.05 to 1.35 g/cm³.

Solution:

Equations 8.1a and 8.1b are used with the given information as shown below. Notice that while Equations 8.1a and 8.1b are presented in terms of unit weight since B is a dimensionless parameter, either unit weight or density can be used in either SI or Imperial units.

Given:

$$\phi = 30, q = 0, H = 20\,\text{m}, H_w = 18\,\text{m}, H_s = 19.4\,\text{m}, \rho_m = 1.8\,\text{g/cm}^3$$

$$\rho_s = 1.05\,\text{g/cm}^3 \text{ to } 1.35\,\text{g/cm}^3$$

Compute:

$$\rho' = 1.8 - 1.0 = 0.8\,\text{g/cm}^3$$

$$m = H_w/H = 18/20 = 0.9 \text{ and } n = H_s/H = 19.4/20 = 0.97$$

B = dimensionless stability index (compute from eq. 12.1a)
F = safety factor (compute from eq. 12.1b)

Discussion: Notice the factor of safety is highly dependent upon the slurry density and is less than one at low slurry densities, similar to the as-mixed slurry density (Figure Ex.8.2).

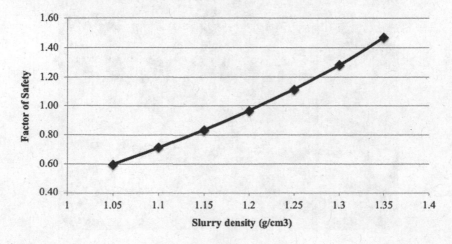

Figure Ex.8.2 Factor of Safety as a function of slurry density.

However, in the field, under most conditions, slurry quickly picks up fines from the soil formation and the density rises to provide an adequate factor of safety. In occasional cases, the slurry becomes so thick and dense that desanding (sand removal) is required.

The assumptions for the solution above are generally applicable to cohesionless soils in long trenches. Additional solutions are available for analysis of the stability of short trenches incorporating 3D effects and for drained analyses for trenches in cohesive soils (Fox 2004). If there is at least a 1.0 m to 1.5 m differential head between the groundwater and the top of the slurry, slurry trenches have been generally observed to be stable in almost all soil types. Trench stability calculations are not commonly performed for short or shallow trenches. Most of the equations developed for the stability of slurry trenches ignore the stabilizing benefit of deep filtration of slurry into the formation soils which can result in a stable trench even when static equilibrium equations show a factor of safety at or near one.

8.2.2 Slurry property measurement

Analysis of trench stability above depended upon the formation of a filter cake and slurry with adequate density. In order to increase the slurry density and to prevent particles suspended in the slurry from falling out of suspension onto the backfill, the slurry also needs to be adequately viscous to keep soil particles in suspension. Tests to measure these characteristics have been adopted from oil drilling practices which also use slurry in order to maintain borehole stability. Some of the oil field drilling tests commonly used for slurry trenching measurements are the *mud balance* used to measure slurry density, the *Marsh funnel* used to measure viscosity, and the *filter press* used to measure filtrate loss and filter cake thickness. These devices are shown in Figures 8.7 and 8.8.

Bentonite-water slurry mixtures generally have no less than 5% bentonite by weight resulting in an as-mixed slurry density of approximately 1.05 g/cm³. The slurry density in the trench increases due to the inclusion of soil formation materials. Some projects limit the maximum slurry density, typically near 1.35 g/cm³. The need for the viscous, fluid-like, backfill to displace the slurry is the reason for an upper limit on slurry density. However,

Figure 8.7 Marsh funnel and mud balance.

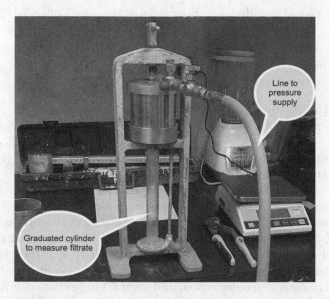

Figure 8.8 Filtrate loss equipment.

rather than specify the density, it is more appropriate to specify the difference in density between the backfill and the slurry (typically about 0.25 g/cm³). The density is typically measured using a mud balance as shown in Figure 8.7. A mud balance is a simple, single-beam bubble balance with a cup and lid at one end that contains a fixed and constant slurry volume. This allows the beam to be calibrated in various units of density or unit weight including g/cm³ and lb/ft³.

The slurry must also have a certain viscosity to keep particles in suspension yet allow excavating equipment to freely operate and allow the backfill to displace the slurry. This viscosity is typically measured with a Marsh funnel, shown in Figure 8.7. In this test, the outlet of the funnel is blocked while slurry is added to the top of the funnel through the screen to remove lumps or other coarse materials. The slurry is then allowed to flow from the funnel outlet into a cup. The time for 973 ml (1 US quart) of slurry to flow is recorded in seconds. This time is the Marsh viscosity. Viscosity specifications typically range from 36 s to 40 s from the Marsh viscosity. For reference, water has a Marsh viscosity of approximately 26 s. Bentonite water slurry is a non-Newtonian fluid. Since the shear rate changes throughout the Marsh funnel test, the test does not yield a true viscosity measurement. A full understanding of the viscosity of a slurry can be measured using a rheometer with viscosity measurements conducted at various and constant shear rates. An approximation of the true viscosity is given by Pitt (2000) in Equation. 8.3:

$$u = \rho(t - 25)$$ (8.3)

where:

u = viscosity (cP)
ρ = density (g/cm³)
t = Marsh viscosity (s)

The ability for the slurry to interact with the adjacent ground and form a filter cake is important to slurry trench stability and is dependent upon both the properties of the slurry and the adjacent soil. For slurry trenches containing suspended particles of silt and fine sand, a filter cake should readily form if the criteria given by Equation 8.4 are met (Filz et al. 1997).

$$D_{15g} \leq (9)D_{85 \, \text{slurry}} \tag{8.4}$$

where:
   $D_{15g} = D_{15}$ of the ground adjacent the trench
   $D_{85 \, \text{slurry}} = D_{85}$ particle size of the slurry

A filter press is used to evaluate the slurry's ability to form a filter cake (Figure 8.8). In this device, fully hydrated slurry is added to a 100 mm diameter vessel that is then pressurized. The pressure forces the slurry through a filter paper. The volume of water passing through the filter paper is measured. Typical criteria limit the water (filtrate) loss to less than 20 cm^3 to 25 cm^3 in 30 min at 695 kPa (100 psi). Factors causing excessive filtrate loss include poor-quality bentonite, slurry that is improperly hydrated, and mixing water that affects slurry hydration. Regarding mixing water, it is common to specify that the pH of the as-mixed slurry be in the range of 6 to 9 and/or the pH of the mixing water be between 6 and 8. Although useful as a means of comparison, and for tracking experience across projects, the filter press tells an incomplete story of the filter cake development. The results should therefore be evaluated in the context of the overall objectives. More information on the use of the filter press in slurry trenching is given in Ruffing et al. (2016).

8.2.3 SB backfill design

The backfill for the SB slurry trench cutoff wall is comprised of a mixture of soil, water, and bentonite. The backfill is prepared using excavated and/or imported soils at natural water content, bentonite-water slurry and, if needed, additional dry bentonite, all blended to produce a material with a viscosity similar to high slump concrete. Low hydraulic conductivity is desired and, frequently, is the only characteristic specified.

Specifications typically include a maximum hydraulic conductivity, such as 1×10^{-7} cm/s. In most projects, the added benefit of the low permeability filter cake is neglected and the hydraulic conductivity of the backfill is taken as the hydraulic conductivity of the cutoff wall. However, this conservative assumption may be counterbalanced by the fact that small-scale laboratory measurements of hydraulic conductivity may underpredict field hydraulic conductivity since the measured value of hydraulic conductivity has been shown to decrease with decreasing sample size (Filz et al. 2003). Specifications may require a special test method, such as a triaxial test with backpressure. These specifications are necessary, but not always sufficient since SB backfill is soft and compressible and the hydraulic conductivity is highly stress dependent. Figure 8.9 shows the relationship between hydraulic conductivity and effective stress for an SB backfill from a project in Pennsylvania, USA. The hydraulic conductivity decreases by more than two orders of magnitude as the effective consolidating stress increases from 8 kPa to 200 kPa. Also shown are error bars revealing greater variability in hydraulic conductivity at lower values of effective stress. These data illustrate three important points. First, the assignment of laboratory permeability tests on SB must specify the effective consolidating pressure or the data produced must be corrected or compared to an appropriate target. Second, to estimate the field hydraulic conductivity from laboratory test results, an estimate of the field effective stress is needed. Third, there is greater variability in hydraulic conductivity near the top of the trench or in shallow trenches or in trenches constructed in cohesive formation soils where the effective stresses are expected to be lowest.

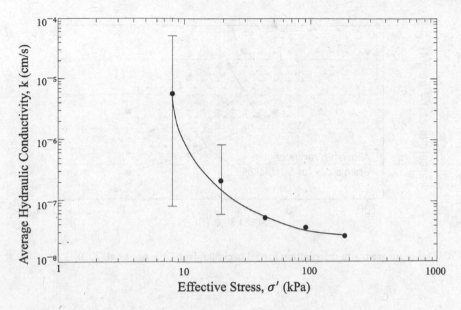

Figure 8.9 Hydraulic conductivity as a function of effective stress (after Ruffing 2009).

In an SB cutoff wall, soft, compressible backfill is placed in a narrow trench with relatively incompressible sidewalls, making the determination of stresses within the wall complex. Recognizing the interaction between the sidewalls of the trench and the more compressible backfill, it was postulated that the stresses in the backfill are less than geostatic (Evans et al. 1985). In the vertical direction, the backfill/trench wall friction impedes the downward movement of the backfill, resulting in load transfer to the formation materials and reducing the vertical stresses, i.e. the backfill experiences *arching* effects. In the transverse direction (perpendicular to the axis of the wall), the transverse stress is equal to the equilibrium lateral earth pressure. In the longitudinal direction (along the axis of the wall), the longitudinal stress is an at-rest pressure relating both the vertical and transverse stresses. Analytical and field studies (Evans et al. 1995; Filz 1996) have shed light on this issue. For shallow trenches (less than 7 m or so), the arching model is recommended for stress estimation. For deep walls, the modified lateral squeezing model (Ruffing et al. 2010) is recommended for the evaluation of transverse stresses within the wall. Where a more complete analysis is required, refer to Ruffing et al. 2010.

The grain size distribution of the SB backfill is important. Desirable attributes of the backfill are low hydraulic conductivity, low compressibility, and longevity. The soil used in the mixture, often termed the base soil, should therefore be well graded with minimum and maximum fines contents of 15% and 80%, respectively. In this way, the soil particle packing will be optimized and contribute to lower compressibility and hydraulic conductivity. An allowable grain size distribution range is presented in Figure 8.10. The preferable grain size distribution is toward the coarse limits of the range. While this figure may be optimum in terms of compressibility, hydraulic conductivity, and longevity, base soils with grain size distributions outside of those in Figure 8.10 have been used successfully. In most cases, the base soil is controlled by what is available at or near the site. Site-specific testing is recommended in all cases.

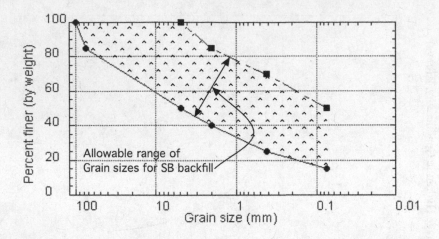

Figure 8.10 Optimum grain size distribution for SB backfill (after LaGrega et al. 2010).

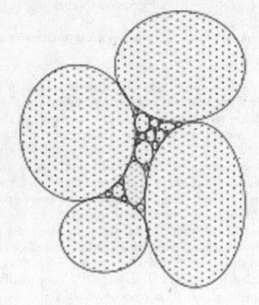

Figure 8.11 Schematic of idealized packing.

A schematic of idealized packing is presented in Figure 8.11 showing the progressive infill of larger voids with progressively smaller particles. Examination of the idealized packing shown in Figure 8.11 shows that the large pore space between the larger particles is filled with progressively smaller particles producing a small pore size. The presence of the swelling bentonite in low quantities is then effective in reducing the void size even further producing a backfill with a low hydraulic conductivity. With this particle size configuration, the impact of environmental conditions that might reduce the swelling of the bentonite is mitigated as the small pore sizes limit the permeability increases that might otherwise arise, even if the bentonite is removed entirely. The recommended particle size distribution also promotes good particle-to-particle contact, reducing the compressibility of the backfill. With reduced compressibility, the settlement of the backfill decreases, as does the downdrag within the trench as the backfill settles. This results in higher vertical stresses in the trench compared with more compressible backfill materials.

Example problem Ex.8.3: Backfill composition

An SB slurry trench cutoff wall is planned to rehabilitate a 7 m high levee. The levee was built on, and of, a 2 m layer of silty clayey sand. Underlying the sand is a thick deposit of silty clay. It is planned to key the wall 1 m into this silty clay layer. Grain size distribution curves for the two materials are shown in Figure Ex.8.2. The water contents for the silty clayey sand and the silty clay are 8% and 19%, respectively. Calculate the grain size distribution and resulting water content (neglecting bentonite addition) for the backfill (Figure Ex.8.3). How does this compare to the information presented in Figure 8.10?

Solution:

The backfill will be comprised of, by volume, 90% silty clayey sand and 10% silty clay as follows:

$$((7\ m + 2\ m) / 10\ m)\ 100 = 90\%$$

$$(1\ m / 10\ m)\ 100 = 10\%$$

Hence using the relative contribution of each material and the given grain size distribution of each layer, the final grain size distribution and moisture content can be calculated as shown in Table Ex.8.3. If density information were available, the relative amounts of each would be adjusted accordingly.

Example problem Ex.8.4: Backfill composition

Continuing with the scenario established in example problem Ex.8.3, backfill will be blended using soils excavated from the slurry trench. Assuming a 5% by weight bentonite-water slurry and target final water content of 30% to achieve the desired slump, what will be the final bentonite content of the backfill by dry weight? What is the solids content by total weight of the mix?

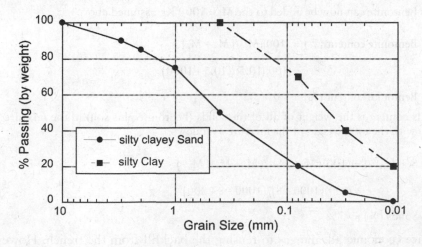

Figure Ex.8.3 Grain size distribution curves for example problem Ex.8.3.

Table Ex.8.3 Grain size distribution results

Grain Size (mm)	Silty Sand (% passing)	Silty Clay (% passing)	Amount from silty sand (%)	Amount from silty clay (%)	Total
10	100		90%	10%	100%
3	90		90%	10%	91%
2	85		90%	10%	87%
1	75		90%	10%	78%
0.4	50	100	90%	10%	55%
0.08	20	70	90%	10%	25%
0.03	5	40	90%	10%	9%
0.011	0	20	90%	10%	2%
Water content (%)	8.0	19.0	90%	10%	9.1%

Solution:

Using the results from Table Ex.8.3, the initial water content is 9.1% and the final water content is 30%. Since

$$w_i(\%) = (M_{wi}/M_s)(100) = 9.1\%$$

$$w_f(\%) = (M_{wf}/M_s)(100) = 30\%$$

So for each 1.0 metric tonne of solid

$$w_i(\%) = (M_{wi}/M_s)(100) \Rightarrow M_{wi} = 91 \text{ kg}$$

$$w_f(\%) = (M_{wf}/M_s)(100) = M_{wf} = 300 \text{ kg}$$

Hence, the additional moisture via slurry is

$$M_{wf} - M_{wi} = 300 \text{ kg} - 91 \text{ kg} = 209 \text{ kg} = \Delta M_w$$

For a 5% bentonite water slurry (by weight of water)

$$5\% = (100)\left(M_b/M_w\right) \text{ where } \Delta M_w = 209 \text{ kg} \Rightarrow M_b = 10.5 \text{ kg}$$

This bentonite can now be added to the M_s = 1000 Kg assumed above

$$\text{Bentonite content } (\%) = (100)(M_b)/\left(M_b + M_s\right)$$

$$= (100)(10.5)/(10.5 + 1000)$$

Bentonite content (%) = 1.0%

Solids content is the weight of all of the solids (bentonite plus soil) in the mixture in a ratio with the total weight of the mixture as follows:

$$\text{Solids } (\%) = (100)\left(M_s + M_b\right)/\left(M_s + M_b + M_{wf}\right)$$

$$= (100)(1000 + 8)/(1000 + 8 + 300)$$

Solids (%) = 77%

There are economic advantages to reusing the backfill from the trench. However, there are times where offsite borrow materials are blended with the excavated soil to improve the gradation of the backfill. For example, in areas of clean sand and where chemical compatibility is particularly important, fine-grained borrow materials can be mixed in to produce a backfill of lower permeability and with improved compatibility. Also imported backfill may be used when the excavated materials are unsuitable for backfill. Unsuitable soils include highly organic and peaty soils, heavily contaminated soils, or soils containing deleterious materials. In some cases, dry bentonite is added to reduce the permeability. The bentonite content is usually near 1% in the backfill simply from the addition of slurry to adjust the water content to achieve the desired slump.

As with most geotechnical projects, construction considerations are on equal footing with design considerations. In this case, the backfill ingredients must be blended into a homogenous, thick, viscous material, free of pockets of unmixed slurry or soil. The blended backfill must then be placed in the trench in a manner that displaces the slurry and does not entrap either slurry or sediment in order to form a continuous, defect-free, cutoff wall. Often backfill is mixed alongside the trench as shown in Figure 8.3. Slurry is added "automatically" as the excavating equipment invariably brings both slurry and excavated soils to the surface mixing area. In order to properly flow into the trench and displace the slurry, the backfill should be of the consistency of high slump concrete (100 mm to150 mm). In the field, slump cone testing equipment developed for concrete (ASTM C143M-15) is used, requiring over 10 kg of material to conduct the test. A miniature cone, one requiring a smaller volume of material for testing, has also been developed for laboratory design mix studies (Malusis et al. 2008). The water content needed to achieve the design slump can be estimated based upon the fines content and the bentonite content of the backfill (Yeo et al. 2005).

By adding an appropriate amount of slurry to produce a backfill having an adequate slump, the process will also add approximately 1% bentonite to the mixture. Depending upon the gradation of the soils used to prepare the backfill and the desired hydraulic conductivity properties, this may be sufficient. If not, an additional 1% to 3% dry bentonite may be added. For dewatering or levee applications, a hydraulic conductivity of 1×10^{-6} cm/s may be sufficient. For environmental applications, 1×10^{-7} cm/s is often required. While it is tempting to simply add more bentonite to the mix to reduce the hydraulic conductivity in these cases, it is also important to note that the bentonite swelling is reversible under certain conditions. Thus, compatibility studies with the backfill and the site contaminants are required. If there is a need to reduce the hydraulic conductivity below that achieved with the soil and slurry, it may be advantageous to add additional fines of low plasticity rather than additional bentonite.

Backfill design must also consider the need to maintain an adequate density difference between the backfill and the slurry such that the backfill can readily and completely displace the slurry. Typically, a density difference of 0.24 g/cm³ (15 pcf) is specified. Should the slurry become too dense, it can be thinned by adding freshly mixed slurry to the trench or by desanding.

For environmental applications, it may be possible to improve compatibility and sorption capacity with variations from traditional mix designs and materials. For example, multi-swellable bentonite (Malusis et al. 2010), zeolite, and attapulgite (Evans et al. 1997; Evans and Prince 1997), and high carbon fly ash (Bergstrom 1989) additives have all been shown to improve performance of SB cutoff wall backfills. Analysis techniques to examine the benefits of additives upon sorption are also available (Khandelwal et al. 1998).

8.2.4 Excavation techniques

Excavation is typically performed with a commonly available backhoe or excavator. However, where deep trenches need to be excavated, contractors may use custom modified

Figure 8.12 Modified excavator capable of digging 30 m in depth.

excavator arms known as long sticks and long booms with backhoes or even clamshells. For long and deep slurry trench excavations, a long stick excavator may work in tandem with one or more clamshells to complete the excavation in stages. Shown in Figure 8.12 shows an excavator modified to excavate to a depth of 30 m. Alternatively, some contractors prefer to excavate 20 m or so with an excavator and then complete the trench with a clamshell. The choice of excavating equipment is largely dictated by the contractor and contained in their presentation of means and methods for the project. That said, a discussion of issues that may arise in deep trenches is in order.

As the excavation depth increases, the length of open trench as measured at the surface increases. This is because the backfill lies on a slope of between 5H:1V and 10H:1V. For example, a trench that is 15 m deep needs up to 150 m of open trench to ensure that the toe of the backfill does not impinge upon the toe of the excavation. For a trench that is 30 m deep, the length of open trench doubles to up to 300 m. Increasing the amount of open trench also increases the likelihood of trench instability as well as sedimentation of material upon the backfill slope. Quality control measures described earlier are particularly important as the depth of trench increases. Another factor to consider when using a long stick as shown in Figure 8.12 is the articulation pattern during excavation. With a shorter stick, the bucket can rise vertically along the excavation face and thus any materials that spill from the bucket fall back into the excavation area. With a long stick and an inexperienced operator, the excavation bucket can rise out of the trench a considerable distance from the excavation face. As a result, it is important to make sure the toe of the backfill is back beyond the excavation bucket as it rises out of the trench. If not, materials that spill from the bucket can land on the slope of the backfill, causing contamination.

8.3 CB SLURRY TRENCH CUTOFF WALLS

CB walls are excavated under a head of slurry made from a mixture of water, bentonite, and cement. The cement may be portland cement or a mixture of portland cement and granulated ground blast furnace slag (GGBFS). The excavating slurry is left to harden in place, hence these walls are sometimes referred to as self-hardening. This is in contrast to the SB cutoff walls where the slurry is displaced by the backfill. The CB wall is therefore a one-phase operation (excavation only) and CB wall properties are largely independent of the site geology. Variations in the nature of the cement and the mix proportions can have a significant effect on the resulting product.

While CB slurries are used to fill the trench as described in this section, it is important to realize that mixtures of bentonite, cement, and water are widely employed in other ground improvement techniques such as grouting, in situ mixed walls, soil mix methods, and sealing of groundwater wells. Thus, the understanding of CB slurries described here underpin the understanding of their use in other ground improvement methods and vice versa. As such, details of CB slurry behavior, presented in the context of slurry wall construction, provide useful instruction on CB slurry behavior used in grouting, in situ mixed cutoff walls, soil mixing, and borehole sealing.

8.3.1 CB mixtures and properties

CB slurry is made by first preparing a bentonite-water slurry identical to that used for SB cutoff walls. Typical bentonite-water slurry is composed of 5% bentonite and 95% water. After mixing, the slurry is typically allowed to hydrate for 24 hours until the viscosity and fluid loss properties stabilize. Once the slurry is properly hydrated, 10% to 25% cement, by total weight, is added and the CB slurry pumped into the trench. The excavation proceeds in and through the CB slurry. When the excavation is complete, the CB is left to harden in place with the initial set (hardening) usually occurring overnight.

As indicated earlier, the cement ingredient may be ordinary portland cement. More commonly, however, the cement is a mixture of PC and GGBFS. Most CB mixtures include 75%pc95% of the PC replaced with GGBFS (sometimes termed slag-CB). Hence the cementitious portion of the CB slurry may be 20% PC and 80% GGBFS. An example recipe (Barker et al. 1997) for a modern CB mixture is:

1. 35 kg bentonite
2. 120 kg GGBFS
3. 30 kg PC
4. 934 l water.

According to the Building Research Establishment (BRE 1994), mix proportions vary as follows:

1. 30 kg to 60 kg bentonite
2. 100 kg to 350 kg cementitious material (mixture of GGBFS and PC)
3. 1,000 l water.

The numbers reveal that the mixture has a very low solids content as compared with SB mixtures. Note that these proportions are as-mixed at the batch plant. Site soils invariably get mixed in during the excavation process. Additional information regarding mix proportions is provided in numerous texts (e.g. Jefferis 2012).

Example problem Ex.8.5: Solids content of CB mixture

What is the solids content of the example CB mixture given in the text (Barker et al. 1997).

Solution:

$$\text{Solids (\%)} = (100)\big(M_b + M_{ggbfs} + M_{Pc}\big)\big/\big(M_b + M_{ggbfs} + M_{Pc} + M_w\big)$$

$$\text{Solids (\%)} = (100)\big(35 + 120 + 30\big)\big/\big(35 + 120 + 30 + 934\big) = \underline{16.5\%}$$

The hydraulic conductivity of hardened CB slurry made with PC is substantially higher than hardened slurry made with a combination of GGBFS and PC (Jefferis 1981). A number of variables influence the hydraulic conductivity of any particular mixture including bentonite source, nature and duration of bentonite hydration, mix water chemistry, source of the GGBFS, mix proportions, mix sequence, and mixing methods. That said, for any combination of site-specific ingredients, the benefits of replacing most of the PC with GGBFS are clear as illustrated by the data in Figure 8.13.

To better understand hardened (or cured) CB slurry, a fundamental understanding of the nature of the slurry components (PC, GGBFS, and bentonite) is helpful. GGBFS is a by-product of iron manufacturing and is composed of oxides of calcium (40%), silicon (30%), aluminum (15%), and magnesium (10%) with small quantities of other oxides and ions. Since there is variation in the chemistry of GGBFS depending upon the source, laboratory testing should be done using the GGBFS from the source expected for the full-scale project and ideally including samples from that source at multiple points in time. GGBFS, when blended with OPC in a ratio of about 80%/20% by weight, not only reduces the hydraulic conductivity as shown in Figure 8.12 but strength is increased, and bleed is decreased. Bleed is the free water that can emerge as PC and GGBFS particles settle in the mixture prior to the initial set. The benefits of PC replacement with GGBFS are believed to be, in part, due to the behavior of the bentonite in the mixture. Upon the addition of PC, the fully hydrated sodium bentonite is overloaded with calcium cations and has a tendency to flocculate, but the slurry made with GGBFS does not exhibit this behavior (Jefferis 1997). The more dispersed nature of a GGBFS slurry also explains the improvement in hydraulic conductivity

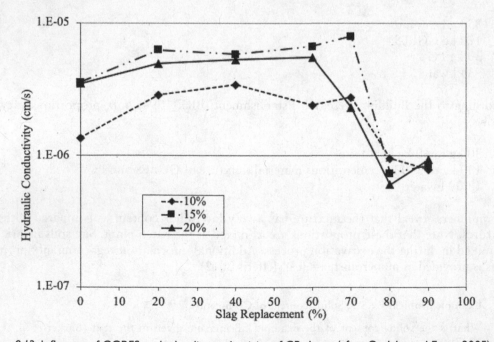

Figure 8.13 Influence of GGBFS on hydraulic conductivity of CB slurry (after Opdyke and Evans 2005).

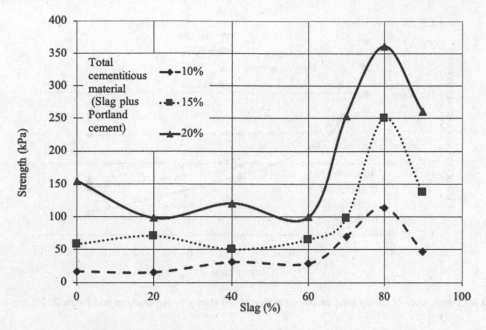

Figure 8.14 Influence of GGBFS on unconfined compressive strength of CB slurry (after Opdyke and Evans 2005).

compared with slurry made with PC. The characteristic dark blue/green color observed for hardened CB made with GGBFS is attributable to the magnesium in the mixture.

CB slurries made with an optimum combination of PC and GGBFS are also stronger than those made only with PC (see Figure 8.14). Notice a minimum 70/30 mix of GGBFS and PC is needed to begin to see the benefits of higher concentrations of GGBFS. Suppliers of cement provide materials to a broad range of clients for many uses and often have 50/50 or 25/75 premixes available for use in conventional reinforced concrete. While these ratios have been used on some slurry wall projects without detriment, there is also little benefit. To achieve the benefit for CB slurries and grouts, a higher GGBFS/PC ratio is needed.

Finally, one of the drawbacks of GGBFS in CB slurries relates to the time rate of curing. The concrete industry has adopted a standard curing time of 28 days for the evaluation of concrete strength. This 28-day time has found its way into ground improvement engineering. The data showing hydraulic conductivity measurements for four CB slurries in Figure 8.15 demonstrates the substantial decrease in hydraulic conductivity that can be expected after the first 28 days. For these data, the hydraulic conductivity declined another one to two orders of magnitude between 1 month and 1 year and other studies have shown improvement beyond 1 year. Similarly, studies have shown that the CB wall strength increases substantially over longer curing periods.

8.3.2 Role of the bentonite in CB mixtures

Much is written and understood about the swelling nature of bentonite and its widespread use in civil engineering applications. However, the role of bentonite in slurry and grout is not well studied. In CB slurry, bentonite is largely a construction expedient to keep the cement particles in suspension until the slurry sets. Neat cement grouts and slurries (mixtures with only cement and water) exhibit significant bleed due to excessive settling of the cement

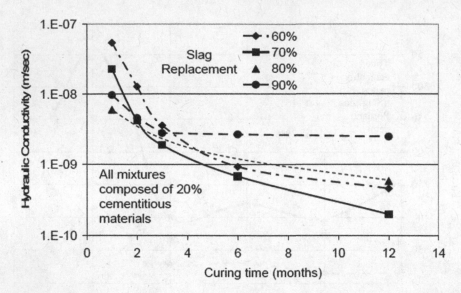

Figure 8.15 Influence of curing time on permeability of CB slurry (after Opdyke and Evans 2005).

particles prior to set. The bentonite significantly helps reduce this bleed by maintaining a dispersion of cement particles. But what becomes of the bentonite as the slurry hydrates? Bentonite is a smectitic clay mineral with a covalently bonded octahedral sheet between two silica tetrahedral sheets forming one layer of clay. Isomorphous substitutions within the sheets give rise to charge deficiencies that are balanced by cations present between successive layers of sheets. As a result, bonding between layers is weak and predominantly controlled by Van der Waal's bonds. While the layer spacing of dry bentonite is 9.6A, in the presence of water the spacing is unlimited, i.e. complete separation is possible (Mitchell and Soga 2005). This attribute (high swelling in the presence of water) is what makes bentonite favorable for many uses in civil engineering. The pore fluid chemistry, and potential changes in such, may have a substantial impact on the relative position of the individual bentonite layers. Hence, mixing PC and GGBFS with bentonite and water produces a markedly different slurry than a mixture created with PC alone. The more dispersed nature of the bentonite in slurry made with GGBFS results in a product with increased strength, reduced compressibility, and reduced permeability compared to slurry made without GGBFS.

This understanding of bentonite behavior in the presence of water is useful in understanding changes in the bentonite that will occur with the addition of PC and GGBFS to the slurry. The behavior of the bentonite at this early time in the life of the mixture has significant influence on the cured properties including hydraulic conductivity, strength, compressibility, and bleed. However, over time, the bentonite (in the form used by civil engineers) is consumed by the hydration reactions and is embodied in the aluminosilicate crystalline structure of the cured CB slurry. Evidence of the demise of the bentonite from a recognizable silicate mineral can be seen in x-ray diffraction studies which reveal the gradual reduction in bentonite peaks over time as the CB slurry cures (Evans et al. 2021). Given that bentonite, after curing of the CB slurry, is no longer present in its original form, it would be incorrect to assume typical hydrated bentonite behavior is similar to its behavior in the hardened CB. For example, low dielectric constant contaminants have been shown to cause increases in hydraulic conductivity in bentonite sand mixtures (Evans et al. 1985) as would be expected from an understanding of pore fluid-bentonite interactions, but similar behavior is not observed in hardened CB (Garvin and Hayles 1999; Evans and Opdyke 2006). Since the active component in SB (bentonite) is not present in fully hydrated CB, the behavior observed for SB does not necessarily apply for CB.

Whereas filtrate loss is measured and controlled for bentonite-water slurries used in SB cutoff wall construction and in CB slurry preparation, filtrate loss is not normally measured on CB slurry. Once the cement, particularly PC, is mixed with the bentonite water slurry, the bentonite flocculation diminishes the ability of the slurry to form a low permeability filter cake. For CB slurry, the bleed is measured and controlled. If the bentonite is not of sufficient quality or quantity, there will be excessive bleed from the slurry. Bleed is measured by filling a 1.0 L graduated cylinder with freshly mixed CB slurry and measuring the volume of solids after set. The UK standard (BRE 1999) requires a minimum of 980 mL of solids at 24 hours out of 1,000 mL of slurry (i.e. 20 mL of bleed water).

8.3.3 Volume change behavior

Volume change behavior of CB slurry is also important to the final behavior of the cured CB cutoff wall. After CB slurry is mixed and pumped into the trench, the settling of solids, combined with the bleed, results in a volume decrease from the as-placed volume. Once the initial set has occurred, no further bleed develops. Shrinkage takes place during the hydration process (termed autogenous shrinkage). As ingredients (water, GGBFS, PC, bentonite) participate in the chemical reactions (hydration reactions), there is an increase in the volume of the solids and a decrease in the volume of water. The net volume change is negative, i.e. there is net shrinkage. The hydration reactions are exothermic and elevate the temperature of the hardened CB slurry. As the material cools to the temperature of the surrounding ground, there is additional shrinkage that can be attributed to this temperature change, i.e. thermal shrinkage. In hot weather or for large concrete pours, the concrete industry has long used chilled water or ice to minimize thermal shrinkage.

For CB cutoff walls, there is the potential for additional shrinkage due to drying of portions of the wall above the water table. For cutoff walls built in arid environments, or for relatively unprotected upper portions of cutoff walls subjected to seasonal wetting/drying, the extent and ramifications of drying shrinkage, including the potential for transverse cracking, must be considered.

Volume change behavior affects the performance of CB slurry. Drawing from experience in the concrete industry, shrinkage of concrete causes cracking. The concrete industry controls the detrimental effects of cracking through mix designs, reinforcement, construction joints, and, in the case of mass concrete, artificial cooling. A concrete pavement, without reinforcement and construction joints, will exhibit substantial cracking. Friction between the pavement and the earth impedes shrinkage and causes tensile stresses to build up. Cracking occurs when these tensile stresses exceed the relatively low tensile strength of the concrete. Similarly, in CB slurry trench cutoff walls, there are frictional forces on both sides and the bottom of the wall that resist movement due to shrinkage. In the transverse direction, shrinkage is the smallest of the three directions as the width of the wall is the smallest in the transverse direction. Further, the transverse stresses exerted on the ground must be equal to the lateral pressures the ground exerts on the CB wall. The adjacent ground can move to balance the CB stresses anywhere between the active and at-rest state limits. As a result, the transverse stress state in the CB cutoff wall will always be positive, i.e. in compression. The downward movement of the material due to gravitational forces partially compensates for shrinkage in the vertical direction. As with SB walls, the downward movement is resisted by friction along the trench sidewalls. Hence, the vertical stresses are likely to remain in compression, although the vertical compression forces are smaller than those in the transverse directions. In cases where there are substantial drying stresses near the surface, the net stress will be tensile and may be of sufficient magnitude to cause cracking.

The decision to adopt SB or CB, or any of the techniques yet to be discussed, depends upon a myriad of factors including site and subsurface conditions, cost, end product properties, and contractor expertise. In the United States, SB is often considered the most economical of the methods and is widely used. In the UK, CB is the method of choice. Table 8.1 presents a comparison of SB and CB methodologies and outcomes.

8.4 STRUCTURAL SLURRY WALLS (DIAPHRAGM WALLS)

Diaphragm walls, or structural slurry walls, or simply slurry walls, are another type of wall quite different than the SB and CB slurry trench cutoff walls already discussed. Since "slurry wall" names a method of construction rather than a final outcome, *diaphragm walls* will be the name applied to this technique. Diaphragm walls are used when substantial structural capacity is needed for earth retention and/or support of the final structure. Similar in function to soldier pile and wood lagging and sheet piles, these are not really ground improvement techniques. Their construction uses some of the same means and methods as other cutoff wall techniques, but their primary aim is to build a retaining structure with substantial moment capacity rather than to modify the behavior of the ground. As a result, treatment of the topic will not be as extensive as other ground improvement techniques. For more information on this technique, the reader is referred to Structural Engineering Institute (2000) and Xanthakos (1979, 1994).

Diaphragm walls are constructed by excavating a panel in the subsurface using bentonite-water slurry to maintain trench stability. The slurry is the same as that used for SB walls described earlier. Analysis of stability may incorporate 3D effects (Tsai and Chang 1996; Fox 2004) since the panels are short (3 m to 6 m) compared to the wall depth (up to 150 m). Widths usually vary from 0.6 m to 1.2 m.

Table 8.1 Comparison of SB and CB practices and outcomes (from Evans and Dawson 1999)

ELEMENT	SB PRACTICE	CB PRACTICE
Barrier composition	Soil-bentonite (SB)	Slag-cement-bentonite (CB)
Hydraulic conductivity	$< 1 \times 10^{-7}$ cm/s	$< 1 \times 10^{-7}$ cm/s after 90 days
Solids content (M_s/M_T)	~ 70%	~20%
Unconfined compressive strength	~0	> 100 kPa @ 28 days
Strain to failure	Plastic (>20%)	Brittle (<5%)
Time dependency	Consolidation: Initial within a few days; continued consolidation & creep for 30 days or more	Initial set: within one day Complete hydration reactions: 90 days or more
Construction stages	Two phase	One phase
Excavation equipment	Backhoe, clamshell	Backhoe, clamshell
Depth, width, length (typical)	20 m, 0.75m, >1km	15 m, 0.6 m, <1 km
Working space needed	Large for slurry plant & backfill mixing	Small for slurry plant & excavation spoil disposal
Material assessment	Hydraulic conductivity, compatibility during design, grain size distribution	Hydraulic conductivity, strength, strain at failure, density

The construction process is illustrated in Figure 8.16. A shallow reinforced guide wall is first constructed to provide the necessary precision to the line and grade of the wall. Excavation of an individual panel under a head of bentonite-water slurry is usually done with a clamshell, hydromill, cutter soil mixer, or similar tool. The individual primary panels are often longer than can be excavated with a single downward pass of the excavation tool and a three-pass sequence as shown in the figure is common. Upon excavating and cleaning the panel to the desired grades and dimensions, end pipes may be used to facilitate a quality cold joint between panels. In other cases, the joint is created using a chain that cuts a key slot into the primary panels during secondary excavations. Once the panel is excavated, the slurry is then desanded so as to optimize the displacement of the slurry with structural concrete and to prevent sand lenses. A reinforcing cage, previously prepared, is then lowered into the bentonite slurry-filled trench. The slurry is replaced with structural concrete poured using the tremie method of construction.

The construction process is further illustrated in the photos presented in Figure 8.17. In the upper left photo, excavation of a panel is proceeding using a hydromill under the head of a bentonite-water slurry. A reinforcing cage prepared for use in the panel is shown on the upper left of the photo. Notice the pipes through the reinforcing cage placed there to later facilitate the installation of tiebacks. The photo on the lower left shows the cage being lowered into the excavated slurry-filled trench. Finally, the lower right photo shows concrete placement using tremie pipes to transport the concrete to the bottom of the trench to allow for concreting from the bottom up.

Diaphragm walls are the pinnacle of vertical barriers in terms of strength. As described above, the slurry is replaced with reinforced concrete and structural steel beams can be

Figure 8.16 Schematic of diaphragm wall construction (courtesy of Keller NA).

Figure 8.17 Photos of diaphragm wall construction.

Figure 8.18 Photo of diaphragm wall in service.

inserted to add structural capacity. Diaphragm walls are also largely watertight permitting excavation in the dry below the water table. Shown in Figure 8.17 is a photo of a deep excavation in New York City with the structural slurry wall exposed after excavation from within the wall perimeter. Because of the significant depth, the photo also shows rakers (the diagonal braces) providing additional support to the structural diaphragm wall. Tiebacks are also commonly used to provide additional lateral support for the diaphragm wall (Figure 8.18).

Many of the materials and methods used to construct diaphragm walls are also used in the construction of other vertical barriers. The functional difference is that diaphragm walls are primarily employed as earth retention systems. That said, diaphragm walls also have a low permeability and, as well as performing their primary function, also control ground-water flow.

8.5 PROBLEMS

8.1 Notice on Figure 8.4 there is a thin, dark, horizontal line extended to the left from the trench near the base of the model. What is indicated by the dark line and what is the likely cause?

8.2 Calculate the percentage of bentonite by total mass of the slurry and by total mass of the water in example problem 12.1.

8.3 Recalculate the amount of bentonite needed to achieve a slurry density of 1.05 assuming the volume of bentonite in the mix is not significant. How much error is introduced? Considering field accuracy of mixing and measuring, does this matter?

8.4 For a bentonite-water slurry with a Marsh viscosity of 40 s and a density of 1.1 g/cm³, estimate the viscosity in cP.

8.5 From a sustainability perspective, which slurry wall type do you believe to be the most sustainable? Provide reasons for you answer.

8.6 A soil-bentonite cutoff wall is being considered for the subsurface section shown below and the in situ soils will be used to prepare the backfill. The wall will key one meter into the clay aquitard. The grain size distributions for the soils are shown below. What will be the percentage of gravel, sand, and silt/clay in the soil-bentonite backfill that would result from the technique (i.e. SB wall construction)? How much slurry (95% water and 5% bentonite) needs to be added to bring the mixture to 40% moisture content? How much bentonite will be in the mix (Figure Pr.8.6)?

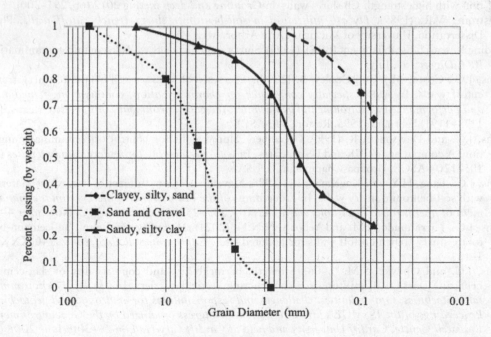

Figure Pr.8.6 Grain size distribution curve for use in Problem 8.6.

Ground Surface	el. 90 m
Clayey Silty Coarse to Fine Sand (w = 18%)	el. 86 m
Sand and Gravel (w = 6%)	el. 82 m
Sandy Silty Clay Aquitard (w = 23%)	

8.7 Calculate the flow rate (in m^3 per day) through a soil-bentonite cutoff wall placed in the center of a levee having the following properties and dimensions:
1.   Elevation of the top of the levee and top of cutoff wall = +30.0 m
2.   Elevation of clay layer and bottom of cutoff wall = +5.0 m
3.   Levee material: Sand
4.   Thickness of cutoff wall = 0.50 m
5.   Hydraulic conductivity of cutoff wall = 1×10^{-7} cm/s
6.   Length of levee = 3000 m
7.   Water elevation on upstream side of levee: +25 m
8.   Water elevation on downstream side of levee: +15 m

8.8 Soil bentonite backfill placed in a slurry trench is relatively compressible compared to the adjacent formation soils. Why does this matter?

8.9 Granulated ground blast furnace slag is often used along with portland cement for slag-CB cutoff walls. What is the principal benefit of slag in the mix?

REFERENCES

ASTM, ASTM. "C143/C143M-15." *Standard test method for slump of hydraulic-cement concrete)*. West Conshohocken, PA: ASTM Standards, ASTM International.

Barker, P., Esnault, A. and Braithwaite, P. (1997). *Containment barrier at pride park, Derby, England* (No. CONF-970208--PROC).

Bellew, G.M., Koirala, A.K., Dillon, J.C. and Mathews, D.L. (2012). Tuttle creek dam seismic remediation with high strength CB slurry walls. In *Grouting and deep mixing 2012* (pp. 291–300).

Bergstrom, W.R. (1989). *Fly ash utilization in soil-bentonite slurry trench cutoff walls*. Ph.D. Dissertation, University of Michigan, Ann Arbor, MI.

Building Research Establishment (BRE). (1994). Slurry trench cut-off walls to contain contamination. *BRE Digest 395*, July.

Evans, J.C., Costa, M.J. and Cooley, B. (2000). The state of stress is soil-bentonite slurry trench cutoff walls. In *ASCE specialty conference on characterization, containment, remediation and performance in environmental geotechnics, the geoenvironment 2000*, New Orleans, LA, 1173–1191, February, 1995. Reston, VA: ASCE.

Evans, J.C. and Dawson, A.R. (1999, December). Slurry walls for control of contaminant migration: A comparison of UK and US practices. In *Geo-engineering for underground facilities* (pp. 105–120). ASCE, Urbana-Champaign, IL USA.

Evans, J.C., Fang, H.Y. and Kugelman, I.J. (1985, November). Containment of hazardous materials with soil-bentonite slurry walls. In *Proceedings of the 6th national conference on the management of uncontrolled hazardous waste sites* (pp. 249–252), Washington, DC, November 4–6.

Evans, J.C., Larrahondo, J.M. and Yeboah, N.N.N. (2021). Fate of bentonite in slag–cement–bentonite slurry trench cut-off walls for polluted sites. *Environmental Geotechnics*, 40(XXXX), 1–13.

Evans, J.C. and Opdyke, S.M. (2006). Strength, permeability, and compatibility of slag-cement-bentonite slurry wall mixtures for constructing vertical barriers. In *5th ICEG environmental geotechnics: Opportunities, challenges and responsibilities for environmental geotechnics: Proceedings of the ISSMGE's fifth international congress organized by the Geoenvironmental Research Centre, Cardiff University and held at Cardiff City Hall on 26–30th June 2006* (pp. 118–125). Thomas Telford Publishing.

Evans, J.C. and Prince, M.J. (1997). Additive Effectiveness in Minerally-Enhanced Slurry Walls. *ASCE Specialty Conference on In Situ Remediation of the Geoenvironment*, Minneapolis, MN, ASCE Geotechnical Special Publication No. 71, October, 1997. Reston, VA: ASCE.

Evans, J.C., Prince, M.J. and Adams, T.L.. (1997). *Metals attenuation in minerally-enhanced slurry walls.* No. CONF-970208--PROC.

Filz, G.M. (1996). Consolidation stresses in soil-bentonite backfilled trenches. In *Environmental geotechnics* (pp. 497–502).

Filz, G.M., Adams, T. and Davidson, R.R. (2004). Stability of long trenches in sand supported by bentonite-water slurry. *Journal of Geotechnical and Geoenvironmental Engineering, 130*(9), 915–921.

Filz, G.M., Boyer, R.D. and Davidson, R.R. (1997). *Bentonite-water slurry rheology and cutoff wall trench stability* (No. CONF-971032-). Reston, VA: American Society of Civil Engineers.

Filz, G.M., Evans, J.C. and Britton, J.P. (2003, June). Soil-bentonite hydraulic conductivity: measurement and variability. In *Proceedings of the 12th Pan American conference on soil mechanics and geotechnical engineering*, Cambridge, MA (pp. 22–26).

Fox, P.J. (2004). Analytical solutions for stability of slurry trench. *Journal of Geotechnical and Geoenvironmental Engineering, 130*(7), 749–758.

Garvin, S.L. and Hayles, C.S. (1999). The chemical compatibility of cement–bentonite cut-off wall material. *Construction and Building Materials, 13*(6), 329–341.

Glass, P. and Stones, C. (2001). Construction of Westminster Station, London. *Proceedings of the Institution of Civil Engineers: Structures and Buildings, 146*(3), 237–252.

Building Research Establishment, Construction Industry Research and Information Association. (1999). *Specification for the construction of slurry trench cut-off walls as barriers to pollution migration*, Thomas Telford.

Jefferis, S. (2012). Cement-bentonite slurry systems. In *Grouting and deep mixing 2012* (pp. 1–24). New Orleans, LA, February 15–18, 2012.

Jefferis, S.A. (1981, June). Bentonite-cement slurries for hydraulic cut-offs. In *Proceedings, tenth international conference on soil mechanics and foundation engineering*, Stockholm, Sweden (Vol. 1, pp. 435–440).

Jefferis, S.A. (1993). In-ground barriers. In *Contaminated land-problems and solutions* (pp. 111–140).

Jefferis, S.A. (1997). *The origins of the slurry trench cut-off and a review of cement-bentonite cut-off walls in the UK.* No. CONF-970208--PROC.

Khandelwal, A., Rabideau, A.J. and Shen, P. (1998). Analysis of diffusion and sorption of organic solutes in soil-bentonite barrier materials. *Environmental Science & Technology, 32*(9), 1333–1339.

LaGrega, M.D., Buckingham, P.L. and Evans, J.C. (2010). *Hazardous waste management.* Waveland Press, Long Grove, IL USA.

Malusis, M.A., Evans, J.C., McLane, M.H. and Woodward, N.R. (2008). A miniature cone for measuring the slump of soil-bentonite cutoff wall backfill. *Geotechnical Testing Journal, 31*(5), 373–380.

Malusis, M.A., McKeehan, M.D. and LaFredo, R.A. (2010). Multiswellable bentonite for soil-bentonite vertical barriers. In *Proceedings of the 6th international congress on environmental geotechnics, 6th ICEG, November* (pp. 8–12), New Delhi.

Mitchell, J.K. and Soga, K. (2005). *Fundamentals of soil behavior* (Vol. 3). New York: John Wiley & Sons.

Nash, K.L. (1974). Stability of trenches filled with fluids. *Journal of the Construction Division, 100*(4), 533–542.

Opdyke, S.M. and Evans, J.C. (2005). Slag-cement-bentonite slurry walls. *Journal of Geotechnical and Geoenvironmental Engineering, 131*(6), 673–681.

Owaidat, L.M., Andromalos, K.B., Sisley, J.L. and Civil Engineer, U. (1999, October). Construction of a soil–cement–bentonite slurry wall for a levee strengthening program. In *Proceedings of the 1999 annual conference of the association of state dam safety officials*, St. Louis, MO (pp. 10–13).

Pitt, M.J. (2000). The Marsh funnel and drilling fluid viscosity: A new equation for field use. *SPE Drilling & Completion, 15*(1), 3–6.

di Cervia, A.R. (1992). History of slurry wall construction. In *Slurry walls: Design, construction, and quality control*. Conshocken, PA: ASTM International. *ASTM Special Technical Publication, 1129*, 3–15.

Ruffing, D.G. (2009). *A reevaluation of the state of stress in soil-bentonite slurry trench cutoff walls* (Master's Thesis, Bucknell University).

Ruffing, D.G. (2012). Personal communication.

Ruffing, D.G. and Evans, J.C. (2010). In situ evaluation of a shallow soil bentonite slurry trench cutoff wall. In *Proceedings of the 6th international congress on environmental geotechnics* (pp. 758–763), New Delhi.

Ruffing, D.G., Evans, J.C. and Malusis, M.A. (2010). Prediction of earth pressures in soil-bentonite cutoff walls. In *GeoFL 2010: Advances in analysis, modeling & design* (pp. 2416–2425). Orlando, FL, February 20–24, 2010.

Ruffing, D.G., Evans, J.C., Spillane, V.A. and Malusis, M.A. (2016). The use of filter press tests in soil-bentonite slurry trench construction. In *Geo-Chicago 2016* (pp. 590–597). Chicago, IL, August 14–18, 2016.

RWE Power International. (2011). Diaphragm containment walls using PFA at Bedfont Lakes, Middlesex, generation aggregates Electron, Windmill Hill Business Park, Whitehill Way, Swindon, Wiltshire, United Kingdom.

Structural Engineering Institute. (2000). *Effective analysis of diaphragm walls*. Reston, VA: American Society of Civil Engineers.

Tamaro, G.J. (2002). World trade center "bathtub": From genesis to armageddon. *Bridge, 32*(1), 11–17.

Tsai, J.S. and Chang, J.C. (1996). Three-dimensional stability analysis for slurry-filled trench wall in cohesionless soil. *Canadian Geotechnical Journal, 33*(5), 798–808.

Xanthakos, P.P. (1979). *Slurry walls*. New York, NY: McGraw-Hill Book Co.

Xanthakos, P.P. (1994). *Slurry walls as structural systems*. McGraw-Hill Book Co.

Yeo, S.S., Shackelford, C.D. and Evans, J.C. (2005). Consolidation and hydraulic conductivity of nine model soil-bentonite backfills. *Journal of Geotechnical and Geoenvironmental Engineering, 131*(10), 1189–1198.

Chapter 9

Ground improvement using geosynthetics

9.1 INTRODUCTION

Mechanical stabilization of soil to improve its properties and performance typically involves adding plastic or metal materials to the soil. Mechanical reinforcement adds tensile strength to soils. The inclusion of these materials allows the engineer to meet client needs faster and more cost-effectively than before.

With increased tensile strength, soil-based projects can be improved. This chapter addresses:

1. strengthened base courses for roads,
2. embankments over soft ground,
3. underfooting reinforcement,
4. general fill strengthening using geofibers, and
5. soil separation.

Soil separation is popular in road and railway construction, requires modest geosynthetic tensile strength, but is not technically a "reinforcement" application.

Geosynthetic reinforcement costs are a small percentage of project costs, making geosynthetic use very attractive. The versatility of the geosynthetics in solving geotechnical problems has led to their rapid adoption by the civil engineering design and construction industry.

9.2 GEOSYNTHETIC GROUND IMPROVEMENT

9.2.1 Introduction

Geosynthetics, introduced in Chapter 3, are, put simply, plastic materials added to soil to create a composite with improved performance relative to the soils alone. Many geosynthetics come in rolls, while others come as discrete fibers, continuous fibers, spray-on materials, or honeycombs. Rolled products are the most popular.

Geosynthetics can perform many soil improvement functions, including:

1. strengthening,
2. filtration,
3. erosion control,
4. drainage, and
5. separation.

There are existing textbooks and design manuals that address all of these functions such as Koerner (2012), Holtz et al. (1998), and FHWA (2008).

Strengthening and separation functions are primarily addressed in this book. Strengthening soil with geosynthetics allows the engineer to increase the soil's bearing capacity. Geosynthetics improve embankment stability, ameliorate settlement (compressibility), and allow faster construction.

The idea of using non-soil materials to reinforce soils is over 2,000 years old, for example, a portion of the Great Wall of China uses mechanically stabilized earth. On a smaller scale, straw has been used for millennia to strengthen clay bricks used in building construction. The famed ziggurats in Iraq and Iran are soil structures with very steep slopes that are reinforced not with the modern plastic or metal, but with plant matter. The idea lay dormant for a few thousand years until the 1960s, when Henri Vidal is credited with bringing the idea into the modern era. Metal reinforcement was used first, followed by the currently widely used plastic.

9.2.2 Geosynthetic types used in ground improvement

The most commonly used ground improvement geosynthetics are *geotextiles* and *geogrids*. These are rolled plastic products with widths of up to 8 m. They have tensile strength commensurate with civil engineering applications.

Geotextiles (cloth) may be woven or nonwoven. The woven textile weave is typically the simple basket weave (Figure 3.1). The fibers/yarns need not be the same, allowing differing strengths in the machine direction (MD) or cross-machine (XD) direction, sometimes called the warp (or weft) and woof directions, respectively. This may be economic when the application only requires strength in one direction (uniaxial). Nonwoven geotextiles are made of many tiny, interlocked fibers and look like felt. Needle punching is the most common production method and is described in Kadolph (2010). Geotextiles are shipped in rolls (Evans et al. 2009).

Geogrids are also rolled plastic products, but have large holes in them, much larger than the holes in geotextiles (Figure 3.4). There are three basic methods of making geogrids: pulled, woven, and heat seamed. Pulled geogrids are made by punching holes in a warm geomembrane sheet and pulling the sheet in one or two directions (giving the geogrid strength in one or two directions). Woven geogrids are made of two strands, attached at the joints by weaving the two strands together or heat seaming them. The two strands may have different strengths. Geogrids are also shipped in rolls.

Geofibers, usually in the form of *staples*, are short (about 10 cm) fibers as shown in Figure 9.1. The geofibers may be individual strands of plastic, or a small accumulation of strands. They are discrete from each other, shipped in bales, and mixed into the soil.

Geocells are a plastic honeycombed structure (Figure 3.5). The cells are formed from plastic strips, set on edge, and welded together at various staggered spacings. Geocells are shipped in the collapsed form, expanded onsite, and filled with soil, producing a composite that is very strong in compression.

9.2.3 Geosynthetic applications in ground improvement

Some applications of geosynthetic ground improvement include soil strengthening, road base improvement, compaction aids, underfooting reinforcement, embankments over soft ground, mechanically stabilized earth (MSE), steepened slopes, and soil separation. Geosynthetics can be used to improve the tensile and compressive strength of soils.

Figure 9.1 Photo of geofibers.

Roads are primarily geotechnical structures, covered with an asphalt or portland cement concrete surface, or wearing course. Figure 9.2 shows parts of road cross section and locations of typical geosynthetics.

The success of the road depends on the success of the base, subbase, and subgrade soils below the surface course. These soils can be improved by interlayering rolled geosynthetics to add tensile strength. Geofibers mixed into these soils also increases compressive strength, while improving drainage.

An asphalt concrete wearing course may be improved by geotextiles or geogrids placed immediately underneath. Pavement structures may be made more water resistant using geotextiles. These topics are discussed in Shukla and Yin (2006) and elsewhere.

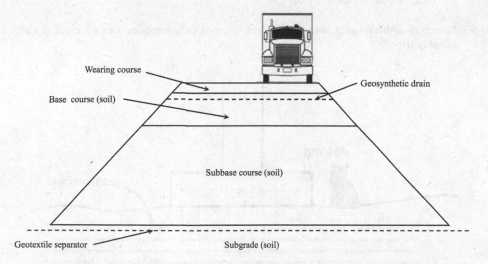

Figure 9.2 Sketch showing subgrade, subbase, base, and wearing course, with locations of geosynthetics. Geosynthetics are used at the interfaces: geotextile separator on the subgrade; geosynthetic drain under the wearing course; geotextile waterproofing, and/or geogrid reinforcement on top of base course (not to scale).

The embankment over soft ground application is given schematically in Figure 9.3. The left side of the figure indicates the possible failure surfaces for a bearing capacity failure whereas the right side indicates a possible slope failure above the subgrade. Geotextiles or geogrids are placed on the subgrade to provide tensile reinforcement at the subgrade and are effective in preventing the failure modes shown in Figure 9.3.

The underfooting reinforcement application is shown in Figure 9.4. This application is economic when the footing is to be placed in an embankment or fill. Layers of geotextiles, geogrids, or geocells are placed in the soil as the embankment is constructed, greatly increasing the bearing capacity of the soil and reducing settlement. Soil strengthening may also be accomplished with the addition of geofibers.

Mechanically stabilized earth (MSE) walls are "changes in grade" that involve stacking layers of soil and geosynthetic reinforcement as shown in Figure 9.5. This combination, when vertical, or near vertical, produces the change in grade to meet project needs. Traditionally, retaining walls were built as large heavy objects that held the soil back and are commonly termed gravity walls. MSE walls use an entirely different principle – adding geosynthetics or metal to the soil so the soil-plus-geosynthetic mass supports itself. This removes the need for a wall made of, say, concrete to retain the soil.

MSE walls often have a nonstructural concrete *facing* to reduce erosion and provide aesthetics such as the example shown in Figure 9.6. MSE walls are a composite structure of soil and geosynthetic, the result being stronger than either. These are the walls of choice (2020)

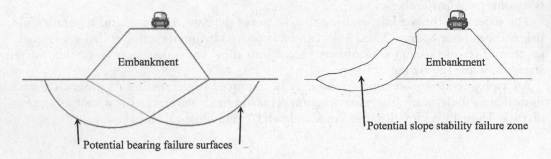

Figure 9.3 Sketch of embankments over soft ground, showing failure modes: bearing capacity (left), slope stability (right).

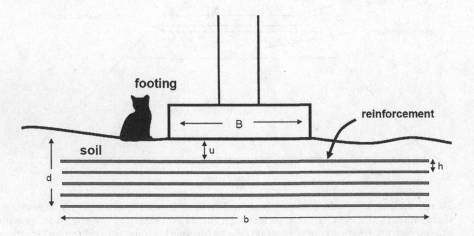

Figure 9.4 Schematic of underfooting reinforcement.

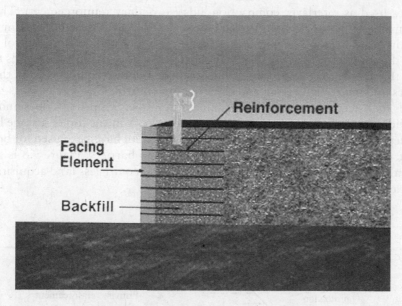

Figure 9.5 Schematic of mechanically stabilized earth wall. (courtesy of GeoStabilization International)

Figure 9.6 Mechanically stabilized earth with block facing.

because they can be constructed easily in many different shapes, are economical and easy to construct. MSE walls are described in greater detail in Chapter 10.

Steepened slopes (embankments) are constructed on the same principle as MSE walls. Technically, *slopes* are naturally occurring; *embankments* are constructed. However, for unknown reasons, these geosynthetic/soil composite embankments are called steepened slopes and that convention is followed here. Unlike MSE walls, steepened slopes do not have facing. Figure 9.7 is a schematic of a mechanically stabilized earth slope, showing geosynthetics used as interlayer compaction aids (secondary reinforcement). These slopes stand at a much steeper angle than the soil alone would stand; hence the term *steepened* slopes. The increase in slope steepness often requires additional erosion control measures. Sometimes, the contractor may use relatively short geosynthetic compaction aids to assist in compacting the soil near the slope face. While steepened slopes require less soil, they require geosynthetics and more engineering and construction control.

MSE walls and steepened slopes create changes in grade. Steepened slopes and walls are needed to widen roadways, create levees, create more space at the top of a slope by making the slope steeper, and for creating foundation preloads. Both have been used for bridge abutments (some even supporting the bridge!) and for sound barriers. Figure 9.8 is a wrapped face MSE wall used as an access ramp. Steepened slopes require less land acquisition for the same crest elevation and location.

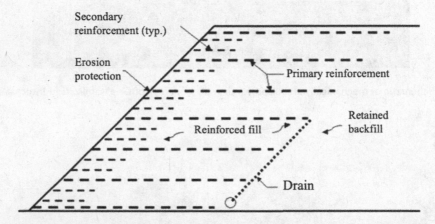

Figure 9.7 Schematic of mechanically stabilized earth slope, showing geosynthetic used as interlayer compaction aids (after Berg et al. 2009).

Figure 9.8 Photo of wrapped face mechanically stabilized earth wall used as an access ramp (Berg et al. 2009).

9.3 PROPERTIES OF GEOSYNTHETICS

9.3.1 Introduction

Geosynthetics are engineering materials, just like concrete, steel, and timber. Their engineering properties are required for design. Chapter 3 discusses the evaluation of permittivity, the basics of interface friction, durability, and survivability, with some discussion of tensile strength. This chapter extends those fundamentals.

9.3.2 Tensile strength

Geosynthetic strength tests largely came from the textile industry. The primary test is the *wide-width* tensile test, discussed in Chapter 3. The test apparatus is shown in Figure 9.9. Typical force-deformation curves from a geotextile wide-width test are given in Figure 9.10. The deformation (and strain) to failure changes significantly with geotextile manufacturing technique. A higher strain to failure indicates lower damage potential. These curves may be used in design.

Geogrid strength, unlike geotextile strength, is evaluated by ASTM D6637-15, the Standard Test Method for Determining Tensile Properties of Geogrids by the Single or Multi-Rib Tensile Method (ASTM D6637, 2015). Here, one or more geogrid ribs are pulled apart in tension. Force and deformation are measured. Geosynthetic strengths are often reported in units of kN/m or lbf/ft.

Geogrids and geotextiles may have different strengths in the machine and cross-machine directions. *Machine direction* refers to the long direction of a roll of geosynthetic. *Cross-machine* is perpendicular to the machine direction.

Geogrids and geotextiles are seamed to provide load transfer. Geotextiles are primarily seamed by stitching or heat welding. The wide-width strength of the seamed geotextile must be considered in design, since seaming weakens the geotextile. Stronger geotextiles suffer

Figure 9.9 Photo of Wide-width test setup, not in failure; sample is eight inches wide. (courtesy of Barry Christopher).

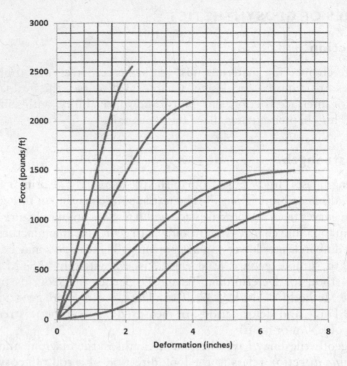

Figure 9.10 Typical force-deformation curves for geotextiles.

greater strength loss, as much as 50%. Seaming methods are given in Qian et al. (2002), Carroll and Chouery-Curtis (1990), and others. Geogrid seams are made using hog rings (or similar), cable ties, or bodkin joints. Bodkin is another word for a large needle. There is no evidence it is named after one of the fourteen tribes of Galway, Ireland. The latter may provide seam efficiencies up to 90%. The seaming method chosen should be tested for capacity.

Geosynthetic ultimate tensile strengths from testing are reduced to the *long-term design strength* (LTDS) for design. The reduction accounts for qualitative, but important, factors that can reduce the strength of the geosynthetic during and after installation. LTDS is:

$$LTDS = \frac{T_{ult}}{RF_{ID}\,RF_{CR}\,RF_{CBD}} \tag{9.1}$$

where
  T_{ult} is the ultimate strength from testing (kN/m or lbf/ft), and
  RF is a reduction factor.

These reduction factors account for installation damage (ID), creep (CR), and chemical/biological degradation (CBD). Typical values vary from 1.0 to 3.0. Tables of RFs for LTDS and other geosynthetic properties are found in Elias et al. (2001), Bonaparte and Berg (1987), and Koerner (2005). The designer may include other reduction factors to account for project-specific concerns. There are other geosynthetic strength tests: grab tensile, puncture, and several tear tests. These are used in geosynthetic survivability specifications, not design specifications.

9.3.3 Interface friction

There are three measures of the shearing resistance that can be developed between a geosynthetic and another material. MSE wall design procedures frequently require these. Interface friction is evaluated by a form of the direct shear test where the shearing occurs on the plane between the two different materials (the soil and the geosynthetic). Interface friction testing is discussed in Chapter 3. Changes in either, or the moisture content of the soil, will change the interface friction angle. ASTM Standard D5321-20 (2020) provides testing guidance.

In testing for the interaction between soils and geosynthetics, the tangent of the interface friction angle is sometimes normalized by the tangent of the angle of internal friction to yield the Coefficient of Direct Sliding (C_{ds}):

$$C_{ds} = \frac{\tan \phi_{\text{interface friction}}}{\tan \phi_{\text{soil}}} \tag{9.3}$$

where ϕ_{soil}, in degrees, is determined from direct shear or other testing of the soil without the geosynthetic.

Pullout testing is used in MSE wall design. ASTM D6706-13 (2013) is the Standard Test Method for Measuring Geosynthetic Pullout Resistance in Soil (ASTM D6706-13, 2013). Here, the geosynthetic is embedded between layers of soil, and, while being squeezed, is pulled out. Figure 9.11 shows the pullout test apparatus.

Unlike the direct shear test, the pullout test shears the geosynthetic on both sides at once, yielding different behavior. The measured result is the pullout force, F. The coefficient of interaction, C_i, is calculated from:

$$C_i = \frac{F}{2 A (\sigma \tan \phi + c)} \tag{9.4}$$

where
   F is the measured pullout force (kN)
   A is the area of the geotextile in contact with the soil (m²)
   σ is the normal stress on the soil during the test (kPa)
   ϕ is the angle of internal friction of the soil (degrees)
   c is the soil cohesion (kPa)

9.3.4 Durability

Durability was introduced in Chapter 3. Geosynthetic *durability* describes how easily geosynthetics degrade in situ. The engineering properties of the geosynthetic must not degrade due to biological or chemical action, or temperature. Geosynthetics are largely made of polypropylene, polyethylene, or polyester. These polymers were chosen for several reasons: low cost, suitable engineering properties, ready availability, and exceptional durability. There are very few natural environmental conditions that affect these polymers. Many laboratory and field studies have shown their extremely durable nature (e.g. Suvorova and Alekseeva 2010; Sprague and Goodrum 1994).

When the designer has concerns about the in situ performance of a geosynthetic, testing should be done. Chemical resistance can be evaluated with ASTM standards D6213-17 (2017) Standard Practice for Tests to Evaluate the Chemical Resistance of Geogrids to Liquids (ASTM D6213, 2017), ASTM D6389-17 (2017) Standard Practice for Tests to Evaluate the Chemical Resistance of Geotextiles to Liquids (ASTM D6389, 2017), and ASTM D5322-17 (2017) Standard Practice for Immersion Procedures for Evaluating the Chemical Resistance of Geosynthetics to Liquids (ASTM D5322, 2017).

Temperature effects on geotextiles may be evaluated with ASTM D4594-96 (2009) Standard Test Method for Effects of Temperature on Stability of Geotextiles (ASTM D 4594, 2009).

9.3.5 Geotextile survivability

Geotextile survivability was introduced in Chapter 3. It refers to the ability of a geotextile to retain its properties during installation. At the time of this writing, the most widely used survivability requirements are given in the American Association of State Highway and Transportation Officials (AASHTO) specification M288-17 (AASHTO M288-17, 2017). The designer evaluates ground and construction equipment condition and selects a survivability class needed. Softer ground and higher equipment pressures require a higher survivability class (Tables 9.1 through 9.5).

The requirements are based on index tests of the geotextile, including grab strength, sewn seam strength, tear strength, and puncture strength. There are ASTM standards for evaluating these. Each geotextile survivability class is divided into two subclasses based on geotextile structure. Nonwoven geotextiles are referred to as "Elongation $\geq 50\%$" while woven geotextiles are referred to as "Elongation $\leq 50\%$."

9.4 ROAD BASE STABILIZATION (CORPS OF ENGINEERS METHODS)

9.4.1 Introduction

This section describes the use of geosynthetics to improve soils underneath asphalt concrete roads. Roads are primarily geotechnical structures. The success of the road depends on the properties of the base and subbase soils, drainage, traffic, and surface course characteristics. The base and subbase soil strength may be improved or preserved using geosynthetics. Soil drainage, the most critical element of successful pavement design (Cedergren 1987, 1994), is covered generally in Chapter 3. Pavement underdrain and edge drain design procedures are given in Christopher et al. (2006).

A typical road cross section is given in Figure 9.2. This figure shows subgrade, subbase base soil, and wearing course. *Subgrade* describes the in situ soils. *Subbase* and *base* soils are imported to support the *wearing course*. These two layers are between the wearing course and the subgrade. The wearing course provides friction for driving, erosion protection from traffic and weather, and some structural support of traffic.

Geosynthetics are economically justified when the subgrade is soft clay, or when the subbase or base soils do not provide the needed support to the wearing course. Because drainage is critical to pavement life, designers prefer to use subbase and base soils that drain well – sands and gravels (granular soil). These soils have high permeability compared to silty or clayey soils. Granular soils on soft, clayey soils will sink into them with time, reducing permeability. With the loss of permeability (and drainage capability), the road begins to fail. Geotextiles are used to separate these soils, prolonging pavement life.

There are many design methods for strengthening base and subbase soils (e.g. Holtz and Sivakugan 1987; Giroud and Noiray 1981). This section covers the US Army Corps of Engineers method (USACE 2003) which is based on the US Forest Service (USFS) method described in Steward et al. (1977). Geotextiles are used for separation; geotextiles and geogrids are used for strengthening in this method. Geofibers and geocells are also used for strengthening. Soil encapsulation (Sale et al. 1973), pile-supported embankments (Han and Gabr 2002), and lightweight fill are other geosynthetic-based methods of improving roads.

Geosynthetics improve soil performance by providing separation of high quality base soil from fine-grained subgrade soils, and by adding tensile strength to the soil.

Table 9.1 Geotextile Strength Property Requirements (from M 288-17 in **Standard Specifications for Transportation Materials and Methods of Sampling and Testing**, by the American Association of State Highway and Transportation Officials, Washington, D.C. Used with permission.)

AASHTO M288-17 – Geotextile Strength Property Requirements

Class[a,b]	Test Methods	Units	Geotextile Class 1		Class 2		Class 3	
			Elongation < 50%[c]	Elongation ≥ 50%[c]	Elongation < 50%[c]	Elongation ≥ 50%[c]	Elongation < 50%[c]	Elongation ≥ 50%[c]
Grab strength	ASTM D 4632	N	1400	900	1100	700	800	500
Sewn seam strength[d]	ASTM D 4632	N	1260	810	990	630	720	450
Tear strength	ASTM D 4533	N	500	350	400[e]	250	300	180
Puncture strength	ASTM D 6241	N	2750	1925	2200	1375	1650	990
Permittivity	ASTM D 4491	sec^{-1}	Minimum property values for permittivity, AOS, and UV stability are based on geotextile application. Refer to Table 9.2 for subsurface drainage, Table 9.3 and Table 9.4 for separation, Table 9.5 for stabilization, and Table 9.6 for permanent erosion control.					
Apparent opening size	ASTM D4751	mm						
Ultraviolet stability (retained strength)	ASTN D 4355	%						

a Required geotextile class is designated in Tables 9.2, 9.3, 9.4, 9.5, or 9.5 for the indicated application. The severity of installation conditions for the application generally dictates the required geotextile class. Class 1 is specified for more severe or harsh installation conditions where there is a greater potential for geotextile damage, and Classes 2 and 3 are specified for less severe conditions.

b All numeric values represent MARV in the weaker principal direction (see section 8.1.2.).

c As measured in accordance with ASTM D 4632.

d When sewn seams are required. Refer to Appendix for overlap seam requirements.

e The required MARV tear strength for woven monofilament geotextiles is 250 N.

Table 9.2 Subsurface Drainage Geotextile Requirements (from M288-17 in *Standard Specifications for Transportation Materials and Methods of Sampling and Testing*, by the American Association of State Highway and Transportation Officials, Washington, D.C. Used with permission.)

AASHTO M288-17 – Subsurface Drainage Geotextile Requirements					
			Requirements		
			Percent in Situ Soil Passing 0.075 mm[a]		
	Test Methods	*Units*	*< 15*	*15–50*	*> 50*
Geotextile class			Class 2 from Table 1[b]		
Permittivity[c,d]	ASTM D 4491	sec^{-1}	0.5	0.2	0.1
Apparent opening size[c,d]	ASTM D 4751	mm	0.43 max avg roll value	0.25 max avg roll value	0.22[e] max avg roll value
Ultraviolet stability (retained strength)	ASTM D 4355	%	50% after 500 h of exposure		

[a] Based on grain size analysis of in situ soil in accordance with T 88.
[b] Default geotextile selection. The engineer may specify a Class 3 geotextile from Table 9.1 for trench drain applications based on one or more of the following:
  1. The engineer has found Class 3 geotextiles to have sufficient survivability based on field experience.
  2. The engineer has found Class 3 geotextiles to have sufficient survivability based on laboratory testing and visual inspection of a geotextile sample removed from a field test section constructed under anticipated field conditions.
  3. Subsurface drain depth is less than 2 m; drain aggregate diameter is less than 30 mm; and compaction requirement is less than 95% of T 99.
[c] These default filtration property values are based on the predominant particle sizes of in situ soil. In addition to the default permittivity value, the engineer may require geotextile permeability and/or performance testing based on engineering design for drainage systems in problematic soil environments.
[d] Site-specific geotextile design should be performed especially if one or more of the following problematic soil environments are encountered: unstable or highly erodible soils such as non-cohesive silts, gap graded soils, alternating sand/silt laminated soils, dispersive clays, and/or rock flour.
[e] For cohesive soils with a plasticity index greater than 7, geotextile maximum average roll value for apparent opening size is 0.30 mm.

Table 9.3 Separation Geotextile Property Requirements (from M 288-17 in *Standard Specifications for Transportation Materials and Methods of Sampling and Testing*, by the American Association of State Highway and Transportation Officials, Washington, D.C. Used with permission.)

AASHTO M288-17 – Separation Geotextile Property Requirements			
	Test Methods	*Units*	*Requirements*
Geotextile class			See Table 9.4
Permittivity	ASTM D 4491	sec^{-1}	0.02[a]
Apparent opening size	ASTM D 4751	mm	0.60 max average roll value
Ultraviolet stability (retained strength)	ASTM D 4355	%	50% after 500 h of exposure

[a] Default value. Permittivity of the geotextile should be greater than that of the soil ($\Psi_g > \Psi_s$).
The engineer may also require the permeability of the geotextile to be greater than that of the soil ($k_g > k_s$).

9.4.2 Unpaved road improvement using geosynthetics

Geosynthetics can improve unpaved roads by providing separation of fine-grained subgrade soils from granular subbase and base materials, and by adding tensile strength to subbase and base materials. The use of geosynthetics reduces the thickness of subbase and base layers, reducing the cost of soil, but increasing the cost of geosynthetics. Fine-grained subgrade

Table 9.4 Required Degree of Survivability as a Function of Subgrade Conditions, Construction Equipment, and Lift Thickness Requirements (from M 288-17 in *Standard Specifications for Transportation Materials and Methods of Sampling and Testing*, by the American Association of State Highway and Transportation Officials, Washington, D.C. Used with permission.)

AASHTO M288-17 – Required Degree of Survivability as a Function of Subgrade Conditions, Construction Equipment, and Lift Thickness (Class 1, 2, and 3 properties are given in Table 9.1; Class 1+ properties are higher than Class 1, but not defined at this time and if used must be specified by the purchaser.)[a]

	Low ground pressure equipment ≤ 25 kPa (3.6 psi)	Medium ground-pressure equipment > 25 to ≤ 50 kPa (> 3.6 to ≤ 7.3 psi)	High ground pressure equipment > 50 kPa (> 7.3 psi)
Subgrade has been cleared of all obstacles except grass, weeds, leaves, and fine wood debris. Surface is smooth and level so that any shallow depressions and humps do not exceed 450 mm (18 in.) in depth or height. All larger depressions are filled. Alternatively, a smooth working table may be placed.	Low (Class 3)	Moderate (Class 2)	High (Class 1)
Subgrade has been cleared of obstacles larger than small to moderate-sized tree limbs and rocks. Tree trunks and stumps should be removed or covered with a partial working table. Depressions and humps should not exceed 450 mm (18 in.) in depth or height. Larger depressions should be filled.	Moderate (Class 2)	High (Class 1)	Very High (Class 1+)
Minimal site preparation is required. Trees may be felled, delimbed, and left in place. Stumps should be cut to project not more than ± 150 mm (± 6 in.) above subgrade. Geotextile may be draped directly over the tree trunks, stumps, large depressions and humps, holes, stream channels, and large boulders. Items should be removed only if placing the geotextile and cover material over them will distort the finished road surface.	High (Class 1)	Very High (Class 1+)	Not recommended

[a] Recommendations are for 150 mm to 300 mm (6 in. to 12 in.) initial lift thickness. For initial lift thicknesses:
300 mm to 450 mm (12 in. to 18 in.): reduce survivability requirement one level;
450 mm to 600 mm (18 in. to 24 in.): reduce survivability requirement two levels;
> 600 mm (24 in.): reduce survivability requirement three levels.
For special construction techniques such as prerutting, increase the geotextile survivability requirement one level. Placement of excessive initial cover material thickness may cause bearing failure of the soft subgrade.

Table 9.5 Stabilization Geotextile Property Requirements (from M 288-17 in *Standard Specifications for Transportation Materials and Methods of Sampling and Testing*, by the American Association of State Highway and Transportation Officials, Washington, D.C. Used with permission.)

AASHTO M288-17 – Stabilization Geotextile Property Requirements			
	Test Methods	*Units*	*Requirements*
Geotextile class			Class 1 from Table 9.1[a]
Permittivity	ASTM D 4491	sec^{-1}	0.05[b]
Apparent opening size	ASTM D 4751	mm	0.43 max avg roll value
Ultraviolet stability (retained strength)	ASTM D 4355	%	50% after 500 h of exposure

[a] Default geotextile selection. The engineer may specify a Class 2 or 3 geotextile from Table 9.1 based on one or more of the following:
  1. The engineer has found the class of geotextile to have sufficient survivability based on field experience.
  2. The engineer had found the class of geotextile to have sufficient survivability based on laboratory testing and visual inspection of a geotextile sample removed from a field test section constructed under anticipated field conditions.

[b] Default value. Permittivity of the geotextile should be greater than that of the soil ($\Psi_g > \Psi_s$). The engineer may also require the permeability of the geotextile to be greater than that of the soil. ($k_g > k_s$).

Figure 9.11 Photo of apparatus for pullout testing of geosynthetics (Cazzuffi et al. 2014).

soil, when moist, may migrate into the base or subbase soils and vice versa (subgrade intrusion). This weakens the subbase and base soils and dramatically reduces drainage, the most important property of subbase and base materials in roads. A little clay goes a long way in reducing permeability. Figure 9.12 shows the effect of soil separation.

Traditionally, subgrade intrusion was accounted for in design by adding extra subbase and base soils that would get contaminated over time. This extra soil was termed *sacrificial aggregate*. Figure 9.13 gives an estimate of the amount of extra subbase or base material needed, based on how soft subgrade clays are. In Figure 9.13, CBR is the California Bearing Ratio, a measure of soil strength, described in ASTM D1883-16 (2016). Note little subgrade intrusion/contamination is expected in soils with CBR > 3%.

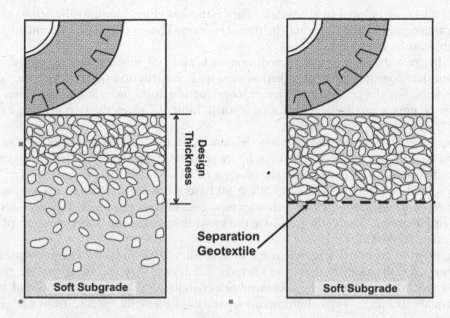

Figure 9.12 Schematic of the effect of geotextile separator on loss of aggregate to a soft subgrade. (courtesy of James Tinjum).

Table 9.6 Subgrade conditions for geosynthetic unpaved road improvement (USACE 2003)

	Subgrade Condition (%)	Geosynthetic Recommendation
Case 1	CBR < 0.5	use geosynthetic separator and geogrid reinforcement; do not reduce base thickness
Case 2	0.5 < CBR < 2	use geosynthetic separator and geogrid reinforcement; reduce base thickness according to a design procedure
Case 3	2 < CBR < 4	use judgment in choosing to use a geosynthetic separator; geogrid not likely to help.

Geosynthetics are useful for roads on clayey subgrades with CBR ≤ 3. Loose, cohesionless subgrades are better improved by vibratory compaction. There are three cases, with attendant design procedures, considered here, based on the CBR of the subgrade clayey soils shown in Table 9.6. Remember, the subgrade condition of interest is the worst condition the road will experience during its lifetime, not necessarily the condition of the subgrade during site exploration or construction.

The design method requires an estimate of the clayey subgrade strength, cohesion (c), at the worst location in the road alignment. This can be estimated from the pocket penetrometer, pocket vane, vane shear test, CBR, or other tests. The Standard Penetration Test (ASTM D1586-18, 2018) is not recommended. Equation 9.5 relates CBR and c

$$c = (1.4)(CBR) \tag{9.5}$$

where
    c is cohesion in pounds per square foot, and
    CBR is the test value in percent.

Figure 9.14 may be used to estimate c. The traffic wheel loading must be estimated. The construction equipment loading may be the worst case. The amount of traffic and allowable rut depth must be estimated.

Typically, two designs are performed: one with and one without the geosynthetic. For the design that does not use geosynthetics, principal construction costs include the costs of the base soils. For the geosynthetic design, construction costs include both the costs of the geosynthetic plus a smaller amount of base soils. Table 9.7 gives the two bearing capacity factors (N_c) for the given conditions.

Figures 9.15–9.17 are used to estimate the depth of CBR $\geq$ 80 base soil, *in 2.5 cm increments*, needed for the road. The choice of figure is based on the vehicle wheel configuration. Enter the figure on the abscissa with the product of c and N_c and use the appropriate wheel load curve and read the thickness of CBR $\geq$ 80 base soil needed. Repeat this for the other N_c. The geosynthetic design yields a thinner base course thickness. The designer selects the less expensive design. The CBR $\geq$ 80 base thickness is adjusted for larger numbers of wheel passes, using Table 9.8.

The geotextile separator's permittivity, survivability, and filtration characteristics must be determined. Filtration is covered in Chapter 3. Survivability specifications are given by AASHTO M288-17 (2017). The geotextile permittivity should exceed 0.1/second to provide adequate drainage. Typical minimum separation geotextile specifications are given in Table 9.9.

A geogrid may be needed, depending on the case given in Table 9.6. Geogrid specifications are given in Table 9.10. The geogrid is placed on top of the geotextile, with seams perpendicular to the roadway alignment, and the high strength direction of the geogrid in the transverse direction of the roadway alignment.

Geosynthetic installation procedures are important to the success of these designs. Sites should be cleared and grubbed. They should be reasonably flat. In some instances, where the ground is very swampy, it may be preferable to leave the plant root mat in place, and

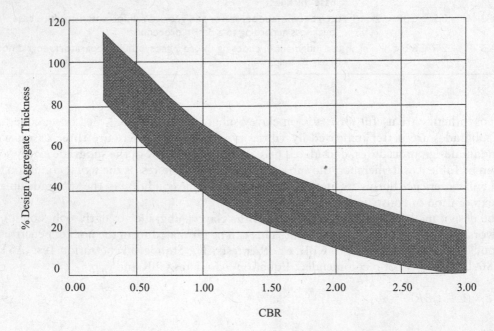

Figure 9.13 Amount of sacrificial aggregate needed to overcome anticipated loss of aggregate to a soft subgrade (FHWA 1989).

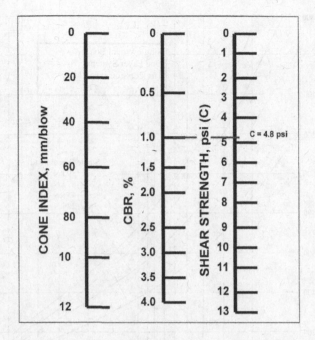

Figure 9.14 Method for estimating cohesion from cone index, or CBR (US Department of Defense 2004).

Table 9.7 Bearing capacity factors for different ruts and traffic conditions both with and without geotextile separators (after Steward et al. 1977)

	Ruts (inch)	Traffic (Passes of 18k Axle Equivalents)	Bearing Capacity (BC) Factor, N_c
Without geotextile	< 1	> 1000	c
	> 4	< 100	3.3
With geotextile	< 1	> 1000	5.0
	> 4	< 100	6.0

account for the expected settlement in design. Removing the root mat may cause the soils to become so soft that construction is hindered at great cost. After the subgrade has been prepared, the geotextile is rolled out by hand, and must not be driven on by any equipment. Since all geosynthetics break down in sunlight, do not leave the geotextile exposed for more than fourteen days before covering with soil. Improving drainage before construction will lengthen the life of the road.

Soil placement is critical. Soil must be placed adjacent to the geotextile and pushed onto the geotextile with low pressure (and, preferably, lightweight) equipment. If the equipment leaves more than 8 cm ruts, reduce the equipment weight and pressure. If the soil is end-dumped onto the geotextile, it is likely to tear the geotextile underneath the pile (out of sight), compromising the separation ability of the geotextile. Soil must not be end dumped on the geotextile. A first lift as thick as 0.6 m may be required with extremely soft sub-grades. If construction equipment leaves ruts deeper than approximately, say, 0.1 m, lower pressure equipment and a thicker first lift should be used. Large rut depths may indicate geo-synthetic rupture. The thicker lift will cause settlement in addition to that caused by the soft subgrade. If construction procedures are questionable, the contractor should construct an

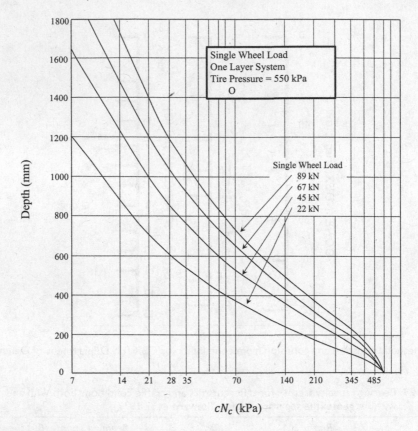

Figure 9.15 Single wheel load plot for determining required aggregate depth (USACE 2003).

on-site *test section* using the actual construction equipment and procedures, then exhume the geosynthetic and examine for damage.

For narrow roads, the width of one geotextile roll, and deploying the geotextile longitudinally, overlap the next geotextile at least 0.6 m in the direction of placement of soil at the end of the roll. At splices, it may be useful to pin the geotextile to the ground prior to soil placement to keep the geotextile from being folded by soil placement equipment. For CBR < 1 subgrades, sewing may be required. Care should be taken to reduce wrinkles. For wider roads, when deploying the geotextile laterally, use the same overlapping criteria and sew the geotextile strips together.

The procedure for placing soil on the geotextile is critical to geotextile survivability. An improper filling procedure may generate a mud wave large enough to rupture the geotextile The general lateral filling procedure is given in Figure 9.18. Occasionally, it may be necessary to wrap the edge of the geotextile around a timber or row of sandbags at the edge of the roadway to form a berm, to keep the initial lift from sliding off the edge of the geotextile. The longitudinal filling procedures for CBR < 1 and CBR > 1 subgrades are given in Figures 9.19 and 9.20. Figure 9.19 is an aerial view of bulldozer advancing soil over a geotextile over CBR < 1 soil. The soil is placed on existing soil (note the three piles waiting to be advanced), not on the geosynthetic. The bulldozer then pushes the soil forward onto the geotextile. Likewise, Figure 9.20 shows the same operation, adjusted for CBR > 1 soils.

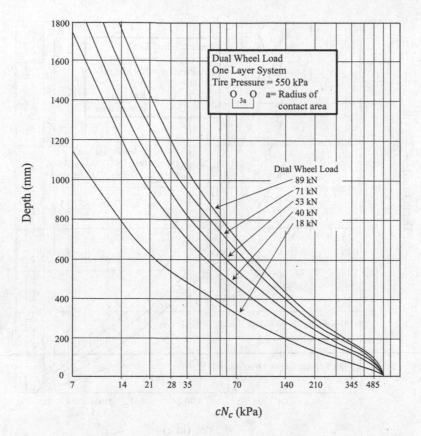

Figure 9.16 Dual wheel load plot for determining required aggregate depth (USACE 2003).

Finally, the following are construction tips. After the first lift, use thinner lifts, typically, 0.3 m to 0.45 m, to minimize mud waving. The mud wave must be kept lower than lift height. Do not place large stones against the geotextile. Consider these best practices: removing large stones that may puncture the geotextile, preventing vehicles from turning on the fill (the tractive force from turning may rupture the geotextile), using inclinometers adjacent to the toe of the embankment to monitor lateral heave, and stockpiling fill off the geosynthetic until the first or second lift is in place to minimize local depressions and reduce risk of geosynthetic rupture. Holtz et al. (1995) provide other advice and a sample specification.

9.4.3 Paved road improvement using geosynthetics

Geosynthetics can improve paved roads by providing separation of clayey subgrade soils from granular subbase and base materials and by adding tensile strength to subbase and base materials. The US Army Corps of Engineers procedure (USACE 2003) is provided. The procedure is summarized in Table 9.11. The properties of the geotextile separator are given in Table 9.9. The properties of the biaxial geogrids are given in Table 9.10.

The "Webster" plot (Figure 9.22) is used to reduce the thickness of the road section when a geogrid is incorporated. The designer compares the savings from a reduced section thickness with the increased cost of the geosynthetics and chooses the less expensive design.

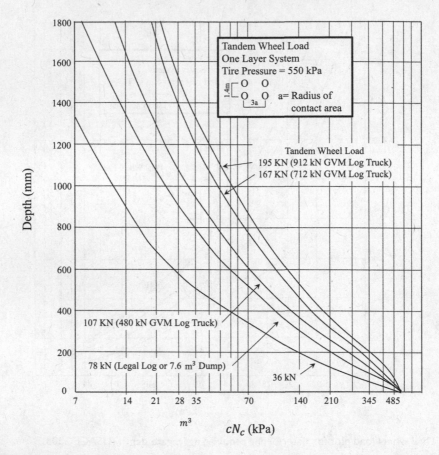

Figure 9.17 Tandem wheel gear weight plot for determining required aggregate depth (USACE 2003).

Table 9.8 Adjustments to base thickness based on axle repetitions (USACE 2003)

Number of repetitions	Amount to increase base thickness taken from Figures 9.15–9.17
2000	10%
5000	20%
heavy equipment	10%

Table 9.9 Typical minimum separation geotextile specification requirements[1] (USACE 2003)

Geotextile Property	ASTM Test Method	Minimum Requirement[2]
Grab strength (lb)	D4632	200
Puncture strength (lb)	D4833	80
Burst strength (psi)	D3786	250
Trapezoid tear (lb)	D4533	80
Apparent opening size (mm)	D 4751	< 0.43
Permittivity (sec^{-1})	D4491	0.05
Ultraviolet degradation (% retained strength @ 500 h)	D4355	50
Polymer type	–	Polyester or polypropylene

[1]This specification is for nonwoven geotextiles, which are recommended for typical separation applications.

[2]Minimum requirements apply to machine and cross-machine directions.

Table 9.10 Minimum biaxial geogrid specification requirements (USACE 2003)

Geogrid Property	ASTM Test Method	Minimum Requirement[1]
Mass per unit area (oz/yd^2)	D5261	9.0
Aperture size – machine direction (in)	Direct Measure	1.0
Aperture size – cross-machine direction (in)	Direct Measure	1.3
Wide width strip tensile strength (lb/ft)%:		
Strength at 5% strain – machine direction		700
Strength at 5% strain – cross-machine direction	D6637	1,200
Ultimate strength – machine direction		1,200
Ultimate strength – cross-machine direction		2,096
Manufacturing process	–	Punched and drawn

[1]Minimum requirements include both machine and cross-machine directions based upon Webster (1993).

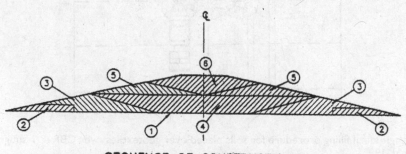

SEQUENCE OF CONSTRUCTION

1. LAY GEOSYNTHETIC IN CONTINUOUS TRAVERSE STRIPS, SEW STRIPS TOGETHER
2. END DUMP ACCESS ROADS
3. CONSTRUCT OUTSIDE SECTIONS TO ANCHOR GEOSYNTHETIC
4. CONSTRUCT INTERIOR SECTIONS TO "SET" GEOSYNTHETIC
5. CONSTRUCT INTERIOR SECTIONS TO TENSION GEOSYNTHETIC
6. CONSTRUCT FINAL CENTER SECTION

Figure 9.18 General lateral filling sequence for soils placed over geotextiles over very soft subgrades (Holtz et al. 1995).

The design steps for designing permanent roads using geosynthetic follow.

1. Use the subgrade CBR to estimate if the geotextile and/or geogrid are useful (see unpaved roads section 9.4.2).
2. If geosynthetics are useful, follow the steps listed below. If not, design the road without geosynthetics.
3. Steps for designing the road without reinforcement:
   a. Get the Road Classification, A to G, from Table 9.12.
   b. Get the Traffic Category, I to IV, from Table 9.13.
   c. Get the Design Index, 1 to 6, from Table 9.14.
   d. Use the subgrade CBR to get a total pavement thickness (base thickness plus asphalt concrete (AC) thickness) from Figure 9.21. This thickness meets the subgrade soil requirement for a good road.
   e. Use Table 9.15 to get the minimum asphalt concrete thickness and minimum base thickness, and the sum of these two numbers.
   f. Subtract the results of (e) from (d) to get the thickness of the base soil needed.

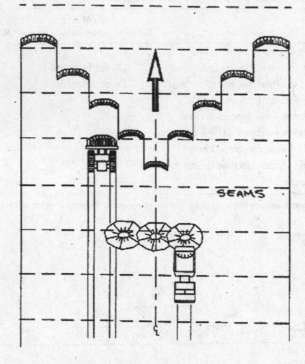

Figure 9.19 Longitudinal filling procedure for soils placed over geotextiles over CBR < 1 subgrades (Holtz et al. 1995).

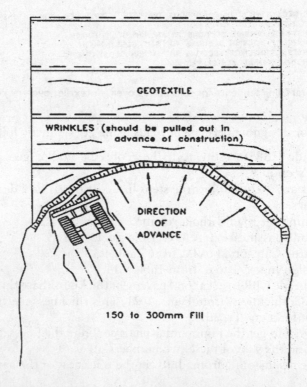

Figure 9.20 Longitudinal filling procedure for soils placed over geotextiles over CBR > 1 subgrades (Christopher et al. 2006).

Table 9.11 Summary of procedure to strengthen paved roads using geosynthetics (USACE 2003)

Case	Advice
CBR < 0.5	use the unpaved roads procedure, above, then use this unpaved road as a subgrade for the paved road design procedure
0.5 < CBR < 4	use a geotextile separator; use a biaxial geogrid; reduce section thickness using the Webster plot (Figure 9.22 Webster's Reinforced Pavement Thickness Design Plot) procedure
4 < CBR < 8	skip the geotextile separator; use biaxial geogrid for base reinforcement; reduce section thickness using the Webster plot (Figure 9.22 Webster's Reinforced Pavement Thickness Design Plot) procedure
CBR > 8	skip the geotextile separator, use a test section to see if geogrid is useful

Table 9.12 Criteria for selecting aggregate surface road class (USACE 2003)

Road Class	Number of Vehicles Per Day
A	10,000
B	8,400–10,000
C	6,300–8,400
D	2,100–6,300
E	210–2,100
F	70–210
G	Under 70

Table 9.13 Traffic composition; Choice of categories (USACE 2003).

Pavement Groups
Group 1. Passenger cars and panel and pickup trucks.
Group 2. Two-axle trucks.
Group 3. Three-, four-, and five-axle trucks.

Category I	Mostly passenger cars, panel and pickup trucks (Group 1 vehicles), and containing not more than 1% two-axle trucks (Group 2 vehicles).
Category II	Mostly passenger cars, panel and pickup trucks (Group 1 vehicles), and containing as much as 10% two-axle trucks (Group 2 vehicles). No trucks having three or more axles (Group 3 vehicles) are permitted in this category.
Category III	Traffic containing as much as 15% trucks, but with not more than 1% of the total traffic composed of trucks having three or more axles (Group 3 vehicles).
Category IV	Traffic containing as much as 25% trucks, but with not more than 10% of the total traffic composed of trucks having three or more axles (Group 3 vehicles).
Category IVA	Traffic containing more than 25% trucks or more than 10% trucks having three or more axles (Group 3 vehicles).

At this intermediate point, one can determine the asphalt concrete thickness from Table 9.15.

The total pavement section thickness is read from Figure 9.21. The base thickness can then be calculated as the difference between the total thickness (d) and asphalt concrete thickness (e).

The pavement thickness can be reduced using geogrid reinforcement in the pavement section. Given the thickness of the road without reinforcement (thickness of base plus AC), enter the ordinate of Webster's plot (Figure 9.22) and read the reduced thickness from the abscissa.

Table 9.14 Design index for pneumatic-tired vehicles (USACE 2003)

	Design Index			
Class	Category I	Category II	Category III	Category IV
A	3	4	5	6
B	3	4	5	6
C	3	4	4	6
D	2	3	4	5
E	1	2	3	4
F	1	1	2	3
G	1	1	1	2

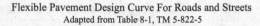

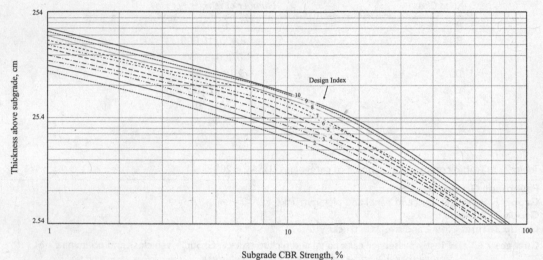

Figure 9.21 Flexible pavement design curves for roads and streets (adapted from Table 8.1 of TM 5-822-5) (USACE 2003).

To use the Webster plot note:

1. Geogrid properties are taken from Table 9.10.
2. Do not reduce the minimum AC thickness determined above.
3. Only use this plot if asphalt concrete thickness is < 8 cm.
4. If base is thicker than 36 cm, put the geogrid in the middle of the base soil.
5. If base is thinner than 36 cm, put the geogrid at the bottom of the base soil.

When CBR ≥ 80 base course material is scarce, or prohibitively expensive, the base course may be subdivided into base course material underlain by less-expensive, weaker subbase material. The US Army Corps of Engineers (USACE 2003) provides a modification to the above design procedure to include subbase materials.

9.4.4 Geofibers in roads

Geofibers, usually in the form of staples, may be used to strengthen soils in road construction. Geofibers are typically 8 cm long, loose plastic fibers mixed with soil. The effect is

Table 9.15 Minimum Pavement Layer Thicknesses (USACE 2003) (extracted from Table 6.1 of USA CoE TM 5-822-5, USACE 2003)[1]

Design Index	Base CBR = 100			Base CBR = 80			Base CBR = 50[2]		
	AC Pavement (inch)	Base (inch)	Total (inch)	AC Pavement (inch)	Base (inch)	Total (inch)	AC Pavement (inch)	Base (inch)	Total (inch)
1	ST[3]	4	4.5[5]	MST[4]	4	4.5[5]	2	4	6
2	MST[4]	4	5[5]	1.5	4	5.5[5]	2.5	4	6.5
3	1.5	4	5.5[5]	1.5	4	5.5[5]	2.5	4	6.5
4	1.5	4	6	2	4	6	3	4	7
5	2	4	6.5	2.5	4	6.5	3.5	4	7.5
6	2.5	4	7	3	4	7	4	4	8
7	2.5	4	7	3	4	7	4	4	8
8	3	4	7.5	3.5	4	7.5	4.5	4	8.5
9	3	4	7.5	3.5	4	7.5	4.5	4	8.5
10	3.5	4	8	4	4	8	5	4	9

[1] Table 6.1 extracted from TM 5-822-5, Chapter 6.
[2] In general, 50 CBR Base Courses are only used for road classes E and F.
[3] Bituminous surface treatment (spray application).
[4] Multiple bituminous surface treatments.
[5] Minimum total pavement thickness for road classes A through D is 6 inches.

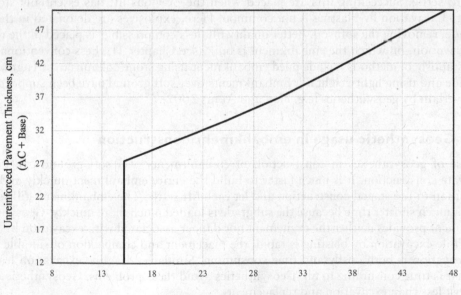

Figure 9.22 Webster's reinforced pavement thickness equivalency chart (USACE 2003).

to strengthen the soil, reduce settlement, and increase permeability. They are the only soil improvement that does all three favorable property changes. Their implementation in road construction is the same as in underfooting reinforcement, described earlier.

9.5 EMBANKMENTS OVER SOFT GROUND

9.5.1 Introduction

Embankments over soft ground refer to the placement of soil over very soft ground. This construction must address several failure modes including bearing capacity, settlement, and slope stability. Geosynthetics can reduce the time to build these embankments, compared to conventional procedures.

Embankments over soft ground are used for road construction and surcharges. Surcharges, here, refer to piles of earth used to compress subgrade soils in advance of construction. The process is, in a nutshell, to first place a pile of soil on the site where a building (or other structure) is to be built. The pile weighs more than the building. Allow the pile to compress the subgrade. When the compression stops, or gets to a calculated amount, remove the pile and replace it with the lighter building. The building will experience minimal recompression settlement because the subgrade has already been compressed by a weight greater than the building. See Chapter 5, "Consolidation," for more information.

9.5.2 Conventional construction of embankments

In the past, embankments over soft ground were constructed by one of two methods: staged construction or excavation of soft subgrade by several methods, including blasting. Building the embankment slowly, in lifts, is termed staged construction. Each lift causes the subgrade to settle. Succeeding lifts are placed when the previous lift has essentially stopped settling. Excavation by blasting is not common. Here, explosives are detonated in the soft subgrade, removing the soft soil. Better quality fill (less compressible) is placed in the resulting excavation, on which the embankment is built (see Chapter 11). Less conventional, but well-established, methods include piled embankment using stone columns to strengthen the subgrade and using lightweight fill. Embankments over soft ground have been supported by piles overlain by geosynthetics (e.g. Han and Akins 2002).

9.5.3 Geosynthetic usage in embankment construction

The use of geosynthetics in construction of embankments over soft ground can greatly accelerate construction. It is much faster to build the entire embankment quickly and let it settle, than to use staged construction and let each lift settle. The embankment still settles, but in a much shorter time because the subgrade is loaded much more quickly. Geosynthetic deployment provides neither the excitement nor danger associated with excavation by blasting. While excavation by blasting is rapid, the placement and compaction of suitable fill in the excavation is both costly and time consuming. Similarly, dragline excavation leaves a hole that is time consuming to fill. Geosynthetics avoid these problems. Geosynthetic materials cost less than excavation and replacement.

The cost of geosynthetics is the primary disadvantage, although this is a tiny percentage of the project cost. The possibility of puncturing the geosynthetic on stumps and debris, negating the separation effect, and contractor unfamiliarity are other disadvantages.

9.5.4 Design procedure

Embankments over soft ground must be designed against these failure modes:

1. slope stability,
2. geotextile tensile failure,
3. sliding of soil on top of geosynthetic,
4. geosynthetic rupture due to sliding,
5. pullout of the geosynthetic,
6. bearing capacity,
7. settlement.

Embankments over soft ground are constructed of soils and geosynthetics. The needed soil properties include shear strength parameters, unit weights, and consolidation characteristics. The needed geosynthetic properties include wide-width tensile strength, seam strength (if seamed), interface friction angle(s), apparent opening size (if a geotextile), and permittivity (if a geotextile). The location of the water table is needed.

The geosynthetic aspects of this design will be covered herein. Methods of obtaining soil shear strength parameters are given in many geotechnical engineering textbooks (e.g. Holtz et al. 2011; Coduto 1999). Rowe and Li (2005) discuss creep in embankments over soft ground.

The critical conditions for this construction are during and immediately after construction. As time passes, the subgrade soils increase in strength. Typically, the *total stress method* is used to evaluate soil strength parameters (Holtz 1989; Rowe and Li 2005).

9.5.4.1 Slope stability

Slope stability of the final embankment cross section is addressed with software. Input soil properties are of the utmost importance in slope stability calculations. Therefore, time and effort must be expended to evaluate the subgrade strength conditions and evaluate the fill soil strength. In this scenario, the subsoil conditions are most likely clay. Undrained strength parameters are often used (Rowe and Li 2005). Modern two-dimensional slope stability software allows the input of a layer of geosynthetic with a user-defined tensile strength, given in force/unit length (e.g. kN/m). Having chosen soil strength parameters, the software analyzes the stability of the slope. If the factor of safety is unacceptable, the engineer may change the slope geometry, the fill material properties, or the strength of the reinforcing geosynthetic. Presuming the fill material and the slope geometry are constrained by the project, the geosynthetic strength is increased until an acceptable factor of safety is reached. This is the geosynthetic strength required for slope stability, $T_{\text{required for slope stability}}$. A factor of safety may be applied to this.

$$T_{\text{adjusted}} = (T_{\text{required for slope stability}}) \, (FS) \tag{9.6}$$

Once a trial geotextile with some ultimate strength, T_{ult}, is selected to specify, its strength is reduced by reduction factors (RF) to get the long-term design strength.

$$LTDS = \frac{T_{\text{ult}}}{RF} \tag{9.7}$$

The reduction factors are given in Koerner (2012) to account for installation damage, chemical/biological degradation, creep, and other nonquantifiable factors so deemed by the designer. The LTDS must be equal to or greater than $T_{adjusted}$. This assures adequate strength against geosynthetic tensile failure.

9.5.4.2 Sliding of soil on top of geosynthetic

Another failure mode to consider addresses the potential for the soil to slide off the geosynthetic. The interface friction between the soil and geosynthetic is lower than the friction between the soil layers. This failure mode, illustrated in Figure 9.23, shows the wedge of soil sliding off the geosynthetic, as embankment soil pushed to the left. The analysis is done by assuming a Rankine lateral earth force is pushing the given wedge of soil across the geosynthetic beneath the wedge.

The driving force, P_A, is given by

$$P_A = \frac{1}{2}\gamma H^2 K_A \tag{9.8}$$

where γ is the unit weight of the driving soil, H is the embankment height, and K_A is the Rankine active lateral earth pressure coefficient for horizontal ground given as:

$$K_A = \tan^2\left(45 - \frac{\phi}{2}\right) \tag{9.9}$$

where
 ϕ is the embankment soil's angle of internal friction.

The resisting force on the upside of the geosynthetic, R_u, is given by

$$R_u = (c_a + \sigma \tan\delta)(L) \tag{9.10}$$

where
 c_a is the embankment soil adhesion on the geosynthetic,
 σ is the average stress on the geotextile under the sliding wedge,
 δ is the embankment soil's angle of interface friction with the geosynthetic, and
 L is the horizontal length of the sliding wedge.

The factor of safety against sliding of soil on top of geosynthetic is

$$FS = \frac{R_u}{P_A} \tag{9.11}$$

9.5.4.3 Geosynthetic rupture due to sliding

The soil sliding over the geosynthetic may cause geosynthetic rupture in addition to the possibility of rupture due to slope stability. The analysis is similar to the sliding analysis above. Here, the resisting to sliding force on the lower side of the geosynthetic is calculated as (if the soil is strictly cohesive) the product of the adhesion (c_a) times the length L of the sliding surface

$$R_l = c_a L \tag{9.12}$$

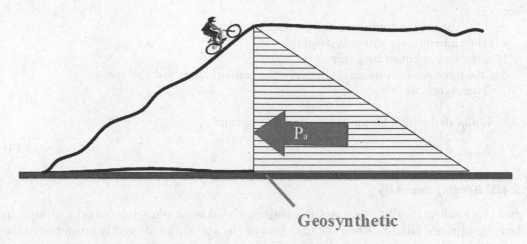

Figure 9.23 Sliding of embankment on geosynthetic.

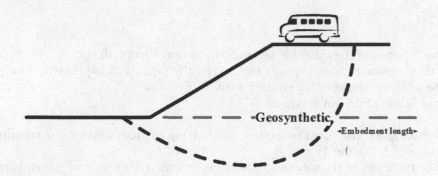

Figure 9.24 Pullout of geosynthetic from embankment.

With the resistances R_u and R_l known, note which is smaller. The smaller one is the one where the failure surface will form. The smaller resistance is then subtracted from P_a. The geosynthetic must have a wide-width tensile strength greater than this difference. The factor of safety is the ratio of the geosynthetic wide-width strength divided by the difference noted above.

9.5.4.4 *Pullout of the geosynthetic*

This failure mode is shown in Figure 9.24. The length of geosynthetic to the right of the failure surface (determined from computer analysis) must be long enough to develop enough force to keep the geosynthetic from pullout to the left.

The resistance to pullout is based on the soil resistance above and below the geosynthetic in the embedment length, L_e, zone (shown in Figure 9.24) and the tensile force in the geosynthetic ($T_{adjusted}$). Typically, the soil above the geosynthetic is frictional and the subgrade is cohesive. Given this is the case, the pullout resistance (right side of equation 9.10) must be greater than $T_{adjusted}$ from the slope stability analysis.

$$T_{adjusted} < \text{lower resistance} + \text{upper resistance} = c_a L_e + (\gamma H \tan \delta \ ) L_e \qquad (9.13)$$

where
  c_a is the subgrade adhesion to the geotextile,
  γ is the embankment soil unit weight,
  H is the embankment height, and
  δ is the interface friction angle between the embankment soil and the
    geosynthetic.

L_e is adjusted with a FS to get an allowable L_e thus

$$L_{e\,\text{allowable}} = (FS)L_e \tag{9.14}$$

9.5.4.5 Bearing capacity

Figure 9.25 schematically illustrates the entire embankment plunging into the ground in a bearing capacity failure. When the thickness of the soft subgrade soil is greater than the height of the embankment, the factor of safety can be approximated from

$$FS = \frac{cN_c}{\gamma H} \tag{9.15}$$

where
  c is the cohesion of the subgrade (assumed to be non-frictional)
  N_c is the cohesion bearing capacity factor typically taken as 5.14, for strip footings
  γ is the soil unit weight of the embankment
  H is the height of the embankment

  If the subgrade has frictional properties, the bearing capacity equation is substituted in the numerator (e.g. Coduto 1999).
  When the thickness of the soft subgrade soil is less than the height of the embankment, the predominant failure mode is soil squeeze. This is addressed by Sowers and Sally (1962), Silvestri (1983), and FHWA (2008).

9.5.4.6 Settlement

The embankment will settle due to internal compression and compression of the subgrade. Settlement calculations are described in Duncan and Buchignani (1987) and in many geotechnical texts. However, the charts for calculating the increase in stress in the subgrade caused by sloping ground (e.g. embankments over soft ground) are given below in Figures 9.26 and 9.27, as they are harder to locate in the open literature. Here, the nondimensional, normalized

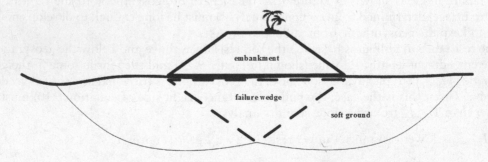

Figure 9.25 Bearing capacity failure of embankment.

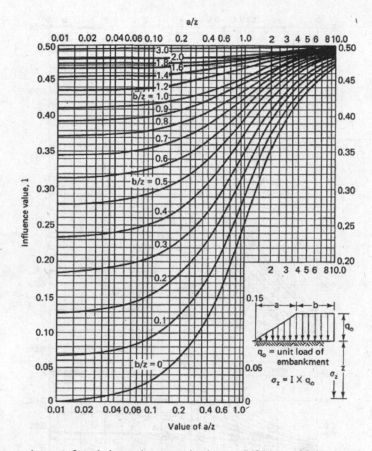

Figure 9.26 Stress under an infinitely long, sloping embankment (US Navy 1986).

figures on the plots are used to determine an Influence Factor (I), which is then used to calculate the stress at depth, using the equations on the figures.

Elastic, primary, and secondary settlement must be calculated. FHWA (1993) provides embankments over soft ground settlement calculation software. Chin and Sew (2000) and Vipulanandan et al. (2009) provide extensive guidance on settlement calculations for embankments over soft ground.

9.5.4.7 Additional checks

The settlement of adjacent structures must be considered. The engineer is reminded that surface and subsurface structures adjacent to the embankment will settle as the embankment settles. Yourman et al. (2006) suggest a method of minimizing this effect.

The geosynthetic must meet the survivability constraints in AASHTO M288-17, regardless of any values computed above. If a geotextile is chosen, it must have a permittivity adequate to provide drainage and must not clog. Geotextile filtration design is given in Chapter 3.

9.5.5 Instrumentation

Instrumentation can reduce uncertainty in settlement and stability of constructed embankments over soft ground. Settlement plates to measure subgrade compression and compression rates are both inexpensive and greatly improve the estimations of the time rate of

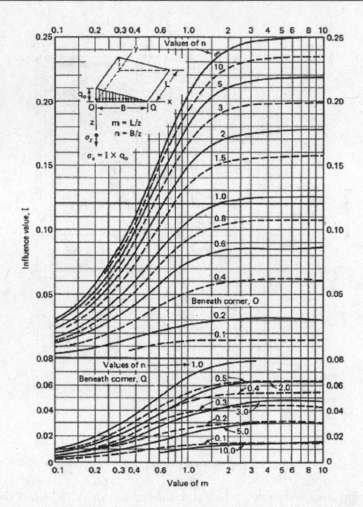

Figure 9.27 Stress under the corners of a sloping embankment of limited length (US Navy 1986).

settlement. Initial predictions are based on approximations or on testing of very small samples that may not model the site very well. During embankment construction, settlement monitoring of what is effectively a full-scale field test allows back calculation of the soil parameters controlling the amount and rate of settlement. This greatly improves total and time rate of settlement predictions.

Pore pressure gages in the subgrade assist in controlling the rate of construction. The measured pore pressure is compared to the pore pressures used in the slope stability analysis. The rate of filling is regulated to keep the measured pore pressures at or below that required for stability. When pore pressures approach the critical value, filling is halted until pore pressures dissipate to a safe value. Alternatively, the measured pore pressure may be used to calculate the pore pressure ratio (r_u), that is, the ratio of pore pressure to total overburden stress. The critical r_u is determined from slope stability analysis. When r_u approaches the critical value, filling is halted until r_u drops to a safe value.

Organics (which decompose to gases) or existing trapped gases in the soil may inflate pore pressure readings and may not drop off as consolidation occurs, as it is difficult for small bodies of trapped gases to escape through groundwater.

The successful use of pore pressure gages lies in the proper location and reading of the gages. Critical locations, based on the slope stability analysis and on experience, must be identified and instrumented. Typically, these locations are on the predicted critical failure surface, at the edges of the embankment, or at discontinuities in the soil profile. The gages must be read frequently and faithfully. Redundancy is recommended.

Surface survey monuments both on and off the embankment, inclinometers, and LiDAR measurements are used to monitor squeeze and settlement. Dunnicliff (1993, 1998) provides instrumentation guidance.

9.5.6 Construction guidance

Construction processes greatly influence the success of embankment over soft ground projects. Proper construction sequencing and instrumentation can decrease the opportunity for failures. The deployment of the geosynthetics and filling procedures, described in section 9.4 "Road Base Stabilization," should be used. Riordan and Seaman (1993) provide other construction advice, including a flowchart to help the designer choose the construction method.

9.5.7 Alternative procedures

Embankments over soft ground may also be constructed by removing and replacing the soft soils under the embankment with stronger, less compressible soil (overexcavation), installing piles with a layer of geosynthetic reinforcement across the piles under the embankment (Han and Akins 2002; Han and Gabr 2002), using a geocell mattress (Bush et al. 1990), flattening or benching the side slopes, or installing columns (e.g. deep soil mixing) to strengthen the subgrade and/or support the embankment. Strip drains, described in Chapter 5, accelerate settlement by allowing groundwater to escape much more quickly. Chapter 7.6 of Christopher et al. (2006) has advice on improving subgrades under roads and embankments. Han et al. (2010) discuss using geocells to strengthen road bases over soft ground, while Madhavi Latha (2011) and Bush et al. (1990) present design procedures for embankments over soft ground using geocells.

9.6 UNDERFOOTING REINFORCEMENT WITH ROLLED GEOSYNTHETICS

9.6.1 Introduction

The bearing capacity of soils underneath spread footings can be increased as much as fourfold by embedding geosynthetic sheets in the soil beneath the footing. In addition, settlement is reduced. The geosynthetic adds tensile strength to the soil. Hence, column loads that might have required deep foundations may now be able to be constructed on much less expensive spread footings. The dramatic improvement in soil strength is easily shown in a simple laboratory experiment, described by Elton (2001). Figure 9.4 is a schematic of underfooting reinforcement using rolled geosynthetics.

Underfooting reinforcement's primary advantage is to save money by replacing expensive deep foundations with less expensive shallow foundations. Additional savings accrue from the speed of construction. The primary disadvantages are the limited experience with the technology (2012) and the uncertainty associated with the amount of settlement. Some contractors may be unfamiliar with geosynthetic installation. Underfooting reinforcement is more economical when installing a fill than installing below grade, which requires soil excavation, then replacement and compaction of the soil with interbedded geosynthetics.

9.6.2 Design procedure

The analysis presented here, and in the literature, is for shallow foundations on horizontal ground, on cohesionless soils (no pore pressure development during shear), loaded statically. The materials are soil and a geosynthetic capable of taking tension. The soil properties needed include effective angles of internal and external (soil/geosynthetic) friction, and soil unit weight. Procedures for the evaluation of these properties are found in geotechnical books (e.g. Holtz et al. 2011; Bowles 1996; Koerner 2012).

The required geosynthetic properties are tensile strength, seam/junction strength, crushing strength (if using geocells) and durability characteristics. In theory, any material with tensile strength will suffice. In practice, geogrids are used almost exclusively because of their availability, cost, large apertures (allowing soil-to-soil interaction and good drainage), and high modulus compared to other geosynthetics. Evaluation of geosynthetic properties is discussed in Chapter 3.

Design procedures for underfooting reinforcement largely come from laboratory studies. Kumar et al. (2007), Kumar and Saran (2003), Yetimoglu et al. (1994), Huang and Tatsuoka (1990), Fragaszy and Lawton (1984), Adams and Collin (1997), Shin et al. (2002), and Das (2009) are among many authors proposing procedures. Most focus on sandy soils, with variation based on the shape of footing and soil layering. The most investigated case is a footing on uniform sandy soil with geogrid reinforcement. All of the procedures to date (2012) are based on laboratory studies and are largely empirical. The procedure below draws primarily from Das (2009) and Wayne et al. (1998). The reader is reminded that this discussion applies to free-draining soils (no soil pore pressure development during loading).

The parameters of interest are in Figure 9.4 and include:

1. depth to first geosynthetic layer, u;
2. depth of deepest layer from the base of the footing, d;
3. footing dimensions B, L in plan (B is the shorter dimension);
4. width of geosynthetic in plan, b (in the B direction of the footing);
5. length of geosynthetic in plan, l (perpendicular to b);
6. vertical spacing of geosynthetic layers, h;
7. number of layers, N;
8. tensile strength of reinforcement.

The improvement in bearing capacity is quantified by the Bearing Capacity Ratio, BCR, thus:

$$BCR = \frac{\text{Improved Bearing Capacity}}{\text{Original Bearing Capacity}} \tag{9.16}$$

Much laboratory research (e.g. Binquet and Lee 1975) has proved the depth to the first layer of reinforcement is important. If the layer is too close to the footing base (u is very small), the upper layer(s) of reinforcement are of little effect. However, if u is too large, the bearing capacity failure occurs above the first layer, negating the effect of the reinforcement. The depth of the first layer of reinforcement below the footing, u, is a function of the footing width, B. Wayne et al. (1998) cite evidence that, for a given bearing capacity ratio, u/B increases with increasing effective angle of internal friction, ϕ. For ϕ greater than 30°, a u/B less than 0.66 will keep the bearing capacity failure from forming above the topmost layer of reinforcement. Increasing the effective angle of internal friction increases the bearing capacity ratio for a given u/B. For footings on sands, on the ground surface, various laboratory experiments suggest optimal values of u/B range from 0.25 to 0.5.

In like manner, the depth of reinforcement, d, affects the improvement in bearing capacity. If d is too shallow, the bearing capacity failure surface penetrates the reinforcement, while, if d is too deep, the lower layers of reinforcement do not affect the improvement in bearing capacity. Omar et al. (1993) recommend the following:

$$\left(\frac{d}{B}\right)_{cr} = 2 - 1.4\left(\frac{B}{L}\right) \text{ for } 0 \le \frac{B}{L} \le 0.5 \tag{9.17}$$

$$\left(\frac{d}{B}\right)_{cr} = 1.43 - 0.26\left(\frac{B}{L}\right) \text{ for } 0.5 \le \frac{B}{L} \le 1 \tag{9.18}$$

where the subscript *cr* stands for the optimal case. These equations suggest the best d/b ratio for square foundations is 1.2B, and about 2B for strip foundations.

The reinforcement width, b, can be estimated from

$$\left(\frac{b}{B}\right)_{cr} = 8 - 3.5\left(\frac{B}{L}\right)^{0.51} \tag{9.19}$$

About 80% of the increase occurs when b/B reaches 2.

The reinforcement length, l, can be estimated from:

$$\left(\frac{1}{B}\right)_{cr} = 3.5\left(\frac{B}{L}\right) + \left(\frac{L}{B}\right) \tag{9.20}$$

The vertical geosynthetic spacing, h, should be less than 0.4B. Typically, four to six layers of reinforcement are used. More layers do not improve the BCR substantially. Wayne et al. (1998) recommend dimensions given in Table 9.16. These tend to be more conservative than those given by Das (2009), cited above. Note the additional criteria of "distance to subgrade from bottom geosynthetic layer."

Table 9.16 Design parameters for underfooting reinforcement using punched and drawn biaxial geogrids (after Table 4, Wayne et al. 1998)

Parameter		Typical Value	Recommended Values Less Than:
u	distance from bottom of footing to top geosynthetic layer	0.15B–0.3B	0.5B
h	spacing between geosynthetic layers	0.15B–0.3B	0.5B
z	distance between top and bottom geosynthetic layers	0.5B–1.0B	2.0B
b	width of geosynthetic layer	2B–3B	4.0B
a	distance to subgrade from bottom geosynthetic layer	0.1B–0.2B	0.3B
l	length of geosynthetic	L+(1.0B–2.0B)	L+4B
N	number of layers	2–4	5

There are two strategies to calculate the bearing capacity for these footings. One is to calculate the ultimate bearing capacity for the unreinforced footing using conventional soil mechanics, and then use a chart of the ultimate BCR_{ult} vs. d/B (Omar et al. 1993) or u/B (Shin and Das 2000) to estimate the reinforced bearing capacity. Given the BCR from the chart, and q_{ult} from calculation, $q_{ult\ reinforced}$ is calculated.

Settlement, rather than bearing capacity, usually governs footing design. Allowable bearing capacity, based on settlement, is addressed by Das (2009). Here, the $BCR_{allowable\ settlement}$ / BCR_{ult} vs. D_f/B is plotted, and $q_{ult\ reinforced}$ is calculated as above. D_f is the distance from the ground surface to the base of the footing.

The second strategy is to calculate an allowable reinforced bearing capacity directly, as done by Wayne et al. (1998). Their equation, developed to account for acceptable settlement as well as bearing capacity, includes two parameters that account for movement of the geosynthetic reinforcement laterally and vertically. They note that underfooting reinforcement reduces settlement. Huang and Tatsuoka (1990) and Huang and Menq (1997) provide equations for special cases of underfooting reinforcement in sand.

Das (2009) provides the following tentative guidelines and equation for the ultimate bearing capacity of footings underlain by reinforcement and on *sand*:

a. u/B should be between 0.25 and 0.33 (see Figure 9.4)
b. h/B should not exceed 0.4
c. b/B should be between 2 and 3
d. N < 5

$$q_{ult\ reinforced} = \left[0.5 - 0.1\left(\frac{B}{L}\right)\right](B + 2d\tan\beta)\gamma N_\gamma + \gamma\left(D_f + d\right)N_q \tag{9.21}$$

where
N_q and N_γ are shallow bearing capacity factors
γ is the unit weight of the soil under the reinforced zone

β is a measure of the projection of the stress from the footing, through the reinforced zone, into the soil below. It can be estimated from

$$\beta = tan^{-1}\left[0.68 - 2.071\left(\frac{h}{B}\right) + 0.743(CR) + 0.03\left(\frac{b}{B}\right)\right] \tag{9.22}$$

where CR is the coverage ratio calculated as the width of a reinforcing strip divided by the horizontal spacing between the strips. For a geotextile (complete coverage), the ratio is one. For metal strips and geogrids, it is less than one. All other terms are defined previously. Das (2009) describes a procedure for reducing $q_{ult\ reinforced}$ to account for settlement.

Other cases have been examined by researchers. Kumar et al. (2007) describe the improvement in bearing capacity when reinforcement is added to a stronger sand layer overlying a weaker sand layer. A three- to fourfold increase in bearing capacity is reported when the reinforcement is added to the upper stronger soil layer. Das (2009) describes a method of estimating the bearing capacity of a footing with underfooting reinforcement subject to cyclic loads, and for eccentrically loaded footings. Mosallanezhada and Nasirib (2015) have research on underfooting reinforcement in clayey soils.

Settlement of footings underlain by reinforcement is reduced from the unreinforced case, often by 50% in laboratory studies. The reinforced zone acts to distribute the footing load over a larger area, reducing stress, and reducing settlement. Wayne et al. (1998) and Das (2009) discuss procedures to estimate settlements. Studies by Akins (2012) indicate that geogrid flexibility enhances the performance by reducing the amount of settlement. Stiffer geogrids tend to cause punching failures, or excessive settlement.

Biaxial geogrids are the geosynthetic of choice for underfooting reinforcement because they have strength in two directions. Typical maximum tensile strengths are 1.36 to 1.90 kN/m. There are no cases of geogrid rupture reported in the literature (2012).

9.6.3 Construction

Construction procedures are critical to success. The subgrade soil must be compacted then covered with a thin, compacted layer of the soil to be placed in the reinforced soil zone. The geogrid must be in good contact with soil. While it is not necessary to tension the reinforcement, it is best practice to remove wrinkles. The geogrid must remain undamaged during placement. Care must be taken to keep construction equipment off the geogrid to avoid damage. Soil should be end dumped or placed only on soil, never on the geogrid. The new soil is then pushed onto the geogrid with lightweight equipment and compacted to about 95% of standard Proctor density (ASTM D698 2012).

The reinforced soil should be granular. Particles must be less than 5 cm in diameter to effect good compaction. The angle of internal friction, ϕ', should be greater than 30°. If gravel is used, it must be well graded.

9.7 UNDERFOOTING REINFORCEMENT WITH GEOCELLS

9.7.1 Introduction

Underfooting reinforcement can be performed with any material that adds tensile strength to the soil. While planar reinforcement is most common (2012), geocells have been used for underfooting reinforcement. Laboratory studies of geocells under shallow foundations show they increase bearing capacity significantly and decrease settlement. Geocells are described in Chapter 3.

Geocells, filled with soil, improve bearing capacity by adding strength to the soil. The geocell confines the soil, keeping a failure surface from forming in the soil, greatly increasing the bearing capacity of the composite soil mass. Geocell placement under a shallow footing may have the effect of increasing the footing depth of embedment, which increases bearing capacity. Several authors (e.g. Rajagopal et al. 1999; Shukla et al. 2009) have found the increase in strength, in laboratory studies, may be characterized by an artificial cohesion value and an enhanced angle of internal friction. Other researchers have found the inclusion of any underfooting reinforcement distributes footing load over a broader area, reducing stress and settlement (e.g. Wayne et al. 1998; Sitharam and Sireesh 2006).

The US Army Corps of Engineers (Webster and Watkins 1977; Webster 1979) did pioneering work on the development of geocells, used for strengthening roadways. It appears Bush et al. (1990) did pioneering work on analysis of underfooting reinforcement using geocells. They used a network of upended geogrids, connected in a cellular pattern, about 1 m deep.

Sitharam and Sireesh (2006) found from laboratory studies that adding a planar layer of reinforcement below the geocells reduced heave and increased bearing capacity, but only when the base of the geocells was within about twice the footing width. This construction above clay subgrades generated only about 20% improvement in bearing capacity, whereas the same construction above cohesionless subgrades generated over 100% improvement.

Moghaddas Tarreshi and Dawson (2010) in laboratory studies found geocells are more efficient, in terms of the weight of polymer used (which reflects cost), than planar geosynthetics in increasing bearing capacity and reducing settlement. They indicate geocells reduce surface heave more than planar reinforcement.

9.7.2 Ultimate load calculation

Koerner (2012) proposed a procedure for calculating the ultimate load sand infilled geocells can support based on the standard bearing capacity equation. An additional term accounts for strength contributed by the geocell. That procedure is modified here.

$$P_{ult} = 2Tt + (B)\left[cN_cS_c + \gamma D_f N_q S_q + 0.5\gamma B N_\gamma S_\gamma\right] \tag{9.23}$$

where

P_{ult} is the ultimate force per unit length that can be placed on the geocell before the subgrade fails

T is the shear stress generated on two walls of the geocell

t is the vertical geocell dimension

B is the width of the applied load

c is subgrade soil cohesion

N_c, N_q, N_γ are bearing capacity factors, found in standard geotechnical texts

S_c, S_q, S_γ are footing shape factors, to account for footing shape (square, rectangular, or strip) found in standard geotechnical texts

The T term is derived from assuming the soil in the geocell shears downward under load. It is calculated from the frictional resistance of soil sliding inside the geocells.

$$T = p\tan^2\left(45 - \frac{\phi}{2}\right)\tan\delta \tag{9.24}$$

where

p is pressure on top of the geocell caused by, say, a footing

δ is the interface friction angle between the infill soil and the geocell

ϕ is the angle of internal friction of the infill soil

This procedure can be used to estimate a factor of safety, FS, of a footing underlain by a geocell thus:

$$FS = \frac{P_{ult}}{P_B} \tag{9.25}$$

where

P_{ult} is the ultimate force per unit length that can be placed on the geocell (capacity)

P_B is the applied force per unit length (demand)

The geocell should extend $\geq 2B$ from the footing edge.

9.7.3 State of practice

No studies have scaled geocell placement under footings to field conditions (Biswas and Krishna 2017). The technology looks hopeful. Dash et al. (2004) found in the laboratory that geocells were more effective than planar geosynthetics for underfooting reinforcement. At very large settlements, exceeding 45% of the laboratory footing width, geocells still didn't fail. These model footings on geocells gave an eightfold increase in bearing capacity.

There are various methods for designing geocells for underfooting reinforcement. The principle is accepted, based on numerous laboratory studies showing increase in bearing capacity and decrease in settlement when geocells are placed under shallow footings (e.g. Han et al. 2010; Dash et al. 2001; Latha and Murthy 2007).

9.7.4 Construction advice

Geocell infill soils should be well compacted. Poor compaction leads to lower bearing capacity and greater settlement and damage to the geocells. Well-graded gravel, while providing excellent strength, may be difficult to compact, while uniform sand, though easy to compact, does not provide as much strength. Compaction is achieved by overfilling the geocells and compacting the cohesionless soil or gravel with a vibrating plate compactor, being careful not to damage the geocells. The geocell should be placed about 0.1B below the footing, where B is the width of the footing.

A geotextile separator should be placed beneath geocells overlying clayey or silty subgrades, to prevent migration of fines upwards into cohesionless infill soil, weakening it. Geotextile separation design is covered in section 9.4.2.

> **Example problem Ex.9.1: Geocell design**
>
> A 2.44 m square footing is planned for a 600 kN column load on a soft, clayey subgrade. The subgrade properties are: $c = 14.4$ kPa, and $\gamma = 18$ kN/m³. The weak subgrade suggested using a 0.2 m thick geocell under the footing. The geocell will be filled with compacted, coarse, $\phi = 30°$ sand which has an interface friction angle of $\delta = 22°$ with the geocell. The footing will rest on a thin layer of sand over the geocell. Comment on the factor of safety of this design.
>
> **Solution:**
>
> Using Koerner's method, calculate
>
> $$T = p \tan^2\left(45 - \frac{\phi}{2}\right)\tan\delta$$
>
> $$T = \left(\frac{600}{(2.4)(2.4)}\right)\left(\tan^2 45 - \frac{30}{2}\right)(\tan 22) = 13.6$$
>
> The Vesic bearing capacity factors for the subgrade ($\phi = 0$) are
>
> $$N_c = 5.14, N_q = 1, \text{ and } N_g = 0$$
>
> The Vesic shape factors are
>
> $$S_c = 1 + (N_q / N_c) = 1 + (1/5.14) = 1.19$$
> $$S_q = 1 + \tan\phi = 1 + \tan(0) = 1$$
> $$S_\gamma = 0.6$$

Calculate the ultimate force per unit length

$$P_{ult} = 2Tt + (B)\left[cN_cS_c + \gamma D_f N_q S_q + 0.5\gamma B N_\gamma S_\gamma\right]$$

$P_{ult} = 2(13.6)(0.2) + 2.4((14.4)(5.14)(1.19) + 18(0)(1)(1) + 0.5\ (18)(2.4)(0)(0.6)) = 216.8$ kN/m
   The capacity of the 2.4 m footing is = 216.8 (2.4) = 529.9 kN
   The factor of safety (FS) is

   FS = capacity/demand = 529.9/600 = 0.89

This is unacceptably low. The footing size should be increased, or the depth of the geocell increased, or a combination of these.

9.8 UNDERFOOTING REINFORCEMENT WITH GEOFIBERS

9.8.1 Introduction

Geofibers (Figure 9.1) improve soil properties including strength, reduced compressibility, and increased permeability. Geofibers, also known as staples, may be very long, or may be short, plastic fibers. Staples are typically five cm long, often fibrillated, and are mixed with soil before compaction. They are described in Chapter 3.

Staples are the most commonly used geofiber (long ones are uncommon). Of these, fibrillated geofibers are preferred, since they provide more soil/geofiber interface strength (Frost and Han 1999). Webster and Santoni (1997) suggest that 5 cm is about the optimal length.

Geofibers are the only geosynthetic that increases strength, decreases compressibility, and increases permeability.

Geofibers have many applications, all related to increased strength with decreased compressibility: embankment stabilization, slope rehabilitation, reinforcement of pavement subgrades, foundation stabilization, underfooting reinforcement, erosion control, and veneer soils used for landfill (or other) cover reinforcement. Consoli et al. (2003), Santoni et al. (2001), Li and Ding (2002), and many others, show the increase in strength due to geofiber dosing. Park and Tan (2005) found geofiber soil behind a retaining wall significantly reduced soil pressure on the wall.

Improvement is closely related to soil type. Park and Tan (2005) found that adding 0.2% fiber by dry weight to a sandy silt increased the friction angle by 3° but had no effect on the cohesion. Consoli et al. (2003) found that adding 0.5% fiber by dry weight to a silty sand increased the cohesion from 23 kN/m^2 to 127 kN/m^2 and had little to no effect on the friction angle. Therefore, it is important to note that fiber inclusion designs are case-specific and require testing.

Geofibers offer these advantages over geotextile or geogrid reinforcement:

1. provide three-dimensional reinforcement with no preferred orientation,
2. increase soil permeability (when high permeability is needed),
3. decrease unit weight (good for embankment over soft ground construction),
4. low cost/unit volume of soil treated,
5. cannot be damaged by construction equipment
6. do not provide a potential failure surface at the geosynthetic – soil interfaces,
7. may be easier to install than geotextiles or geogrids,
8. reduce swell potential,

9. reduce crack formation,
10. can be used to strengthen onsite soils, reducing need to import soils, thereby reducing cost, and
11. stresses in geofibers are not cumulative, as they are discrete geosynthetics.

Geofibers offer these disadvantages over geotextile or geogrid reinforcement:

1. cannot be deployed on windy days,
2. increase soil permeability (when low permeability is needed),
3. mixing uniformity may be difficult in clayey soils, and
4. application may be slower then rolled geosynthetics.

9.8.2 Design procedure for strength increase

Geofibers are used primarily for increasing soil strength. Increase in geofiber-treated soil strength is related to soil type, compaction water content, compaction method, geofiber properties, and geofiber dosage rate. The soil type and geofibers are usually fixed for a given project. Too few or too many geofibers are not effective. Too few geofibers result in minimal strength increase. Too many geofibers may result in strength lower than the untreated soil (e.g. Chen et al. 2010).

The procedure for estimating the strength increase from geofibers begins with the compaction test, described in Chapter 4. The optimum moisture content of the soil is determined with Proctor testing. The optimum percentage of geofibers is determined experimentally. Trial dosages of geofibers are added to the soil at optimum moisture content, which is compacted to a standard or modified Proctor density, as specified by the engineer. Typical geofiber dosage rates are 0.2% to 1.0% by dry weight. The compacted soil samples are extruded from the Proctor molds and strength tested, typically in unconfined compression (ASTM D698-12e1 2012). The unconfined compressive strength is reported and used in conventional design procedures. The effect of the geofibers that have tensile capacity is to add a cohesive component to the strength of the soil. If the application warrants further testing, appropriate triaxial testing is specified. The results will show an increased cohesion and angle of internal friction, which are then used in conventional design procedures. Proctor mold-sized specimens are recommended for triaxial testing, since it is difficult to trim geofiber enhanced samples. The process is repeated at increasing dosages, often at 0.2% intervals, and plotted, until the optimum dosage is found.

Roadway applications may require parameters other than strength, such as resilient modulus or R-value (California Test 301 2000). These parameters will also vary with geofiber dosage. The same experimental process, described above, is used to find the optimum dosage.

9.8.3 Construction advice

Mixing polymer fibers into soil is straightforward. The candidate soil is placed lifts approximately 0.2 m thick. Geofibers are distributed on top of the lift by hand, or with a blower. This cannot be done effectively on windy days as the lightweight geofibers will be scattered by the wind, resulting in uneven dosages. A rotary soil mixer (or blender) combines the soil and geofibers. One pass of the mixer is typically adequate. It is easier to blend geofibers into sandy soils than clayey soils. For clayey soils, the lift is then compacted with a sheepsfoot roller with long enough feet to penetrate the full lift. If the geofiber-enhanced soil is particularly sandy, a vibratory roller will be more effective.

9.9 SOIL SEPARATION

9.9.1 Introduction

Cohesionless soils placed over cohesive soils mix together over time. When this happens, cohesionless soils lose much of their strength and permeability, and become more compressible when mixed with even small amounts of cohesive soils. This phenomenon gives rise to the geotechnical aphorism "a little clay goes a long way," meaning that even small amounts of clay in a cohesionless soil cause the entire mass to act as a cohesive soil. Cohesive soils drain poorly and may have lower strength (recall the discussion on effective stress in Chapter 2). The presence of clay makes the soil mass sensitive to loss of strength with increasing water content, unlike cohesionless soils (up to saturation, that is). Cohesive soils are more difficult to compact than cohesionless ones. So, when a cohesionless soil is placed over a cohesive soil, it is desirable to separate them, so that the cohesionless soil may maintain its properties. Geotextiles separate soils. Figure 9.12 shows the problem and solution schematically. Figure 9.2 shows the road construction application of a separator.

Clayey subgrade intrusion into overlying cohesionless soils can occur due to gravity leading the cohesionless soils downward, or, if the soils are cyclically loaded, as in roadways, by pumping of the clayey soils upwards. The effect is the same.

The savings in using a geotextile separator can come from:

a. reduced amount of base soil needed (for roads), lowering costs. This is described in section 9.3.5.
b. reduced maintenance and longer project life.

Roads built on clayey subgrades require separation of the granular base soils, or, for railroads, ballast, from the subgrade. Keeping clay from infiltrating the clean base soils greatly helps retain strength, low compressibility, and good permeability. Historically, roads over clayey subgrades were built with more clean base material than was needed for the pavement structural design. The extra base soil, termed sacrificial aggregate, could be estimated from Figure 9.13. The percentage of extra aggregate needed is a function of the subgrade California Bearing Ratio. Parking lots, loading dock ramps, unpaved roads, construction access roads, sports fields, and embankments over soft, clayey ground use geotextile separators.

9.9.2 Design procedures

The geotextile separator requires certain engineering properties to function. The properties are in two categories: those for construction survivability and those for filtration and drainage. Construction survivability refers to the ability of the geotextile to survive the installation process. Geotextiles are fragile, compared to most civil engineering materials. The American Association of State Highway and Transportation Officials (AASHTO) standard M288-17 addresses survivability. Table 9.17, from Caltrans (2009), is based on the 1996 version of M288. The Table 9.17 values are based on minimum average roll value (MARV) in the weaker principle direction, except for apparent opening size, which is based on maximum average roll value.

FHWA (2008) provides Table 9.18. This table includes *degree of survivability*, which is evaluated by the engineer. Sites with very soft soils (say, CBR < 2) required a geotextile with

Table 9.17 Geotextile properties for survivability (after Caltrans 2009)

Separation Geotextile Property		ASTM specification		
	Woven			Nonwoven
Elongation at break, %		D4632	<50	≥50
Grab tensile strength, lb minimum		D4632	250	160
Wide width tensile strength at 5% strain, lb/ft minimum		D4595	–	–
Wide width tensile strength at ultimate strain, lb/ft minimum		D4595	–	–
Tear strength, lb minimum		D4533	90	60
Puncture strength, lb minimum		D6241	500	310
Permittivity, sec^{-1} minimum		D4491	0.05	0.05
Apparent opening size (maximum), inch		D4751	0.012	0.012
Ultraviolet stability (retained strength after 500 hours exposure), % minimum		D4355	70	70

Table 9.18 Geotextile survivability properties (based on FHWA 2008, and AASHTO 2017)

Property	ASTM Test Method	Units	Required Degree of Survivability		
			Very high woven/nonwoven	High woven/nonwoven	Moderate woven/nonwoven
Grab strength	D4632	lb	300/225	250/150	180/110
Tear strength	D4533	lb	110/80	90/50	70/40
Puncture strength	D6241	lb	620/430	500/310	370/225

very high survivability. Sites with tree stumps or large rocks require very high-survivability geotextile separators.

The design procedure is as follows. Estimate the degree of survivability needed, then choose the geotextile that will meet this criterion. If water is expected, the geotextile must be designed as a filter (Chapter 3). The geotextile permeability should be at least ten times greater than the soil permeability.

9.9.3 Construction advice

The success of geotextile separators is a strong function of the installation procedures. Care must be taken to keep the geotextile intact during construction operations. Many designers prefer staple nonwoven geotextiles, since they can stretch more than others, reducing likelihood of tearing.

The following installation advice will reduce geotextile damage:

1. When the subgrade CBR < 3, have at least 20 cm of soil on the geotextile before trafficking; this may be reduced to 15 cm for CBR≥ 3 subgrades.
2. If equipment makes ruts deeper than 5 cm, lessen the equipment pressure or increase the thickness of cover soil.
3. Use drum or rubber-tired rollers on first lift, not sheepsfoot.
4. Equipment should drive on/back off. Do not allow turning, especially with tracked vehicles. Avoid turning with tired vehicles.
5. Remove wrinkles as much as possible.
6. Don't leave geotextiles in the sun for more than two weeks.
7. Extremely angular aggregate should be avoided for the first lift. When using angular aggregate, use a 10 cm sand layer between angular aggregate and the geotextile.

8. Based on subgrade CBR, overlap the geotextile at least 0.3 m for soils with CBR ~3 to at least 0.6 m for soils with CBR ~1. Soils with CBR < 1 may require sewn seams.
9. When layout goes around curves, cut the geotextile in pieces and overlap in the direction of soil placement.
10. Never allow soil to be end dumped on the geotextile. This procedure will likely tear the geotextile beneath the pile of soil due to large shear stresses as the soil is dumped. Rather, place soil on existing soil and spread onto the geotextile with equipment.
11. After the first lift is placed, limit the pile height of end-dumped soils, especially on the first lift. Soft subgrades may fail under large piles. A rule of thumb is to limit the pile height to the height of the final road grade.
12. When using a geotextile separator beneath railroad ballast, be sure the geotextile is buried deeper than the length of the tamping tines used to compact the ballast to avoid geotextile puncture by the tines.
13. Close inspection before covering with soil is recommended. If tears are found, cover the tears with a geotextile before covering with soil. The cover piece should be pinned to the ground to prevent uplift by the covering equipment spreading the soil.

FHWA (2008) provides more details on design, installation procedures, and specifications.

9.10 PROBLEMS

9.0 What are geosynthetics?
9.1 Name three civil engineering projects that could use geosynthetics, and the geosynthetics involved.
9.2 There are two basic ways of fabricating geotextiles – name both. Describe the structure of each.
9.3 Geogrids are sold based on strength properties. Are there geotextiles with similar strengths? Other than strength, what would be an advantage of geogrids over geotextiles?
9.4 Given the geotextile force-deformation data in Figure 9.10, and that the specimen has a 10 cm gage length, estimate the following for the top curve.
   a. peak strength (in units of force/length)
   b. strength at 2%, 5%, and 10% strain (in units of force/length)
9.5 For the geotextile force-deformation data in Figure 9.10, given that the specimen has a 10 cm gage length, estimate the 2% strain LTDS for the following reduction factors for the top curve:
   installation damage   1.5
   creep   1.2
   chemical/biological degradation   1.1
9.6 A 5 mm thick geotextile will be used to filter a drain behind a retaining wall. Tests show the geotextile allows 0.11 m^3 of water to flow through a 0.09 m^2 piece of geotextile in a minute at a hydraulic gradient of one. What is the geotextile's permittivity? (Note: Darcy's law is needed to solve this.)
9.7 A pullout test is run on a geotextile embedded in a sandy soil with an angle of internal friction of 30°. The peak pullout force on the 0.3 m wide by 0.6 m long geotextile was 9.34 kN. The normal stress on the geotextile was 4.79 kPa. Estimate the coefficient of direct sliding.
9.8 Sketch and name the major parts of a road cross section. What are the functions of each part?

9.9 What is "sacrificial aggregate"? Why was/is it needed?

9.10 A roadway alignment primarily traverses cohesive soil in a wet climate. You specified CBR tests along the alignment to use in roadway design. The lowest CBR value was 2.5%. Estimate the soil's cohesion, in kPa.

9.11 The US Forest Service method is used to select a geotextile separator. What is the separator used for?

9.12 The US Army Corps of Engineers has a roadway design method that incorporates geosynthetics. What two types of geosynthetics may be specified in this method? Describe each. What are the functions of each geosynthetic?

9.13 Design problem. An unpaved road is planned in an area with a clayey subgrade. The minimum CBR ≈ 1 (very soft!). The road will carry about 2,000 passes of 167 kN tandem axles of 245 kN trucks. 50 mm ruts are allowable. Recommend thicknesses of CBR ≥ 80 base material for two designs: using geosynthetics and not using geosynthetics. Give the geosynthetic specifications.

9.14 Describe the placement and compaction of soil over a geotextile placed on a soft subgrade.

9.15 A paved road is planned over a CBR ≈ 6 subgrade. The design parameters are: Class E traffic; category IVA; base CBR ≥ 80; 20,000 tandem axle passes make up 40% of the traffic. Estimate the required thickness of this base and recommended asphalt concrete thickness.

9.16 An oil company wishes to construct a very wide embankment, 9.2 m high, on a very soft site in Louisiana, USA. The embankment slopes are 31° from horizontal. The fill soil has an internal angle of friction of 22° and a mild cohesion of 4.79 kPa. The fill will be compacted to 18.9 kN/m³. The very wet, soft, clay subgrade has a unit weight of 15.4 kN/m³. Your site investigation, using a CPT, shows c=28.7 kPa. Since any failure would be rapid, you estimate $\phi = 0°$. To create the embankment in the shortest time, you choose to use geosynthetics – an embankment over soft ground. Your geosynthetic has $\delta_u = 21°$ and $\delta_L = 0°$, $c_{a\ upper} = 3.83$ kPa, and $c_{a\ lower} = 19.2$ kPa.

a.   A geosynthetic with wide-width strength of 1,459 kN/m produces a slope stability FS = 1.5 Estimate the length of geosynthetic needed beyond the failure surface for FS = 1.5. Then, estimate the factors of safety for sliding and breakage.

b.   What other failures modes must be checked to complete the design?

9.17 What are the advantages of using a geogrid instead of a geotextile to reinforce an embankment over soft ground?

9.18 When deploying geosynthetic reinforcement over very soft ground, it may be necessary to leave organic debris on the site, to avoid reducing the strength of the soils to the point where construction is impossible. What concerns do you have about this?

9.19 What is stage construction? Why is it done? What problems have to be solved for successful stage construction?

9.20 A 369 kN column load is to be carried by a 1.52 m square footing. The $\gamma = 16.5$ kN/m³, $\phi = 28°$ subgrade soil will not carry this load without failure. Rather than use soil improvement, or deep foundations, you choose to use five layers of geogrid underfooting reinforcement. In keeping with good practice, the footing will be embedded 0.3 m in the ground. The infill soil is well-graded crushed rock, with 5 cm maximum size. A $FS_{bearing\ capacity} \geq 3$ is required.

Prepare the final design, with sketches showing dimensions. Estimate the BCR.

9.21 A 5338 kN column load is to be carried by a square footing on sandy soil. The subgrade has $\gamma = 18.2$ kN/m³, $\phi = 25°$. A footing with no underfooting reinforcement, embedded 0.3 m, would be about 5.5 m square, for FS=2. 5.5 m is judged too large. Prepare an underfooting reinforcement design to carry this load with FS ≥ 2.

Prepare the final design, with sketches showing dimensions. Comment on this design.

9.22 A 2 m square footing is planned for a 133 kN column load on a soft, clayey subgrade. The subgrade properties are: $c = 15.3$ kPa, and $\gamma = 14.9$ kN/m³. The weak subgrade suggested using a 0.15 m thick geocell under the footing. The geocell will be filled with compacted, coarse, $\phi = 30°$ sand which has an interface friction angle of $\delta = 25°$ with the geocell. The footing will rest on a thin layer of sand over the geocell. Estimate the factor of safety of this design. The footing is not embedded in the ground.

9.23 What polymer(s) are most common in geocells? Why do you think this is?

9.24 Some geocells are made of textiles, rather than polymeric sheets. What are the advantages and disadvantages of these geocells over the polymeric sheet geocells?

9.25 Geofibers increase soil strength. Describe how this works.

9.26 Geofibers increase soil permeability, while increasing strength. Explain why this may be desirable.

9.27 Describe a construction project where soil separation would be important.

9.28 What properties of geotextiles are important in separation applications?

REFERENCES

AASHTO M288-17. (2017) *Geotextile Specifications and Highway Applications – In Standard Specification for Transportation Materials and Methods of Sampling and Testing.* American Association of State Highway and Transportation Officials, Washington, DC.

Adams, M.T., Collin, J.G. (1997) Large model spread footing load tests on geosynthetic reinforced soil foundations. *Journal of Geotechnical and Geoenvironmental Engineering,* 123(1), 66–72.

Akins, K. (2012) Personal communication. Tensar Corporation, Atlanta, Georgia.

ASTM D1586-18. (2018) *Standard Test Method for Standard Penetration Test (SPT) and Split-Barrel Sampling of Soils.* American Society for Testing and Materials, West Conshohocken, PA.

ASTM D1883. (2016) *Standard Test Method for CBR (California Bearing Ratio) of Laboratory-Compacted Soils.* American Society for Testing and Materials, West Conshohocken, PA.

ASTM D4594. (2009) *Standard Test Method for Effects of Temperature on Stability of Geotextiles.* American Society for Testing and Materials, West Conshohocken, PA.

ASTM D4595. (2017) *Standard Test Method for Tensile Properties of Geotextiles by the Wide-Width Strip Method.* American Society for Testing and Materials, West Conshohocken, PA.

ASTM D5322-17. (2017) *Standard Practice for Immersion Procedures for Evaluating the Chemical Resistance of Geosynthetics to Liquids.* American Society for Testing and Materials, West Conshohocken, PA.

ASTM D6213-17. (2017) *Standard Practice for Tests to Evaluate the Chemical Resistance of Geogrids to Liquids.* American Society for Testing and Materials, West Conshohocken, PA.

ASTM D6389-17. (2017) *Standard Practice for Tests to Evaluate the Chemical Resistance of Geotextiles to Liquids.* American Society for Testing and Materials, West Conshohocken, PA.

ASTM D6637. (2015) *Standard Test Method for Determining Tensile Properties of Geogrids by the Single of Multi-Rib Tensile Method.* American Society for Testing and Materials, West Conshohocken, PA.

ASTM D6706-13. (2013) *Standard Test Method for Measuring Geosynthetic Pullout Resistance in Soil.* American Society for Testing and Materials, West Conshohocken, PA.

ASTM D698-12e1. (2012) *Standard Test Methods for Laboratory Compaction Characteristics of Soil Using Standard Effort.* American Society for Testing and Materials, West Conshohocken, PA.

ASTM D5321-20. (2020) *Standard Test Method for Determining the Coefficient of Soil-Geosynthetic or Geosynthetic-Geosynthetic Friction by the Direct Shear Method.* American Society for Testing and Materials, West Conshohocken, PA.

Berg, R.R., Christopher, B.R, Samtani, N.C. (2009) *Design and Construction of Mechanically Stabilized Earth Walls and Reinforced Soil Slopes – Volume I.* NHI Publication No. FHWA-NHI-10-024. National Highway Institute, Washington, DC.

Binquet, J., Lee, K.L. (1975) Bearing capacity analysis of reinforced earth slabs. *Journal of Geotechnical Engineering Division*, American Society of Civil Engineers, *101*(12), 1257–1276.

Biswas, A. and Krishna, A.M. (2017) Geocell-reinforced foundation systems: a critical review. *International Journal of Geosynthetics and Ground Engineering*, *3*(2), 17.

Bonaparte, R., Berg, R. (1987) Long-term allowable design loads for geosynthetic soil reinforcement. In *Proceedings of Geosynthetics '87, IFAI* (pp. 181–192), Vol. 1, New Orleans, LA.

Bowles, J.E. (1996) *Foundation Analysis and Design*, 5th ed.. McGraw-Hill, New York.

Bush, D.I., Jenner, C.G., Bassett, R.H. (1990) The design and construction of geocell foundation mattresses supporting embankments over soft grounds. *Geotextiles and Geomembranes*, *9*(1), 83–98.

California Test 301. (2000) *Method for Determining the Resistance "R" Value of Treated and Untreated Bases, Subbases, and Basement Soils by the Stabilometer*. California Department of Transportation, Sacramento, CA.

Caltrans. (2009) *Guide for Designing Subgrade Enhancement Geotextiles*. California Department of Transportation, Sacramento, CA.

Carroll Jr., R.G., Chouery-Curtis, V. (1990) Geogrid connections. *Geotextiles and Geomembranes*, *9*(4–6), 515–530, Conference: Seaming of Geosynthetics.

Cazzuffi, D., Moraci, N., Calvarano, L.S., Cardile, G., Gioffrà, D., Recalcati, P. (2014) European experience in pullout tests: Part 2 - The influence of vertical effective stress and of geogrid length on interface behaviour under pullout conditions. *Geosynthetics Magazine*, *32*(2), 40–50.

Cedergren, H. (1987) Undrained pavements: A costly blunder. *ASCE Civil Engineering*, *57*, 6.

Cedergren, H. (1994) America's pavements: World's longest bathtubs. *ASCE Civil Engineering*, *64*, 56–58.

Chen, R.-H., Chi, P.-C., Wu, T.-C., Ho, C.-C. (2010) Shear strength of continuous-filament reinforced sand. *Journal of Geo Engineering*, *6*(2), 99–107.

Chin, T.Y., Sew, G.S. (2000) Design and construction control of embankment over soft cohesive soils. In *SOGISC-Seminar on Ground Improvement-Soft Clay*, 23–24 August 2000, Kuala Lampur, Malasia.

Christopher, B.R., Schwartz, C., Boudreau, R. (2006) *Geotechnical Aspects of Pavements*. FHWA Report NHI-05-037. National Highway Institute, Federal Highway Administration, U.S. Department of Transportation, Washington, DC.

Coduto, D. (1999) *Geotechnical Engineering*. Prentice Hall, Upper Saddle River, NJ.

Consoli, N.C., Casagrande, M.D.T., Prietto, P.D.M., Thome, A. (2003) Plate load test on fiber-reinforced soil. *Journal of Geotechnical and Geoenvironmental Engineering*, *129*(Part 10), 951–955.

Das, B.M. (2009) *Shallow Foundations, Bearing Capacity and Settlement*, 2nd ed. CRC Press/Taylor and Francis, Boca Raton, FL.

Dash, S.K., Krishnaswamy, N.R., Rajagopal, K. (2001) Bearing capacity of strip footings supported on geocell-reinforced sand. *Geotextiles and Geomembranes*, *19*, 235–256.

Dash, S.K, Rajagopal, K., Krishnaswamy, N.R. (2004) Performance of different geosynthetic reinforcement materials in sand foundations. *Geosynthetics International*, *11*(1), 35–42.

Duncan, J.M, Buchignani, A.L. (1987) *An Engineering Manual for Settlement Studies*. University of California - Berkeley, Berkeley, California, USA, 188p.

Dunnicliff, J. (1993) *Geotechnical Instrumentation for Monitoring Field Performance*. NCHRP Synthesis 89. Transportation Research Board, Washington, DC. USA.

Dunnicliff, J. (1998) *Geotechnical Instrumentation Reference Manual*. NHI Course No. 13241, Module 11. FHWA-HI-98-034. Federal Highway Administration, U.S. Department of Transportation, Washington, DC, USA.

Elias, V., Christopher, B.R., Berg, R. (2001) *Mechanically Stabilized Earth Walls and Reinforced Soil Slopes*. U.S. Department of Transportation, Publication No. FHWA-NHI-00-043, Federal Highway Administration, NHI Course No. 132042, Washington, DC, USA.

Elton, D.J. (2001) *Soils Magic*. Geotechnical special publication 114, American Society of Civil Engineers, Reston, VA.

Evans, J., Elton, D., Ruffing, D. (2009) Interior spaces to accommodate geotechnical engineering personnel. *Journal of Universal Rejection*, 9(12), 45–51.

FHWA. (1989) *Geotextile Design Examples*. Contract no. DTFH-86-R-00102. Geoservices, Inc. Report to the Federal Highway Administration, Washington, DC.

FHWA. (1993) *"EMBANK," Computer Program*. User's Manual Publication No. FHWA-SA-92-045. Federal Highway Administration, Washington, DC.

FHWA. (2008) *Geosynthetic Design and Construction Guidelines*. NHI Course No. 132013. Publication No. FHWA NHI-06-116, US Federal Highway Administration, Washington, DC, USA.

Fragaszy, R., Lawton, E. (1984) Bearing capacity of reinforced sand subgrades. *Journal of Geotechnical and Geoenvironmental Engineering*, ASCE, 110(10), 1500–1511.

Frost, J.D., Han, J. (1999) Behavior of interfaces between fiber-reinforced polymers and sands. *Journal of Geotechnical and Geoenvironmental Engineering*, 112(8), 804–820.

Giroud, J.P., Noiray, L. (1981) Geotextile-reinforced unpaved road design. *Journal of the Geotechnical Engineering Division*, ASCE, 107(GT9), 1233–1254, and discussions (108(GT12), 1654–1665).

Han, J., Akins, K. (2002) Use of geogrid-reinforced and pile-supported earth structures. *Geotechnical Special Publication No. 116*, ASCE, 668–679.

Han, J., Gabr, M.A. (2002) Numerical analysis of geosynthetic-reinforced and pile-supported earth platforms over soft soil. *Journal of Geotechnical and Geoenvironmental Engineering*, 128(1), 44–53.

Han, J., Pokharel, S.K., Parsons, R.L., Leshchinsky, D., Halahmi, I. (2010) Effect of infill material on the performance of geocell-reinforced bases. In *Geosynthetics for a Challenging World*, E.M. Palmeira, D.M. Vidal, A.S.J.F. Sayao, M. Ehrlich (eds.), *Proceedings of the 9th International Conference on Geosynthetics*, Brazil, May 23–27, 2010, pp. 1503–150.

Holtz, R.D., Christopher, B.R., Berg, R.R. (1995) *Geosynthetic Design and Construction Guidelines*. FHWA HI-95-038. Federal Highway Administration, Washington, DC.

Holtz, R.D, Christopher, B.R., Berg, R.B. (1998) *Geosynthetic Design & Construction Guidelines: Participant Notebook*. National Highway Institute, U.S. Dept. of Transportation, Federal Highway Administration, National Highway Institute., Washington, DC, USA

Holtz, R.D., Kovacs, W.D., Sheahan, T. (2011) *An Introduction to Geotechnical Engineering*, 2nd edition. Prentice Hall, Englewood Cliffs, NJ.

Holtz, R.D., Sivakugan, N. (1987) Design charts for roads with geotextiles. *Geotextiles and Geomembranes*, 5, 191–199.

Holtz, R.R. (1989) Treatment of problem foundations for highway embankments, *Synthesis of Highway Practice 147*. NCHRP, TRB, Washington, DC, 72pp.

Huang, C.C., Menq, F.Y. (1997) Deep-footing and wide-slab effects in reinforced sandy ground. *Journal of Geotechnical and Geoenvironmental Engineering*, 123(1), 30–36.

Huang, C.C., Tatsuoka, F. (1990) Bearing capacity of reinforced horizontal sandy ground. *Geotextiles and Geomembranes*, 9, 51–82.

Kadolph, S.J. (2010) *Textiles*, 11th edition. Prentice Hall, Upper Saddle River, NJ.

Koerner, R.M. (2005) *Designing with Geosynthetics*, 5th edition. Van Nostrand Reinhold, New York, NY.

Koerner, R.M. (2012) *Designing with Geosynthetics*, 6th edition. Xlibris Corporation, www.xlibris.com.

Kumar, A., Ohri, M., Bansal, R. (2007) Bearing capacity tests of strip footings on reinforced layered soil. *Geotechnical and Geological Engineering*, 25(2), 139–150.

Kumar, A., Saran, S. (2003) Bearing capacity of rectangular footing on reinforced soil. *Journal of Geotechnical and Geoenvironmental Engineering*, 21(3), 201–224, Kluwer Academic Publishers, The Netherlands.

Latha, G.M., Murthy, V.S. (2007) Effects of reinforcement form on the behavior of geosynthetic reinforced sand. *Geotextiles and Geomembranes*, 25(2007), 23–32.

Li, J., Ding, D.W. (2002) Nonlinear elastic behavior of fiber-reinforced soil under cyclic loading. *Soil Dynamics and Earthquake Engineering*, 22(9–12), 977–983.

Madhavi Latha, G. (2011) Design of geocell reinforcement for supporting embankments on soft ground. *Geomechanics and Engineering*, *3*(2), 117–130.

Moghaddas Tafreshia, S.N., Dawson, A. (2010) Comparison of bearing capacity of a strip footing on sand with geocell and with planar forms of geotextile reinforcement. *Geotextiles and Geomembranes*, *28*(1), 72–84.

Mosallanezhad, M., Nasiri, I. (2015) A novel reinforcement to improve the bearing capacity of soil investigation. *International Journal of Engineering & Technology Sciences*, *3*, 123–134.

Omar, M.T., Das, B.M, Yen, S.C., Puri, V.K., Cook, E.E. (1993) Ultimate bearing capacity of rectangular foundations on geogrid-reinforced sand. *ASTM Geotechnical Testing Journal*, *16*(2), 246–252.

Park, T., Tan, S.A. (2005) Enhanced performance of reinforced soil walls by the inclusion of short fiber. *Geotextiles and Geomembranes*, *23*, 348–361.

Qian, X., Koerner, R.M., Gray, D.H. (2002) *Geotechnical Aspects of Landfill Design and Construction*. Prentice Hall, Upper Saddle River, NJ.

Rajagopal, K., Krishnaswamy, N.R., Madhavi Latha, G. (1999) Behaviour of sand confined with single and multiple geocells. *Geotextiles and Geomembranes*, *17*(3), 171–184.

Riordan, N.J., Seaman, J.W. (1993) *Highway Embankments over Soft Compressible Alluvial Deposits: Guidelines for Design and Construction*. Contractor Report 341. Transport Research Laboratory, Department of Transport, London, UK.

Rowe, R.K., Li, L.L. (2005) Geosynthetic-reinforced embankments over soft foundations. *Geosynthetics International*, Special Issue on the Giroud Lectures, *12*(1), 50–85.

Sale, J.P., Parker, F. Jr., Barker, W. (1973) Membrane encapsulated soil layers. *Journal of the Soil Mechanics and Foundations Division*, *99*(12), 1077–1090.

Santoni, R.L., Tingle, J.S., Webster, S.L. (2001) Engineering properties of sand-fiber mixtures for road construction. *Journal of Geotechnical and Geoenvironmental Engineering*, *127*(3), 258–268.

Shin, E.C., Das, B.M. (2000) Experimental study of bearing capacity of a strip foundation on geogrid-reinforced sand. *Geosynthetics International*, *7*(1), 59ff.

Shin, E.C., Das, B.M., Lee, E.S., Atalar, C. (2002) Bearing capacity of strip foundation on geogrid-reinforced sand. *Geotechnical and Geological Engineering*, Kluwer, *20*, 169–180.

Shukla, S.K., Sivakugan, N., Das, B.J. (2009) Fundamental concepts of soil reinforcement – An overview. *ASCE Journal of Geotechnical and Geoenvironmental Engineering*, *3*, 329–342.

Shukla, S.K, Yin, J-H. (2006) *Fundamentals of Geosynthetic Engineering*. Taylor and Francis, London.

Silvestri, V. (1983) Bearing capacity of dykes and fills founded on soft soils of limited thickness. *Canadian Geotechnical Journal*, *20*(3), 428–436.

Sitharam, T.G., Sireesh, S. (2006) Effects of base geogrid on geocell-reinforced foundation beds. *Geomechanics and Geoengineering*, *1*(3), 207–216.

Sowers, G.F., Sally, H.L. (1962) *Earth and Rockfill Dam Engineering*. Asia Publishing, New York.

Sprague, C.J., Goodrum, R.A. (1994) Selecting standard test methods for experimental evaluation of geosynthetic durability. *Transportation Research Record*, *1439*, 32–40, National Research Council, Washington, DC.

Steward, J.E., Williamson, R., Mohney, J. (1977) *Guidelines for Use of Fabrics in Construction and Maintenance of Low-Volume Roads*. FHWA Report No FHWA-TS-78-205, Federal Highway Administration, Washington, DC.

Suvorova, Y.V., Alekseeva, S.I. (2010) Experimental and analytical methods for estimating durability of geosynthetic materials. *Journal of Machinery Manufacture and Reliability*, *39*(4), 391–395.

TM 5-822-5. (1992) *Pavement Design for Roads, Streets, Walks, and Open Storage Areas*. TM 5-822-5/AFM 88-7, Chapter 1, Joint Departments of the Army and the Air Force, Washington, DC.

US Department of Defense. (2004) *Engineering Use of Geotextiles*. UFC 3-220-08FA 16. US Department of Defense, Washington, DC (also US Army (1995) *Engineering use of Geotextiles*. ARMY TM 5-818-8. US Department of Defense, Washington, DC).

US Navy. (1986) *Design Manual 7.01*. Naval Facilities Engineering Command, Alexandria, VA.

USACE. (2003) *Use of Geogrids in Pavement Construction*. ETL 1110-1-189. Department of the Army, US Army Corps of Engineers, Washington, DC.

Vipulanandan, C., Bilgin, Ö., Ahossin Guezo, Y.J., Vembu, K., Erten, M.B. (2009) *Prediction of Embankment Settlement Over Soft Soils*. Report 0-5530-1, FHWA/TX-09/0-5530-1. Texas Department of Transportation Research and Technology Implementation Office, Austin, TX.

Wayne, M.H., Han, J., Akins, K. (1998) The design of geosynthetic reinforced foundations. In *Design and Construction of Retaining Systems, ASCE Geo-Institute Geotechnical Special Publication, No. 76*, edited by John J. Bowders et al. pp. 1–18.

Webster, S.L. (1979) *Investigation of Construction Concepts across Soft Ground*. Mississippi Report S-79-20. US Army Waterways Experiment Station, Vicksburg, MS.

Webster, S.L. (1993) *Geogrid Reinforced Base Courses for Flexible Pavements for Light Aircraft: Test Section Construction, Behavior under Traffic, Laboratory Tests, and Design Criteria*. Technical Report GL-93-6. U.S. Army Corps of Engineers Waterways Experiment Station, Vicksburg, MS.

Webster, S.L., Santoni, R.L. (1997) *Contingency Airfield and Road Construction using Geosynthetic Fiber Stabilization of Sands*. Tech. Rep. GL-97-4. U.S. Army Corps of Engineers Waterways Experiment Station, Vicksburg, MS.

Webster, S.L., Watkins, J.E. (1977) *Investigation of Construction Techniques for Tactical Bridge Approach Roads across Soft Ground*. Mississippi Report S-77-1. US Army Corps of Engineers Waterways Experiment Station, Vicksburg, MS.

Yetimoglu, T., Wu, J.T.H., Saglamer, A. (1994) Bearing capacity of rectangular footings on geogrid-reinforced sand. *Journal of Geotechnical Engineering*, ASCE, 120(12), 2083–2099.

Yourman, A.M. Jr., Diaz, C.M., Gilbert, G.K. (2006) Jet grouted settlement isolation wall at the Henry Ford Avenue grade separation. In *GeoCongress 2006, Geotechnical Engineering in the Information Technology Age*, eds., DeGroot, D.J, DeJong, J.T, Frost, D., Baise, L.G., Atlanta, Georgia, USA.

Chapter 10

Reinforcement in walls, embankments on stiff ground, and soil nailing

10.1 INTRODUCTION

The idea of and ability to add tensile capacity to soil caused a paradigm shift in geotechnical engineering and construction. Broadly, reinforcement of soils allows many projects to be completed in a more cost-effective manner than previously possible. In this chapter, three areas of earth reinforcement are examined, including:

1. mechanically stabilized earth walls,
2. embankments over stiff ground, and
3. soil nailing.

Mechanically stabilized earth walls and embankments over stiff ground often involve the use of geosynthetics although metal reinforcement is also used. Soil nailing uses metal exclusively.

10.2 MECHANICALLY STABILIZED EARTH WALLS

10.2.1 Introduction

Retaining walls are broadly defined as structures having a hard facing that retain soil on one side and that create steep changes in grade on the other. For thousands of years, retaining walls were massive, heavy objects (gravity walls) that supported the load of the soil behind them without significant displacement. For gravity walls, the weight of the wall ensures stability. In the 20th and 21st centuries, these conventional walls were/are often made of concrete. Figure 10.1 shows a gravity wall and a cantilever wall (a variation on the gravity wall) with the parts labeled. The wall proper, backfill, and foundation are the primary parts, but properly constructed retaining walls also include drain systems to relieve water pressure.

A large heavy object, a retaining wall, can be more efficiently created using a reinforced soil mass. Figure 10.2 is a schematic of this approach. This reinforced (composite) mass resists displacement. This class of walls is called mechanically stabilized earth (MSE), mechanically stabilized backfill (MSB), or geosynthetic reinforced soil (GRS). While there are some design differences, the term MSE will be used here. Early versions of these types of structures were installed thousands of years ago, e.g. parts of the Great Wall of China, ~2,300 years ago and the ziggurats in the former Babylonia, over 4,000 years ago. These ancient structures use plant matter as the reinforcement mixed with soil. Berg et al. (2009) provide a history of modern MSE structures that use geotextiles, geogrids, or metal. These materials add tensile capacity to soil, allowing the composite mass to stand vertically.

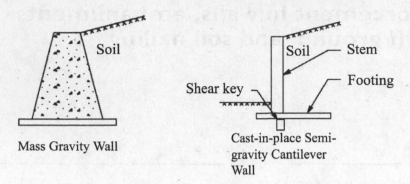

Figure 10.1 Mass concrete and cantilever retaining walls (after Wisconsin DOT 2010, Bridge Manual, Madison, WI).

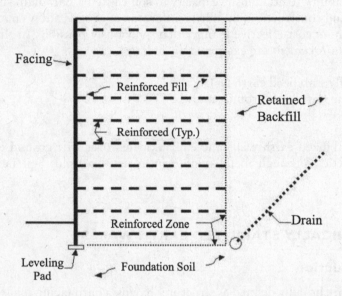

Figure 10.2 Schematic of typical MSE wall. Note: reinforcement lengths may vary across section (after Berg et al. 2009).

Today, geosynthetic and metal MSE retaining (and other) walls are very widely used. Their design is significantly different from the gravity walls that formed the basis of wall design for centuries. MSE walls are quick to construct, use a wider variety of backfill soils, are inexpensive, and can be easily formed into shapes attractive to architects and clients.

Mechanically stabilized earth construction consists of three parts: soil, tensile reinforcement, and facing material. Unlike gravity wall systems, where the soil is backfill (taking up space), the soil in a mechanically stabilized earth system is a part of the structural system. That is, the soil strength is required for success.

Mechanically stabilized earth walls have many applications. Examples include raising elevations of buildings, bridge abutments, true bridge abutments (supporting the bridge deck), streambank erosion control, lateral expansion of roadway right of way, tank confinement berms, reducing the footprint of soil preloads, blast/sound walls, transfer stations, and stabilizing slopes and dams.

10.2.2 Design philosophy

Mechanically stabilized earth walls function differently than gravity walls. Gravity walls depend on self-weight of the wall for stability. MSE walls depend on the internal strength of the soil and geosynthetic mass for stability. The layers of geosynthetics and soil form a composite mass that has high internal strength and, thus, the reinforced soil mass becomes equivalent to the mass of a conventional gravity wall. The strength is derived from the strength of the soil, the strength of the geosynthetic, and the interaction between them. Some of this interaction is due to confinement of the soil by the geosynthetic, and some by soil arching between the geosynthetic layers. While the controlling phenomena are complex, there are simplified and accepted design procedures that are widely used.

10.2.3 Advantages and disadvantages of MSE walls

MSE walls offer many advantages over gravity walls, including:

1. flexibility (more likely to sustain settlement without visible damage),
2. ease of construction (no concrete formwork or steel work is needed, allowing for reduced labor costs),
3. less expensive to create unusually shaped walls,
4. lower cost and lighter construction equipment needed,
5. less site preparation,
6. speed of construction, especially with higher walls (no formwork or concrete curing time needed),
7. highly resistant to seismic loads,
8. do not require a footing,
9. sustain high lateral displacement without structural failure,
10. are economic for very large heights (over 30 m),
11. wider variety of backfill soils can be used leading to more frequent use of onsite soils with consequent cost reduction,
12. phenomenal improvement in backfill strength allowing heavier structure to be placed closer to the wall face,
13. often require less construction space in front of the wall,
14. may be built on softer foundations since MSE is much more tolerant of settlement,
15. requires less foundation preparation, and
16. aesthetics (although beauty is in the eye of the beholder).

Mechanically stabilized earth walls have a few disadvantages over gravity/cantilever walls. These include:

1. minimum space behind the wall for construction is greater to accommodate the reinforcement,
2. backfill excavation after construction is not possible as excavation that severs the reinforcement will cause failure, and
3. only useful for active earth pressures.

10.2.4 Design using geosynthetics

Figure 10.2 shows the components of mechanically stabilized earth systems including the facing, reinforcement, retained soil, backfill soil, and foundation soil. There are several accepted methods of designing MSE walls (e.g. FHWA 2018; Koerner 2012a; Miyata and Bathhurst 2007). Two are presented in this chapter. The first is based on NCMA (1993). Design methods must address multiple failure modes including reinforcement breakage, reinforcement pullout, sliding of the reinforced mass, block stability, bearing capacity, overall stability, and settlement. Figures 10.3 and 10.4 schematically illustrate the modes unique to MSE construction: geosynthetic breakage and geosynthetic pullout modes. In Figure 10.3, the failure of the geosynthetic in tensile loading is illustrated. In Figure 10.4, failure of the

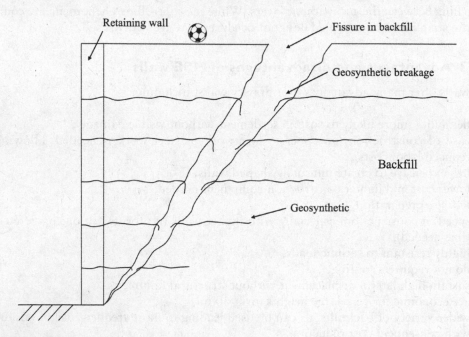

Figures 10.3 Schematic illustration of geosynthetic breakage.

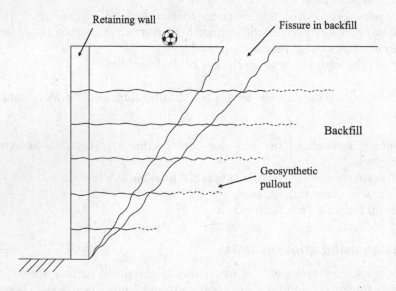

Figures 10.4 Schematic illustration of geosynthetic pullout.

geosynthetic by pullout is illustrated. These two failure modes are also applicable to MSE walls reinforced with metal strips.

The second approach is based on the US Federal Highway Administration Geosynthetic Reinforced Soil - Integrated Bridge System method (FHWA 2018). This method, developed for walls supporting bridges, can also be used for walls not supporting bridges. This method offers a simplicity uncommon to other MSE design methods.

10.2.4.1 Sliding of the reinforced mass

The backfill soil generates an active earth force tending to move the reinforced mass. The geosynthetic/soil interface at the base of the wall represents the weakest sliding plane. Resistance to sliding comes from friction between the soil and the geosynthetic. The resisting frictional force is characterized as

$$R_{\text{sliding}} = (W \tan \delta)(L) \tag{10.1}$$

where

$R_{sliding}$ is the resisting frictional force
W is the weight per meter parallel to the wall acting on the geosynthetic
δ is the interface friction angle between the retained soil and the geosynthetic
L is the length of the geosynthetic

δ may be estimated or be the result of testing (ASTM D5321-08, 2008). W includes the weight of the soil plus any permanent surface loads. Note that the critical sliding case may occur during construction.

The active earth pressure, P_a, is the driving force and is calculated using Coulomb earth pressure theory. For a granular backfill soil, which is recommended because it drains well:

$$P_a = 0.5 K_a \sigma_{\text{avg vertical}} H \tag{10.2}$$

where

$\sigma_{avg\,vertical}$ is the average vertical stress on the layer of geosynthetic in question
H is the average depth to the geosynthetic layer being checked for sliding
K_a is the Coulomb lateral earth pressure coefficient for granular soils

The active earth pressure coefficient can be determined from tables in various texts or from Equation 10.3 as follows:

$$K_a = \frac{\sin^2(\alpha + \phi)}{\sin^2 \alpha \, \sin(\alpha - \delta)\left[1 + \sqrt{\dfrac{\sin(\phi + \delta)\sin(\phi - \beta)}{\sin(\alpha - \delta)\sin(\alpha + \beta)}}\right]} \tag{10.3}$$

where

ϕ is the backfill soil angle of internal friction
α is the angle formed by the rear edges of the reinforcement measured clockwise from horizontal for left-facing walls
β is the angle of the surface of the retained soil above the horizontal
δ is the lower of the internal friction angles of the retained fill (that between layers of geosynthetic) or the backfill

To design an MSE wall, it is first necessary to choose a trial number of layers of geosynthetic and to assume the length, L, of a given layer of geosynthetic reinforcement being analyzed. The geosynthetic lengths need not be the same and are often longer at the top. For long and tall walls, there may be significant savings by making the lowest layers shorter than the topmost layers. The factor of safety, in sliding, for a given layer is

$$FS_{\text{sliding}} = \frac{R_{\text{sliding}}}{P_a} \tag{10.4}$$

where

P_a is calculated using the average depth, H, to the geosynthetic layer being analyzed for sliding

The FS calculation is repeated for each layer of reinforcement. If the FS_{sliding} is unacceptably low, the geosynthetic length L can be increased. Maintaining equal FS_{sliding} in each reinforcement layer may require a different minimum length of reinforcement for each layer. For ease of construction, usually all layers are the same length.

The number of layers of geosynthetic reinforcement, n, must then be determined. In theory, a few layers of geosynthetic will work. In practice, it's better to use many layers, both to provide redundancy, and because the geosynthetic/soil system forms a composite only if the layers are close enough to cause the soil to arch. If modular concrete blocks are used as facing, it is easiest to place a layer of geosynthetic between every course of blocks. This greatly encourages the contractor to compact every lift of soil and compaction is critical to success.

10.2.4.2 Reinforcement breakage

The geosynthetic reinforcement must be strong enough to retain the soil without breaking. Breakage analysis is done by comparing soil forces stressing the geosynthetic against the tensile strength of the geosynthetic. The forces stressing the geosynthetic are due to the active earth force, P_a, calculated above. Each layer of geosynthetic must carry its tributary area of the pressure diagram, a function of the vertical spacing (S_v) of the reinforcement. Figure 10.5 schematically shows this vertical spacing concept.

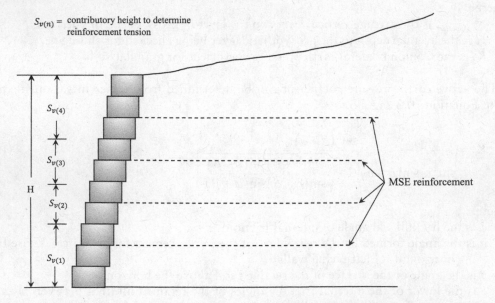

Figure 10.5 The tributary area, of the pressure diagram, contributing to pullout force (after Berg et al. 2009).

The $FS_{breakage}$ for each geosynthetic layer is calculated from

$$FS_{\text{breakage for a layer}} = \frac{LTDS}{P_{a \text{ contributing to that layer}}} \tag{10.5}$$

where
    $LTDS$ is the long-term design strength, described in Chapter 3.

The factor of safety for breakage for each geosynthetic layer must be checked.

10.2.4.3 Reinforcement pullout

Each layer of reinforcement must be long enough to keep it from sliding out from between layers of reinforced soil. The tributary part of P_a, above, is considered the driving force. The resisting force is due to friction on both sides of the geosynthetic. The resisting force is:

$$FS_{\text{resisting pullout}} = \frac{AC}{P_{a \text{ contributing to that layer}}} \tag{10.6}$$

where
    AC is the anchorage capacity
    $P_{a \text{ contributing to that layer}}$ is the active earth force contributing to that layer

The anchorage capacity is calculated from:

$$AC = 2\,(\sigma \tan\phi)C_i\,L_a \tag{10.7}$$

where
    AC is the anchorage capacity
    σ is the average stress on the layer of geosynthetic (dead loads only)
    ϕ is the angle of internal friction of the retained soil
    C_i is the coefficient of interaction between the geosynthetic and the retained soil (may estimate, or evaluate by testing)
    L_a is the anchorage length. L_a is the length of reinforcement between the failure surface and the free end of the reinforcement.

The failure surface is taken as a line extending above the horizontal from the back side of the base of the wall face to the ground surface at an angle equal to $45° + \phi/2$, where ϕ is the angle of internal friction of the retained soil. Each geosynthetic layer must be evaluated separately.

10.2.4.4 Other failure modes

Block stability refers to the stability of modular block facings. Blocks may bulge if there is inadequate friction between blocks. Closer reinforcement spacing can be used to reduce this. NCMA (1993) provides details of this analysis. As a construction note, bulging usually occurs from a failure to compact the backfill soil touching the rear face of the block. Environmental degradation of the concrete facing is addressed in the specification of the concrete used. While concrete block is not required for wall system stability, the blocks may be needed for erosion mitigation and aesthetics. To illustrate that the facing blocks are not needed for stability, see Figure 10.6 which shows a stable MSE wall without facing.

Figure 10.6 Photo of unfaced MSE wall used for buttress for failing bridge abutment (courtesy of GeoStabilization International).

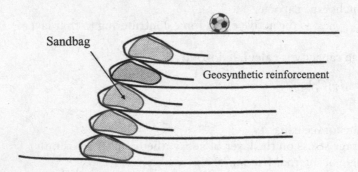

Figure 10.7 Schematic of a geosynthetic wrapped face wall.

Other facing types include precast panels, timbers, metallic facing, gabions, geosynthetic wraparound, shotcrete, cast-in-place concrete, wire baskets/mesh, and even organically stabilized soil materials (plants). Each facing block type has individual stability and durability considerations. Figures 10.7 shows a schematic of geosynthetic wrapped face often used where it is desirable for the wall to blend into the natural environment. Figure 10.8 is a photograph of a shotcrete face.

10.2.5 Design of internal components

The internal stability of the wall must be addressed, that is, the soil strength and the geosynthetic strength. The FHWA Geosynthetic Reinforced Soil - Integrated Bridge System (GRS-IBS) design method is presented, taken from FHWA (2018). The procedure is for walls supporting bridge decks but is presented here for unloaded walls. External stability concerns mirror those for the NCMA (1993) procedures. Required features of this method are modular block facing; close reinforcement spacing (30 cm); geosynthetic/modular block connection failure is not a failure state; frictional connection of geosynthetic to the modular blocks is adequate (no connection calculation needed).

Figure 10.8 Photo of a MSE wall with shotcrete face (courtesy of Bob Barrett).

Most of the design procedures including dimensions of reinforcement, reinforcement spacing, and wall dimensions are based on testing and experience. Equation (10.8) is used to calculate the required reinforcement strength, T_{req}, at given depths:

$$T_{req} = \left[\frac{\sigma_h}{0.7^{\left(\frac{S_v}{6d_{max}}\right)}} \right] S_v \qquad (10.8)$$

where
  S_v is the vertical reinforcement spacing
  d_{max} is the maximum grain size in the backfill
  σ_h is the total Rankine horizontal stress in the reinforced zone at a given depth, all in
      consistent units

Note that σ_h changes with depth; hence the required reinforcement strength changes with depth. T_{req} is calculated for each layer. To avoid jobsite confusion, it is typical to specify the same (largest) of these T_{req} for each layer to avoid using several geosynthetics, each with a different strength. The horizontal stress, σ_h, at any given depth is a result of the weight of the materials above that depth. These may include the pressures due to reinforced soil self-weight, road base, and non-areal loads on the backfill (e.g. bridge girders). Methods of calculating horizontal stress from non-areal surface loads are given by US Navy (1986) and many geotechnical textbooks.

The allowable tensile strength, T_{allow}, considers the lesser of T_{req} (Equation 10.8) and $T_{2\%\,strain}$, the tensile force needed to produce 2% strain (to limit deformations). The latter is provided by testing. The lesser is called T_f. T_{allow} is calculated

$$T_{allow} = \frac{T_f}{FS_{reinf}} \qquad (10.8)$$

T_f is chosen so that T_{allow} is greater than T_{req}. FHWA (2018) recommends $T_f \geq 70$ kN/m for abutments supporting bridges. For walls supporting bridge decks, a reinforcement Factor of Safety (FS_{reinf}) of 3.5 is recommended. For non-load-carrying walls, lower FS_{reinf} may be selected, based on experience.

The reinforcement lengths vary from shorter at the bottom to longer at the top. The minimum base width B (see Figure 10.2) is the greater of 2 m, or 0.3H, where H is the height of the wall. The reinforcement length increases to 0.7H at the top of the wall. Longer lengths may be used.

The first layer of reinforcement (and, thus, the first layer of facing blocks) is underlain by a granular bearing pad, shown in Figure 10.9. Note the reinforcement lengths need not be the same. The depth of the pad is 0.25B and the width of the pad is 1.25B as shown in Figure 10.9. If wall toe scour is anticipated, wrap the bearing pad in a geotextile with a tensile strength of at least 70kN/m. A construction photo is shown in Figure 10.10.

Backfill soil properties are critical to the success of the design and the long-term performance of the system. Soil permeability, strength, durability, and compressibility must be considered. Open-graded soils are recommended for walls that may see any degree of submergence. The ideal soil should be durable aggregates, made of angular particles, and free of organics. Well-graded crushed aggregates, ranging in size from 5 cm to minus 200 material with PI ≤ 6, are recommended for typical applications. For flood-prone areas, where backfill drainage is more critical, the open graded AASHTO no. 89 gradation is suggested. The bearing pad should be well-graded material, which compacts particularly well, reducing opportunity for settlement.

The top three courses of concrete facing blocks are hollow, filled with grout and one piece of light reinforcing steel to hold them together, reducing the opportunity for displacement by vandals. If coping is used, it should be attached with a construction adhesive.

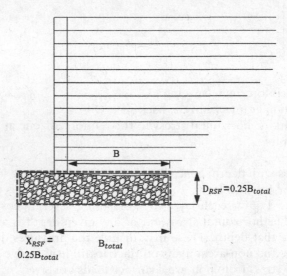

Figure 10.9 Typical granular bearing pad dimensions (after FHWA 2018).

Figure 10.10 Photo of placement of the granular bearing pad for first layer (courtesy of Defiance County, Ohio, Engineer).

Figure 10.11 Photo of bridge in Defiance County, OH, USA, resting on MSE abutments (courtesy of Defiance County, Ohio, Engineer).

The FHWA (2018) procedure describes design of walls with significant vertical loads, such as bridge girders. Also, bearing capacity, settlement, and overall stability must be checked. Block stability (e.g. bulging of the wall facing blocks) is not an issue. Figure 10.11 provides an excellent example of an MSE wall used as a bridge abutment for a highway bridge in Defiance County, Ohio, USA.

10.2.6 External stability

Slope stability analysis must be done for walls on sloping ground. Potential slope failure surfaces through and behind the reinforced zone must be considered. In addition, walls on slopes may change the groundwater and surface water runoff patterns, which must also be considered.

Bearing capacity of the reinforced mass must be checked. The width of the "footing" is taken as the length of the lowest layer of reinforcement. Live and dead loads are considered to act on this footing, and conventional bearing-capacity analysis is applied. Some designers consider P_a to cause an eccentricity in the resulting force on the footing, and design for this. Others consider the footing so flexible (soil on soil) that no eccentricity can develop. Bearing-capacity analysis is covered in, for example, Coduto et al. (2001) and Das (2015).

Settlement of the foundation soil and its effect on adjacent and underground structures must be considered. MSE walls are quite flexible and unlikely to fail structurally due to settlement. However, structures on the reinforced and backfill soils, adjacent to the wall, and under the wall may be damaged by wall settlement. Settlement may change the backfill and slope drainage patterns, leading to foundation and retained soil saturation. Foundation settlement calculations are found in many references (e.g. Duncan et al. 1987).

Some designers and clients require overturning analysis, as it is for rigid retaining walls. Since mechanically stabilized earth is extremely flexible, and has virtually no tensile strength, this analysis is not relevant. Nonetheless, if the client requires it, consider the reinforced soil zone rigid and perform the conventional analysis.

10.2.7 Typical factors of safety

Typical factors of safety with respect to failure modes are (Elias et al. 2001; Elias and Christopher 1996):

1. Sliding: 1.5
2. Bearing capacity: 2.5
3. Deep-seated stability: 1.3
4. Compound stability: 1.3
5. Seismic stability: 75% of static factor of safety
6. Pullout resistance: 1.5

10.2.8 Inclusions in the backfill

Horizontal and vertical inclusions in the backfill, such as guard rail posts and drop inlets, may be incorporated in the backfill during construction. The integrity of the reinforcement must be assured. Horizontal inclusions are easily accommodated by laying the inclusion between layers of reinforcement during construction. This may require altering the vertical reinforcement alignment. Berg et al. (2009) provide advice on construction of MSE walls incorporating backfill inclusions. Figure 10.12 shows a large vertical inclusion. Notice the metal reinforcing in Figure 10.12.

10.2.9 Drainage

Since water reduces soil strength, all retaining walls must incorporate drainage. All walls should have adequate drainage, internal and external. Surface drainage should remove rainfall and snowmelt before it soaks into the retained fill. The backfill must have adequate internal drainage to remove groundwater, perched water, and water from flooding and tidal action. This is also important for conventional gravity walls, but more so for MSE walls, where the soil is a structural component.

Three drain locations are available: on the backfill surface, immediately behind the facing, and on the subgrade. Drains behind the facing serve to reduce seepage through the wall

Figure 10.12 Photo of MSE wall under construction with large vertical inclusion (Berg et al. 2009).

face, reducing staining. FHWA (1975), USACE (1993), Terzaghi et al. (1996), and Holtz et al. (2011) provide general design procedures for filters and drains. Elias et al. (2001) give design procedures for MSE wall drains. Surface drainage is provided by sloping the backslope and foreslope and providing ditches to carry surface water away from the wall. It is important to avoid infiltration at the base of the wall, an often overlooked area.

Retaining walls built on slopes may change the drainage patterns on the slope, potentially leading to backfill saturation and failure. Sloping the backfill and including drainage swales at the top of the wall and at the foot of the wall are particularly recommended.

All soil drains require a filter. The filter must not clog or blind. When the filter clogs or blinds, porewater pressure increases, leading to loss of soil strength. The hydraulic gradient increases at the soil/drain interface, and, hence increases the seepage force which moves small soil particles into the drain. If unfiltered, these small particles will blind or clog the drain. Filter design using soils (graded granular filters) is described in USACE (1993). Filter design using geotextiles is described in Chapter 3.

10.2.10 Other considerations

Design procedures for walls with corners, seismic considerations, limited backfill distance to rock, face penetrations, or used as integral bridge abutments are given in Elias et al. (2001) with an update by Berg et al. (2009). Also note that MSE walls are usually designed using software.

10.2.11 Construction guidelines

Construction procedures are essential to MSE success. Over 75% of MSE wall failures are due to poor construction (Koerner 2012b). MSE walls are much more susceptible to failure due to poor construction than gravity walls. The general construction procedures are:

1. site preparation,
2. construction of subgrade and base of wall drainage features,
3. placement of initial reinforcement layer and facing,

4. soil placement and compaction,
5. repetition of 3 and 4 until desired height is reached, and
6. completion of surface drainage features.

Inspection is needed at all stages of construction.

MSE walls require good soil compaction of the retained soil (that between the layers of geosynthetics) and especially next to the facing. The primary cause of MSE failure is poor compaction immediately behind the facing. Contractors are reluctant to compact immediately behind the facing for fear of displacing the facing, leading to unsightliness. Thick lifts (due to wide reinforcement spacing) invite negligence. For modular block walls, using geosynthetic reinforcement between every course virtually requires the contractor compact in lifts no thicker than the block height.

The facing foundation and the backfill foundation should be prepared equally, to equalize settlement. If a modular block facing rests on an improved foundation, and the backfill does not, differential settlement will occur, with the potential to tear (shear) the geosynthetic across the back edge of the blocks. Blocks with rounded edges on the horizontal surface are recommended, as they reduce stress on the geosynthetic against the block should there be differential settlement.

The geosynthetic must be attached to the blocks. For modular block facings, it is suggested that the geosynthetic be placed across the full width of the block (towards the front of the blocks) before placing the next layer of blocks. This frictional connection will be adequate, for, if the backfill soil is properly compacted, there is very little soil pressure on the blocks. If precast panel facing is used, a positive connection, such as a bodkin joint, is required. Failure to do this results in facing detachment.

Geosynthetic placement is critical. Avoid wrinkles, which do not develop tensile strength needed to hold retained soil together. While geogrids don't usually wrinkle, geotextile reinforcement does. Anisotropic geosynthetics must be placed with the strong direction perpendicular to the wall face. Strip geosynthetics should be placed perpendicular to the wall. Seams in the geosynthetic should be avoided. If required, seam strength (which is less than the geosynthetic strength for geotextiles) must be as large as the design width strength. When the wall face is curved, the geosynthetic reinforcement will overlap. Overlapping geosynthetics should be separated by a soil layer to generate friction. A few centimeters of soil will suffice.

As in all construction projects, ongoing communication between the designer, owner, and contractor reduces problems.

10.3 MECHANICALLY STABILIZED EARTH WALLS USING METAL REINFORCEMENT

10.3.1 Introduction

Metal strips have long been used to stabilize earth masses to create walls and steepened slopes. This procedure is very similar to the use of geosynthetics, with the same outcome. Instead of plastic sheets, strips of metal, often about 7.6 cm wide, are embedded in the soil during construction. Like geosynthetic reinforced structures, metal-reinforced structures are built from the ground up. Having begun in the 1960s, metal reinforcement predates geosynthetic reinforcement. While metal mesh is also used, it is not included in this coverage. NYS DoT (2007) provides some background.

Metal reinforcement serves the same purpose as geosynthetic reinforcement – providing internal tensile strength to the soil. Figure 10.12 shows the arrangement using metal instead of geosynthetic. While the strips may appear to be tiebacks, they are not. Rather, they develop friction between the soil and the strip, transferring some of the weight of the soil to shear stress on the strip. If the strips are long enough and have enough friction, they resist pullout, and if they are strong enough, they resist breakage – the two main failure modes. Modern use of metal reinforcement is attributed to Henri Vidal (Vidal 1969).

10.3.2 Differences between metal and geosynthetic reinforcement

Metal strip reinforcement, unlike geosynthetic reinforcement, must be connected solidly to the wall facing elements. Bolts are commonly used. A typical bolted connection is shown in Figure 10.13. Failure to make this connection properly leads to facing failure. Because the strips are not laterally continuous, the soil confinement effect provided by geosynthetic reinforcement is not present. Hence, the design procedures are different, involving bilinear earth pressure diagrams and three-dimensional tributary areas for estimating the force in the metal strips. That is, each strip carries the soil force on the tributary area of the facing between it and the adjacent strips.

Metal strip reinforcement walls often use large (2 m² to 3 m²) concrete panels with embedded tabs to connect the metal strips, since connection strength is critical to success in these designs. Loose or missing bolts, and bolts made of dissimilar materials from the connecting tab have resulted in corrosion and caused facing failures (Been 2011). Other facings are available, but still require positive connection of the strips to the facing. Welded wire mesh and bar mats can also be used as reinforcement. The mats may be attached to precast concrete facing elements, or cast-in-place facing, or may be formed up to create the facing. The latter offers extreme flexibility, attractive to contractors, and useful when settlement is expected, as it doesn't crack. Figure 10.14 shows an MSE wall with wire mesh facing.

All steel used in MSE is susceptible to corrosion. Coatings, including galvanization, protect steel from chemicals and water in soil, extending life. Engineers should avoid low pH backfill soils. Unlike tieback construction, sacrificial anodes are not used to reduce corrosion. Elias et al. (2009) provide corrosion advice.

10.3.3 Failure modes and typical factors of safety

Metal reinforced MSE walls must be checked for these failure modes (similar to geosynthetic MSE wall failure modes): reinforcement pullout, reinforcement breakage, seismic

Figure 10.13 Photo of bolted connection for metal reinforcement (Berg et al. 2009).

Figure 10.14 Photo of MSE wall with wire mesh facing (courtesy of Tensar International Corporation).

loading, overall (slope) stability, settlement, sliding, and bearing capacity. Overturning is not relevant, as the reinforced mass is too flexible to overturn. The all-important internal and external drainage considerations for geosynthetic MSE walls must be applied to metal-reinforced MSE walls.

Typical factors of safety with respect to failure modes are the same as for geosynthetic reinforced walls with the addition of these (Elias et al. 2001):

1. steel strip reinforcement strength = (0.55) (steel yield strength), and
2. steel grid reinforcement = (0.48) (steel yield strength) when connected to concrete panels or blocks.

10.3.4 Inclusions in the backfill

Horizontal and vertical inclusions may be incorporated in the backfill during construction, as in geosynthetic reinforced MSE. Splaying the reinforcement around inclusions, or attaching the reinforcement to vertical inclusions, are possible solutions. Berg et al. (2009) FHWA NHI-10.024 provide advice on constructing MSE walls incorporating backfill inclusions. Figure 10.12 shows a large, vertical inclusion in a metal reinforced backfill.

10.3.5 Construction guidelines

The general construction procedures for metal reinforced MSE walls follow those for geosynthetic reinforced walls. In addition, construction guidelines for metal reinforced MSE walls include attachment of the metal strips to the facing. As in all construction projects, ongoing communication between the designer, owner and contractor reduces problems.

10.4 REINFORCED SOIL EMBANKMENTS ON FIRM FOUNDATIONS USING GEOSYNTHETIC AND METAL REINFORCEMENT

10.4.1 Introduction

Slopes are non-vertical, non-horizontal ground surfaces. There are two types: cut slopes, made by removing soil, and fill slopes (embankments), made by adding soil. This section discusses embankments built with the addition of mechanical reinforcement, metal or geosynthetic, which greatly improves the composite soil strength. The resulting structures are called *steepened slopes* because they stand steeper than unreinforced slopes. The increase in strength allows the slope to be much steeper than if it were constructed of soil alone. Most soil slopes can safely stand at ~30° from the horizontal. Reinforced slopes may safely stand at ~75° from the horizontal, or steeper. There are many advantages to this, such as creating space at the top of a hill, for, say, lane widening for roads. Figure 10.15 shows parts of a steepened slope.

Slopes constructed with reinforcement create a steeper slope face, which creates more space at the crest of the slope, for a given toe location. This extra space may be used for buildings, parking lots, storage, lane widening (roads), or other construction. If the reinforcement extends beneath a building, the reinforcement enhances bearing capacity (under-footing reinforcement effect). See Sections 9.6, 9.7, and 9.8.

Similarly, for a given crest location, a steepened slope requires less property because the embankment footprint is smaller. When used for roadway lane widening, steepened slopes avoid purchasing additional right-of-way.

Steepened slopes are useful when surcharging because they reduce the embankment footprint. This is particularly useful when property lines are near the edge of the surcharge (e.g. Shah et al. 2008). Earth dams are occasionally retrofitted with berms to improve downstream stability. Berms built with steepened slopes allow a greater weight to be placed in the same footprint. This is particularly attractive in environmentally sensitive areas. Similarly, dam slopes themselves can be constructed steeply using reinforcement (e.g. Dewey 1989; Engemoen and Hensley 1989).

It is desirable to use reinforcement when reconstructing failed slopes. This often allows reuse of onsite soils, at considerable savings. Here, layers of reinforcement are embedded in the soil during the re-placement and compaction operation.

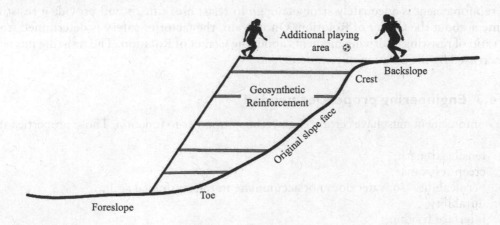

Figure 10.15 Schematic of slope, showing how to create space at the top of a slope with using steepened slope technology.

Any reinforcement material with tensile strength may be used. The most commonly used materials are geogrids, geotextiles, metal strips (often about 7.6 cm wide), and welded wire mesh. Chain link fencing and tire sidewalls (wired together) have been used with success (Barrett 2007). In principle, any continuous material with tensile strength can work. Geotextiles and welded wire mesh are most popular (2012).

Steepened slopes offer the following advantages over conventional embankments:

1. less fill required for a given crest location,
2. more horizontal ground created at crest or toe (or both),
3. faster construction since compaction is faster (less soil to compact),
4. better compaction of slope face because reinforcement allows heavier compactors closer to the crest,
5. less right-of-way needed for given crest location, and
6. less environmental impact due to smaller footprint.

Disadvantages include:

1. more engineering analysis needed,
2. reinforcement cost (often minimal),
3. more attention to fill placement needed, since filling is interrupted by reinforcement placement,
4. construction inspector has additional duties (reinforcement inspection), and
5. excavations in the fill that sever the reinforcement cannot be made after construction.

10.4.2 Philosophy of how reinforcement for steepened slopes works

Steepened slopes stand because the reinforcement adds tensile strength to the soil. The strengthened soil resists slope failure. Conventional circular arc slope stability analysis analyzes stability based on the sum of the moments about a Center of Rotation. The weight of the soil mass creates a driving moment, while the shear strength of the soil creates a resisting moment. The slope factor of safety is the ratio of these moments (resisting/driving). Reinforcement creates additional resisting moments. Figure 10.16 shows layers of reinforcement creating resisting moments. If the reinforcement extends beyond the failure surface, and this length of reinforcement is adequately long enough to resist pullout, and the reinforcement is adequately strong enough to resist breaking, it will provide a resisting moment about the Center of Rotation. Once again, the factor of safety is determined from the ratio of resisting to driving moments about the Center of Rotation. The resisting moment is increased by the addition of reinforcement.

10.4.3 Engineering properties needed

The reinforcement must have certain engineering properties to function. Those properties are:

1. tensile strength,
2. creep resistance,
3. permeability (so water does not accumulate in the reinforced soil),
4. durability,
5. interface friction,
6. flexibility, so the reinforcement conforms to the ground, and
7. ability to survive installation.

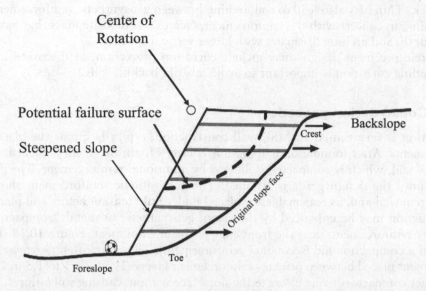

Figure 10.16 Layers of reinforcement creating resisting moments.

Figure 10.17 Photo of terraced, steepened slope with metal wire facing (Berg et al. 2009).

Some reinforcements (particularly steel) may be angled upwards at the slope face to assist in slope face stability. The resulting slope may have a terraced face. Figure 10.17 is an example.

10.4.4 Design notes

Steepened slopes are large compaction projects. The better the compaction, the better the chances of success. Designers are encouraged to make the spacing between reinforcement layers about 0.3 m. This requires the contractor to compact lifts that are never more than

0.3 m thick. Thin lifts also lead to soil arching between geosynthetic reinforcement layers. This arching, in concert with the reinforcement, creates a composite mass that has tremendous strength. Soil arching dissipates with larger vertical spacings.

Metal reinforcement designs may include corrosion protection, as discussed in section 10.3. Avoiding corrosion is important so avoid low pH backfill soils.

10.4.5 Construction procedure

Construction is very similar to MSE wall construction, typically minus the placement of facing elements. After foundation preparation, a layer of reinforcement is placed, followed by a lift of soil, which is compacted, followed by additional reinforcement. The process is repeated until the design grade is attained. The geosynthetic reinforcement should have wrinkles removed and, as reasonable, be placed under mild tension before soil placement.

Construction may be enhanced by the use of geosynthetic or metal "compaction aids" (secondary reinforcement) near the front edge of the embankment. Figure 10.18 shows the concept of a compaction aid (secondary reinforcement). These are often narrower strips of reinforcement placed between primary reinforcement layers. Typically 1 to 1.5 m long, they allow larger compactors to get closer to the slope face without causing soil failures, resulting in better compaction. Without them, lighter compactors must be used in the typical 1.5 m setback from the face. The use of heavy compactors near a fill slope face accounts for the large incidence of surficial slope failures. Geofibers (Chapter 3) are also effective compaction aids.

The slope face may be faced by welded wire mesh, geogrids, or geotextiles. The wire mesh is folded, vertically or sloping, to help retain the surface soils. Occasionally sod is placed, facing out, immediately behind the mesh to assist in erosion control and aesthetics. Similarly, to improve face stability, geotextiles may be wrapped over the succeeding soil lift, and tucked in. Figure 10.7 shows this schematically. Figure 10.19 is an example. The sandbags are optional, but speed construction.

10.4.6 Inclusions in the backfill

As with MSE walls, horizontal and vertical inclusions may be incorporated in the backfill during construction. Splaying the reinforcement around inclusions or attaching the

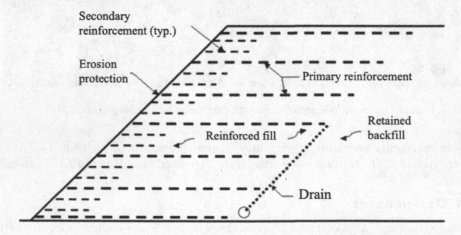

Figure 10.18 Schematic of mechanically stabilized earth slope, showing geosynthetic used as interlayer compaction aids (after Berg et al. 2009).

Figure 10.19 Photo of wrapped face slope using sandbags for formwork.

reinforcement to vertical inclusions are possible solutions. Design details for vertical inclusions can be found in Berg et al. (2009).

10.4.7 Internal stability: pullout and breakage, internal slope stability

Pullout and breakage analysis must be done. The reinforcement must be strong enough to resist breakage while developing enough soil friction to resist pullout. These mechanisms are described in section 10.2.4.2 and 10.2.4.3.

The slope must be analyzed for internal slope stability. Conventional slope stability software is used, accounting for the reinforcement, which acts to strengthen the soil. Facial stability requires engineering judgment.

10.4.8 External stability: bearing capacity, sliding, and settlement

All geotechnical construction requires global stability evaluations. These include bearing capacity, sliding, and settlement. Bearing capacity is checked with conventional bearing-capacity analysis, using the length of the reinforcement as the width of the footing. Some designers include the lateral earth force from the embankment soil behind the tail of the reinforcement. This force is considered to cause an eccentricity in the base of the reinforced zone, which effect is accounted for in conventional bearing-capacity calculations. Other designers consider the embankment far too flexible to sustain an eccentricity. Sliding analysis follows the procedure for MSE sliding analysis (see section 10.2.4.1).

Settlement analysis, perhaps the most overlooked and the most troublesome, follows conventional analysis. The stress distribution beneath sloped surfaces, required for analysis, is given in section 9.4.5.6. Settlement of the embankment disrupts drainage, which may lead to failure in the embankment. Surface and subsurface structures may be affected by embankment settlement. Leakage from disrupted pipes in the embankment can easily cause slope failure.

10.4.9 Slope face stability: veneer instability, erosion control, and wrapped faces

Steepened slope faces have an increased propensity for erosion and veneer instability. The faster runoff increases erosion. A steepened slope has greater shear forces on the face, leading to veneer instability. Geosynthetic or metallic wrapped faces may be used to abet erosion, in addition to conventional erosion control practices. Many erosion control products assist in veneer stability, particularly turf reinforcement mats, which are buried slightly below the surface during construction. Veneer stability is particularly difficult to assess because of the low confining stresses on the slope face. The very steepest faces may require concrete facing.

10.4.10 Drainage

Drainage for steepened slopes should follow the same advice given in section 10.2.9 concerning drainage for MSE walls. Like all geotechnical structures, drainage is essential for success as soil saturation reduces soil strength, and, thus, structural stability.

10.5 SOIL NAILING

10.5.1 Introduction

Soil nailing is a method of stabilizing soil by inserting steel rods or cables in the in situ soil. This creates a reinforced soil mass strong enough that it won't move. Soil nailing is used for top-down slope and wall support, for remedial, temporary and permanent construction of slopes and walls. As an excavation proceeds, soil nails are installed to stabilize the soil. Soil nails, unlike tiebacks (Schnabel 2002), are not prestressed, and, hence, are called passive reinforcement, a scheme used in tunneling since the 1960s (e.g. Lang 1972; Elias and Juran 1991). The essence of the soil nailing technique is adding tensile strength to the soil. The tensile strength holds the soil mass intact. The steel reinforcement develops tension at required small soil deformations, allowing the soil to be strengthened without distress to the overall project. Figure 10.20 is a schematic of typical soil nail installation.

In principle, MSE walls and soil nail walls function the same way. That is, both reinforce the soil to add tensile capacity. The difference is that MSE walls are constructed from the bottom-up and the reinforced soil is fill, whereas soil nail walls are constructed from the top-down and the reinforced soil is in situ soil. Soil nailing is also used for temporary support of excavation. Tiebacks can appear similar to soils nails but they function in a fundamentally different manner. Tiebacks transfer load deep behind the face whereas soil nails are a soil reinforcement method.

A permanent wall or slope facing is required for permanent installations. Typically, shotcrete (Bernard 2010) is used for a facing. Harder, more attractive, facings, such as masonry or concrete, may overlie shotcrete facing. The concrete may be cast in place, or precast.

Compared to tieback installations, soil nails are closely spaced. The soil nails are typically sloped 10° to 20° below horizontal for wall and excavation support. When soil nails are used for landslide repair, the nails may be at any angle, with the intent of orienting the nails

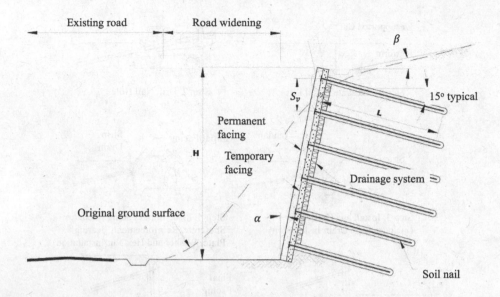

Figure 10.20 Typical soil nail installation for road widening (after Lazarte et al. 2015).

Figure 10.21 Photo of soil nail wall (courtesy of JC Baldwin Construction).

perpendicular to the sliding surface to resist shear. Micropiles are also used in this application. Figure 10.21 shows a photo of a soil nail wall with a shotcrete facing and illustrates the spacing of the soil nails.

To construct a soil nailed wall, the excavation is done in stages as illustrated on Figure 10.22. As shown, first a near-vertical cut is made to a depth that is safe and stable. Next, the soil nails are installed by drilling and grouting. As shown on the figure, a strip drain is installed and then covered with shotcrete as an initial facing. The process repeats itself to the final grade and, if required, a final facing is installed.

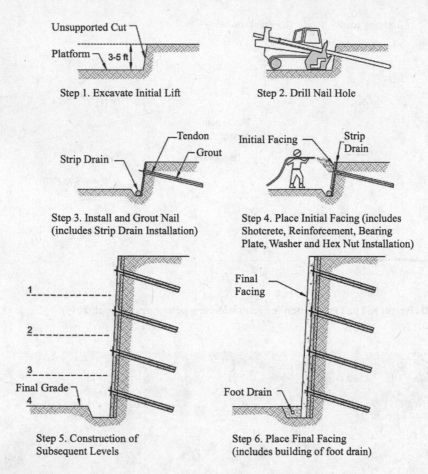

Figure 10.22 Typical sequence of construction operations for soil nail wall (after Lazarte et al. 2015).

10.5.2 Applications

The earlier soil nailing applications (1980s) were primarily temporary excavation support. Soil nailing filled the need to hold a slope in place temporarily while a permanent retention system (such as a wall) was installed, or until construction in the excavation was complete and the need for support ended. Currently, soil nails are used for permanent wall construction and slope stabilization. Cut slopes and failed slopes may be stabilized with soil nails. The US Federal Highway Administration published design and inspector's manuals for soil nailing (FHWA 1998, 1999).

Fang (1990) lists these advantages and applications of soil nailed cut slopes:

1. reduce excavation volumes for concrete work near footings,
2. reduce amount of cutback, construction disturbance, and right-of-way acquisition, by providing steeper slopes which is important in areas that are built, or expensive to acquire,
3. less concrete is needed for the wall facing due to reduced soil pressures,
4. eliminate need for backfill behind walls,
5. provide shorter construction time compared to gravity or tieback walls due to reduction in material volumes and soil nails do not require post-tensioning,

6. may be used for temporary excavation support while constructing the permanent facility,
7. used behind (or above) a retaining wall as they reduce the pressure on the wall thereby reducing need for wall foundation piles to support the wall structure (particularly when there is a steep backslope), and
8. can be used in highly sensitive soils, as less soil disturbance occurs than with conventional excavation.

Soil nails offer other advantages over conventional construction. In temporary construction excavation, soil nails remove the need for cumbersome bracing that obstructs workspace. Soil nails are cheaper than conventional bracing in cuts of more than 4.6 m to 6.1 m deep and/or widths of greater than 18.3m. Finally, there is an increase in public safety since less construction area is required (PileBuck 1990).

Soil nails walls are not the solution for all situations. Some disadvantages include:

1. They typically cost more than conventional walls.
2. In clays, groundwater drainage systems may be hard to build and maintain.
3. In soft, clayey soils, creep may allow long-term displacements.
4. Soft, clayey soils may require long nail lengths to achieve adequate pullout capacity of the nails.
5. The metal nails may degrade in corrosive soils.
6. Adjacent construction may relieve soil stresses, loosening soil nails.
7. There may be the need for underground easements behind the slope or wall face.
8. Soil nails may interfere with future utility installation.
9. Backfill trenches, which sever the nails, can't be installed after construction.
10. Some surface soil deformation will occur, however small.

Soil nailing is widely used for temporary support of excavation applications. A popular use is temporary ground support in building foundation excavations requiring soil support before permanent ground support such as by a basement wall.

The use of soil nailing for permanent applications has been increasing as expertise and experience with the method increases. Support of retaining walls is a major use of soil nails in permanent applications. Although soil nails are not usually used to support building foundation walls since underground easements are required from adjacent property owners, there are cases such as a steeply sloping site which make the use of soil nails attractive. Soil nails may be used to prevent a landslide or to stabilize slopes which have experienced movement. Fractured rock may be stabilized using soil nails. Anchors are used to support marine structures such as walls at harbors. Finally, soil nails may be used to repair or alter existing walls and abutments.

There are alternatives to soil nails such as tiebacks. Typically, tiebacks (cables or rods) are inserted into sloping drilled holes. The bottom of the tieback is grouted in place. The cables or rods are tensioned and tied off at the surface against a bearing plate, putting the soil in compression, increasing its strength (Schnabel 2002). Other alternatives include relocating the project or grouting the slope so it can be cut more steeply, or in the case of failing slopes, stop movement.

10.5.3 Applicable sites

Soil nails may be used economically only at some sites. Sites with soils containing obstructions, or rocks that must be drilled through, or near surface bedrock are not candidates for soil nailing. Excavation support applications require soils with adequate stand-up time for a one meter or more face (approximately) to install the nails. It is preferable that holes drilled in the soil remain open and dry for enough time to insert and grout the soil nail. Temporary casing may be used to assist grouting in collapsible soils. Dry, loose sands, for example, rarely exhibit adequate stand-up time. It is preferable to perform a field trial before construction to estimate the stand-up time.

Stand-up time, the length of time an unsupported soil will stand vertically, is critical to success. The soil must stand up long enough for the excavation to be made, nails installed and shotcrete (or other) facing applied. Stand-up time is evaluated in the field, by testing, a trench is dug, and the time to slumping observed. While the surface soil may have adequate stand-up time, deeper layers may not. These must be identified in the site investigation. Similarly, a soil with adequate stand-up time above the water table may be unstable below the water table. Dry, uniform sands typically do not have a stand-up time; similarly, saturated uniform sands. These are called running sands, or fluidized soils. Stand-up times can be increased by grouting or soil freezing. Soils with minimal stand-up times may be shotcreted immediately after excavation, and before drilling and installation of soil nails.

Groundwater conditions affect soil nail construction. High groundwater velocities will affect the grout by washing it away from the soil nail before the grout can set. Very porous soils, with high groundwater flows, are particularly to be avoided. Similarly, the grouted soil nails must not restrict normal groundwater flow, lest high pore pressures and low soil strength ensue. Normally, the grout does not affect groundwater flow, though high grout takes are cause for concern (HCL 2007).

There are other considerations. Every excavation site must be evaluated for base heave (Terzaghi et al. 1996) and seepage problems. Soft clay sites may creep excessively, causing wall damage or unacceptable slope displacement. Such sites may require a large number of exceedingly long soil nails. Large amounts of seepage will interfere with stand-up time and will present shotcreting, seepage controls, and ground control problems during and after construction. Sites with high water tables require dewatering with the attendant problems of settlement and slope instability. Buried structures can interfere with successful soil nailing. Backfilled trenches present potential failure planes if poorly backfilled. Subterranean utilities or nearby surface construction sensitive to movement negate the use of soil nails, as some movement during soil nailing will occur. The site must have adequate room for the soil nailing equipment. A minimum of five meter benches is recommended. Underground easements are required if the soil nails underlie someone else's property. Highly corrosive soils may interfere with soil nail durability. Extremely loose, cohesionless soils may consolidate due to construction operations, greatly increasing the loads on the nails and adding potential distress to adjacent structures.

Owners must have permission to use the underground space associated with soil nailing. Adjacent underground parking garages, basements, and buried utilities may preclude use of soil nails.

10.5.4 Components of a soil nail system

Conventional soil nail systems consist of a steel bar, grout to hold it in the ground, a nail head, and drainage media. Launched soil nails systems, which consist solely of the nail, are addressed in section 10.5.11. The steel is typically 20 mm to 35 mm in diameter. The borehole is typically 10 cm in diameter. Steepened slopes and walls may have drainage media on the face, covered with shotcrete. Soil nail walls require the nail head and associated hardware to attach to the wall facing. Figure 10.23 is a schematic showing the component parts. All systems should include corrosion protection. While mild steel, susceptible to corrosion,

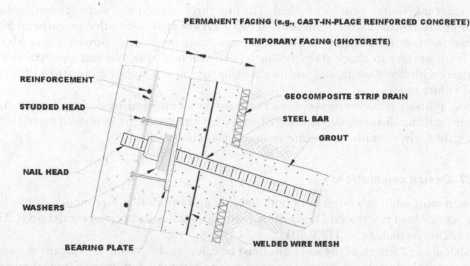

PERMANENT FACING (e.g., CAST-IN-PLACE REINFORCED CONCRETE)

TEMPORARY FACING (SHOTCRETE)

REINFORCEMENT

STUDDED HEAD

GEOCOMPOSITE STRIP DRAIN

STEEL BAR

GROUT

NAIL HEAD

WASHERS

BEARING PLATE

WELDED WIRE MESH

Figure 10.23 Parts of a soil nail system including steel bar, grout, nail head, drainage media (Lazarte et al. 2003).

is typically used, stainless steel and polymeric reinforcement have been used (e.g. Cheung and Lo 2005).

10.5.5 Methods of installing soil nails

There are two basic methods of installing soil nails: drilling and grouting, and driving. For the former, holes are drilled, the soil nail is inserted and grouted. Alternatively, soil nails may be driven into the ground using hammers (or vibrators) or using compressed air nail guns called nail launchers (section 10.5.11).

Conventional (drilling and grouting) soil nailing consists of the nail and the grout. The hole is drilled in the candidate soil; the nail (cable or bar) is inserted and full-length grouted in place. Typically, wire mesh is attached to the bare soil face, a head is attached to the nail and mesh, and the area is shotcreted.

An alternate procedure is to drive the nail in, using a percussion or vibratory hammer, depending on soil conditions and contractor choice. Debris laden soils, rocky soils and soils with particularly stiff layers make driving difficulty uncertain, and risk nail damage. These methods are rapid, do not use grout, and have limited lengths due to the energy limitations of the installation equipment. There is more uncertainty about the final nail location than with drilling and grouting, as nails may bend/displace during installation.

Self-drilled soil nails are hollow, pervious tubes that serve as the drill bit and drill stem, and then as the soil nail after installation. Grouting is done during drilling through the per-forated, hollow self-drilled nails.

10.5.6 Design of soil nailed walls

10.5.6.1 Failure modes

Soil nail design must account for several basic *internal* failure modes: local failure around the soil nails, bearing failure under soil nail head, breakage of the soil nails (including due to corrosion), pullout of the soil nails, bending or shear failure of the soil nails, connection failure at the soil nail head, drainage system failures, and combined structural and connec-tion failure at the facing. Some are shown in FHWA (2003).

Four external failures must be checked. The first three (settlement, slope (global) stability (circular arc and block sliding), and bearing capacity) are analyzed with conventional methods found in many introductory foundation engineering books (e.g. Bowles 1996; Murthy 2003). It's necessary to check stability during construction also, not just the final design, particularly with tiered walls, and with walls using, or on, soft soils. Figure 10.24 illustrates internal failure modes.

Seismic analysis is not reviewed here but must be considered in design. Liquefaction, caused by seismic disturbance, is covered in many references, including Kramer (1996). FHWA (2003) gives details on seismic analysis of soil nail walls.

10.5.6.2 Design calculations

The design must address a number of both internal and external failure modes. While limit state or service load designs (SLD) are often used, load and resistance factor design (LRFD) procedures are available (NCHRP 2011).

The spacing and length of the soil nails must be calculated. Spacing is typically estimated by the geotechnical engineer, based on experience and ground conditions. Typical horizontal spacing is 1.5 m to 2 m in staggered rows; typical vertical spacing is 1.5 m. Uniformity of spacing aids constructability. Staggered rows provide better face stability during construction but may make installation of drains more difficult.

Soil nails slope 5° to 20° below horizontal (Hong Kong 2008), with 15° being typical. The downward slope is required to keep non-pressurized grout from flowing out of the hole. Steeper inclinations may be needed to avoid intercepting near-surface buried utilities. Steep inclinations should be avoided, as nail capacity reduces non-linearly with inclination.

While software is typically used to calculate nail length, a preliminary estimate of the length may be had using the manual procedure, below. Typically, soil nail lengths vary from

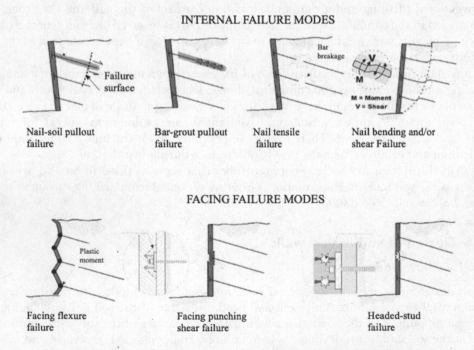

Figure 10.24 Internal and facing failure modes of a soil nail system (Lazarte et al. 2003).

60% to 100% of the wall height (FHWA 1998). Shorter nails may be used at the base, based on design. Uniform nail lengths are more typical and ease construction problems. FHWA (1998, p. 122) provides guidance on staggered nail lengths.

FHWA (1998) provides charts for preliminary design of soil nail lengths and required soil nail strengths for given face batter values and backslope angles. Figures 10.25, 10.26, 10.27, and 10.28 are used first. The results from these Figures are then used with Figures 10.29, 10.30, 10.31, and 10.32, respectively, to get the soil nail length.

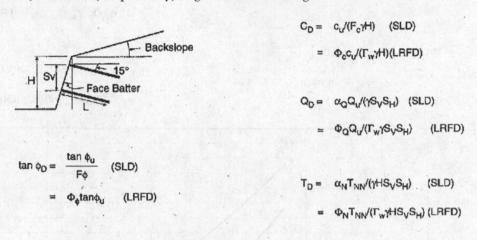

$$C_D = c_u/(F_c \gamma H) \quad \text{(SLD)}$$

$$= \Phi_c c_u/(\Gamma_w \gamma H) \text{(LRFD)}$$

$$Q_D = \alpha_Q Q_u/(\gamma S_V S_H) \quad \text{(SLD)}$$

$$= \Phi_Q Q_u/(\Gamma_w \gamma S_V S_H) \quad \text{(LRFD)}$$

$$\tan \phi_D = \frac{\tan \phi_u}{F_\phi} \quad \text{(SLD)}$$

$$= \Phi_\phi \tan \phi_u \quad \text{(LRFD)}$$

$$T_D = \alpha_N T_{NN}/(\gamma H S_V S_H) \quad \text{(SLD)}$$

$$= \Phi_N T_{NN}/(\Gamma_w \gamma H S_V S_H) \text{(LRFD)}$$

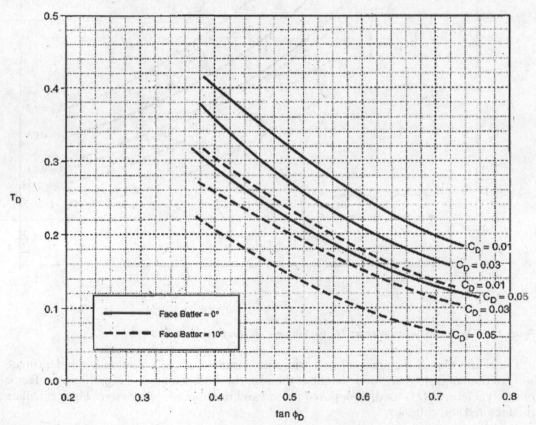

Figure 10.25 Preliminary design chart 1A, backslope = 0° (Lazarte et al. 2003).

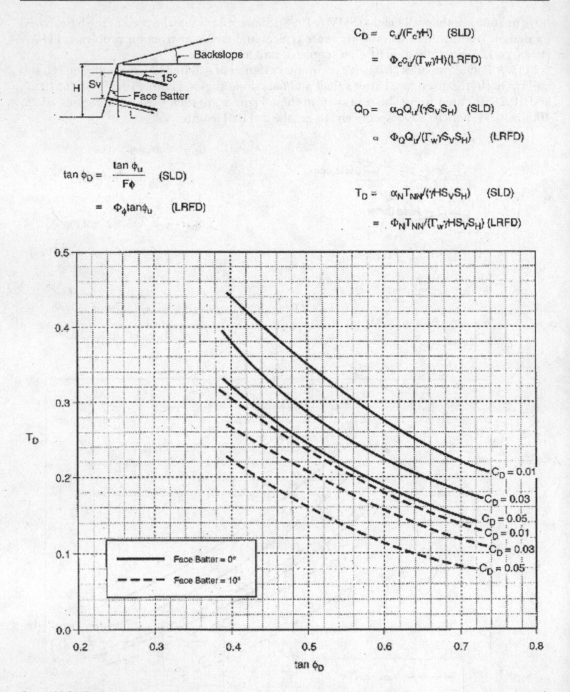

$$C_D = c_u/(F_c \gamma H) \quad \text{(SLD)}$$
$$= \Phi_c c_u/(\Gamma_w \gamma H) \text{(LRFD)}$$

$$Q_D = \alpha_Q Q_u/(\gamma S_V S_H) \quad \text{(SLD)}$$
$$= \Phi_Q Q_u/(\Gamma_w \gamma S_V S_H) \quad \text{(LRFD)}$$

$$\tan \phi_D = \frac{\tan \phi_u}{F\phi} \quad \text{(SLD)}$$
$$= \Phi_\phi \tan \phi_u \quad \text{(LRFD)}$$

$$T_D = \alpha_N T_{NN}/(\gamma H S_V S_H) \quad \text{(SLD)}$$
$$= \Phi_N T_{NN}/(\Gamma_w \gamma H S_V S_H) \text{(LRFD)}$$

Figure 10.26 Preliminary design chart 2A, backslope = 10° (Lazarte et al. 2003).

The charts require factored soil strength values for ultimate soil cohesion (c_u) and ultimate soil friction angle (ϕ_u), the unit weight of the soil, the wall height, and a preliminary factor of safety. Here, SLD is used, as opposed to load and resistance factor design. First, calculate the factored soil cohesion

$$C_D = c_u / FS \gamma H \tag{10.9}$$

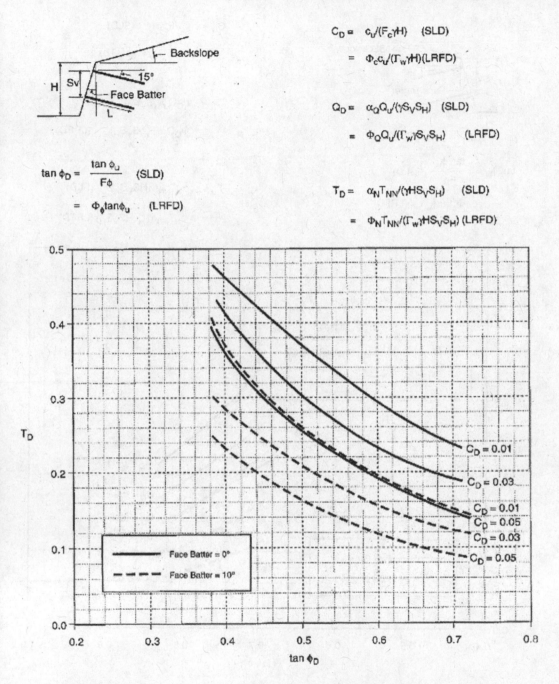

$C_D = c_u/(F_c\gamma H)$ (SLD)

$= \Phi_c c_u/(\Gamma_w\gamma H)$ (LRFD)

$Q_D = \alpha_Q Q_u/(\gamma S_V S_H)$ (SLD)

$= \Phi_Q Q_u/(\Gamma_w)S_V S_H)$ (LRFD)

$\tan\phi_D = \dfrac{\tan\phi_u}{F_\phi}$ (SLD)

$= \Phi_\phi\tan\phi_u$ (LRFD)

$T_D = \alpha_N T_{NN}/(\gamma H S_V S_H)$ (SLD)

$= \Phi_N T_{NN}/(\Gamma_w\gamma H S_V S_H)$ (LRFD)

Figure 10.27 Preliminary design chart 3A, backslope = 20° (Lazarte et al. 2003).

where
C_D = factored soil cohesion
H = full height of the wall
c_u = ultimate shear strength of a cohesive soil
FS = factor of safety (1.35 is a typical value for preliminary designs)
γ = unit weight of the soil in the wall

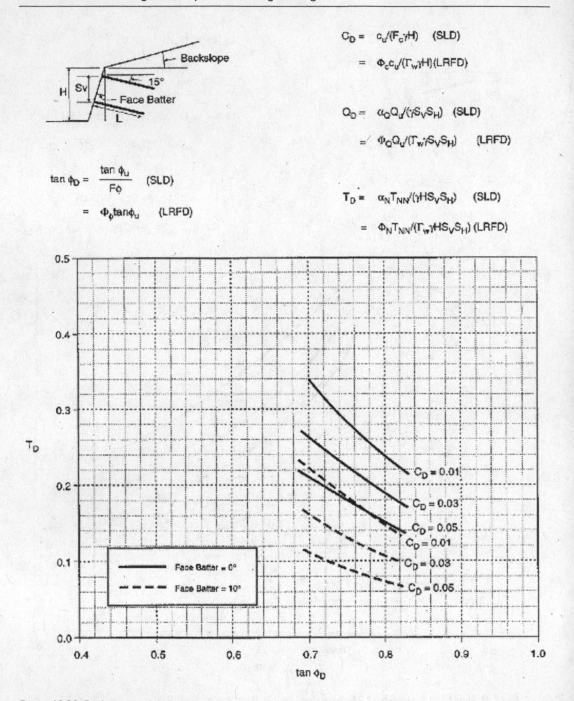

$$C_D = c_u/(F_c \gamma H) \quad \text{(SLD)}$$
$$= \Phi_c c_u/(\Gamma_w \gamma H) \quad \text{(LRFD)}$$

$$Q_D = \alpha_Q Q_d/(\gamma S_V S_H) \quad \text{(SLD)}$$
$$= \Phi_Q Q_d/(\Gamma_w \gamma S_V S_H) \quad \text{(LRFD)}$$

$$\tan \phi_D = \frac{\tan \phi_u}{F_\phi} \quad \text{(SLD)}$$
$$= \Phi_\phi \tan \phi_u \quad \text{(LRFD)}$$

$$T_D = \alpha_N T_{NN}/(\gamma H S_V S_H) \quad \text{(SLD)}$$
$$= \Phi_N T_{NN}/(\Gamma_w \gamma H S_V S_H) \quad \text{(LRFD)}$$

Figure 10.28 Preliminary design chart 4A, backslope = 34° (Lazarte et al. 2003).

The factored soil friction angle is

$$\phi_D = \arctan\left(\frac{\tan \phi_u}{FS}\right) \tag{10.10}$$

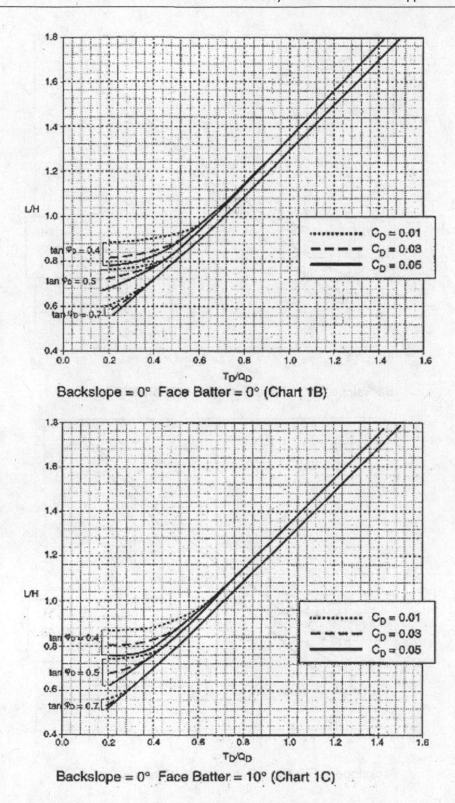

Figure 10.29 Preliminary design chart 1B and 1C (Lazarte et al. 2003).

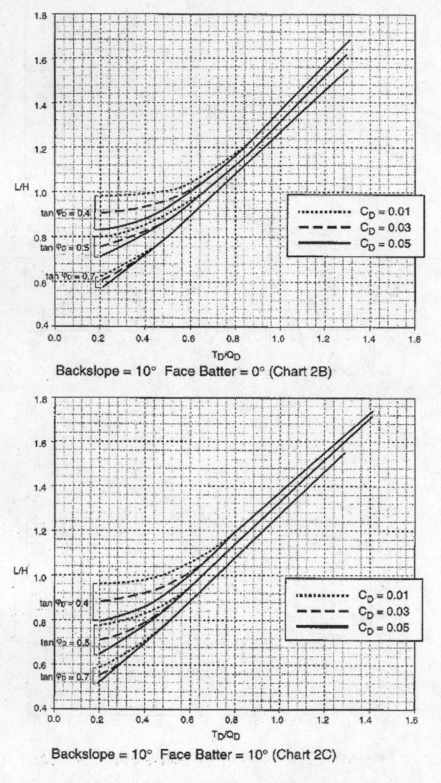

Figure 10.30 Preliminary design chart 2B and 2C (Lazarte et al. 2003).

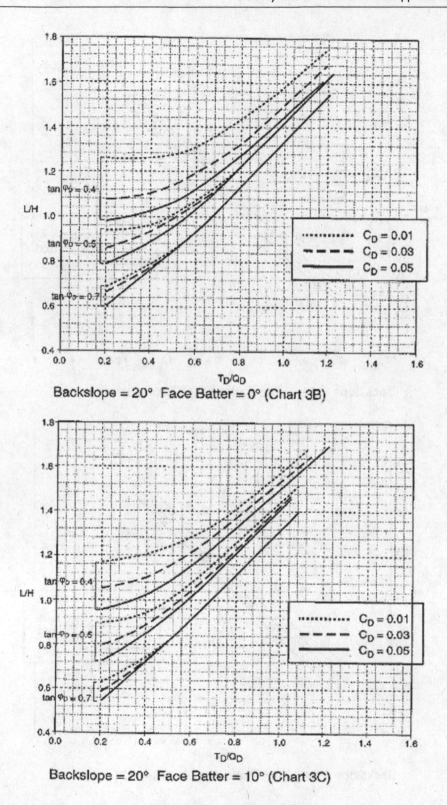

Backslope = 20° Face Batter = 0° (Chart 3B)

Backslope = 20° Face Batter = 10° (Chart 3C)

Figure 10.31 Preliminary design chart 3B and 3C (Lazarte et al. 2003).

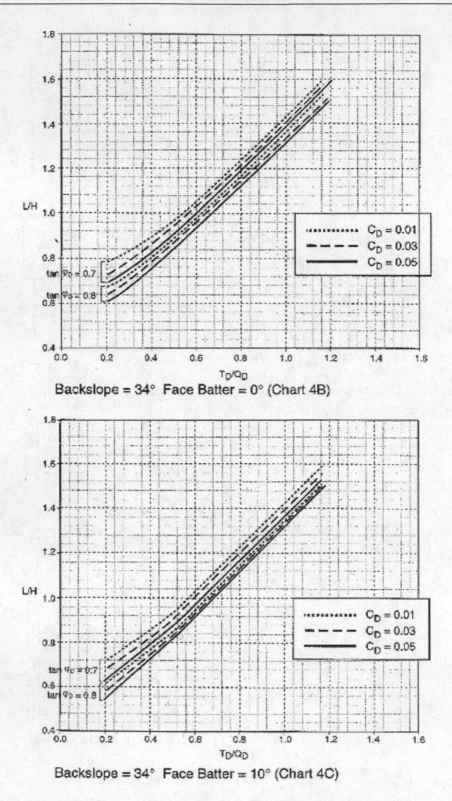

Figure 10.32 Preliminary design chart 4B and 4C (Lazarte et al. 2003).

where

ϕ_D = factored soil angle of friction

ϕ_u = ultimate friction angle

FS = factor of safety. 1.35 is a typical value for preliminary designs.

With C_D and ϕ_D known, and a chosen face batter angle and given backslope angle, one of the Figures 10.24, 10.25, 10.26, or 10.27 is then utilized. To use the figure, take the tangent of ϕ_D and enter on the abscissa, reading upwards to the appropriate curve corresponding to either 0° or 10° face batter and the calculated C_D. Then, the dimensionless nail tensile capacity, T_D, is read on the ordinate. From this, the required nominal nail strength (T_{NN}) is calculated from

$$T_{NN} = \left(T_D S_V S_H \gamma H\right) / \alpha_N \tag{10.11}$$

where

α_N = nail tendon strength factor, typically taken as 0.55

S_V = vertical nail spacing

S_H = horizontal nail spacing

γ = soil unit weight

H = wall height

The cross-sectional area of the soil nail can be estimated from the required nail strength, T_{NN}, and the yield strength of the steel used for the nail. The required cross-sectional area of the nail (bar) is calculated from

$$A_b = T_{NN} / F_y \tag{10.12}$$

where

A_b is the area of bar

F_y is the yield strength of the bar

For US standard steel bars, the proper bar is then chosen from Table 10.1.

Knowing T_D, the length of the soil nail can be estimated from the next series of figures. Calculate the dimensionless nail pullout resistance, Q_D

Table 10.1 Standard bar sizes (FHWA 1998)

Bar Designation No.	Nominal Diameter, in. [mm]	Nominal Area, in² [mm²]
3 [10]	0.375 [9.6]	0.11 [71]
4 [13]	0.500 [12.7]	0.20 [129]
5 [16]	0.625 [15.9]	0.31 [199]
6 [19]	0.750 [19.1]	0.44 [284]
7 [22]	0.875 [22.2]	0.60 [387]
8 [25]	1.000 [25.4]	0.79 [510]
9 [29]	1.128 [28.7]	1.00 [645]
10 [32]	1.270 [32.3]	1.27 [819]
11 [36]	1.410 [35.8]	1.56 [1006]
14 [43]	1.693 [43.0]	2.25 [1452]
18 [57]	2.257 [57.3]	4.00 [2581]

*Soft metric bar designation numbers, nominal diameters and areas are the values enclosed within brackets. Bar designation numbers approximate the number of millimeters of the nominal diameter of the bar.

$$Q_D = (\alpha_Q Q_u) / (\gamma S_V S_H) \qquad (10.13)$$

where

α_Q is the nail pullout resistance strength factor, often taken as 0.5.
Q_u is the ultimate unit pullout resistance of the grout against the soil in the borehole.

The ultimate unit pullout resistance is estimated from:

$$Q_u = (\text{bond strength of the soil})(\text{circumference of the borehole}) \qquad (10.14)$$

Typical bond strengths of soils are given in Tables 10.2, 10.3, and 10.4. Typical boreholes are 10 to 30 cm in diameter; the circumference is π times the borehole diameter. Q_u has units of force/length.

Table 10.2 Estimated bond strength of soil nails in soil and rock (Elias and Juran 1991)

Material	Construction Method	Soil/Rock Type	Ultimate Bond Strength, q_u (kPa)
Rock	Rotary Drilled	Marl/limestone	300–400
		Phyllite	100–300
		Chalk	500–600
		Soft dolomite	400–600
		Fissured dolomite	600–1000
		Weathered sandstone	200–300
		Weathered shale	100–150
		Weathered schist	100–175
		Basalt	500–600
		Slate/Hard shale	300–400
Cohesionless soils	Rotary Drilled	Sand/gravel	100–100
		Silty sand	100–150
		Silt	60–75
		Piedmont residual	40–120
		Fine colluvium	75–150
	Driven Casing	Sand/gravel	190–240
		Low overburden	280–430
		High overburden	380–480
		Dense moraine colluvium	100–180
	Augered	Silty sand fill	20–40
		Silty fine sand	55–90
		Silty clayey sand	60–140
	Jet Grouted	Sand	380
		Sand/gravel	700
Fine-Grained Soils	Rotary Drilled	Silty clay	35–50
	Driven Casing	Clayey silt	90–140
	Augered	Loess	25–75
		Soft clay	20–30
		Stiff clay	40–60
		Stiff clayey silt	40–100
		Calcareous sandy clay	90–140

Elias, V. and Juran, I. (1991). "Soil Nailing for Stabilization of Highway Slopes and Excavations," Publication FHWA-RD-89-198, Federal Highway Administration, Washington, D.C.
Note: Convert values in kPa to psf by multiplying by 20.9.
Convert values in kPa to psi by multiplying by 0.145.

Table 10.3 Ultimate bond stress for cohesionless soils (FHWA 1998)

Construction Method	Soil Type	Unit Ultimate Bond Stress kN/m² (psi)
Open Hole	Non-plastic silt	20–30 (3.0–4.5)
	Medium dense sand and silty sand/sandy silt	50–75 (7.0–11.0)
	Dense silty sand and gravel	80–100 (11.5–14.5)
	Very dense silty sand and gravel	120–240 (17.5–34.5)
	Loess	25–75 (3.5–11.0)

Table 10.4 Ultimate bond stress for cohesive soils (FHWA 1998)

Construction Method	Soil Type	Unit Ultimate Bond Stress kN/m² (psi)
Open Hole	Stiff clay	40–60 (6.0–8.5)
	Stiff clayey silt	40–100 (6.0–14.5)
	Stiff sandy clay	100–200 (16.5–29.0)

Next, knowing the ratio of T_D and Q_D, and knowing C_D and ϕ_D, enter the appropriate figure (either Figures 10.29, 10.30, 10.31, or 10.32) on the abscissa. Choose the appropriate curve on the figure, based on ϕ_D and C_D and read the L/H value on the ordinate. With H (wall height) known, calculate the required nail length, L = (H)(the L/H value from the ordinate of the figure).

In theory, the length of the soil nail, L, can be reduced by increasing the borehole diameter. A very large borehole diameter can reduce L to a very small number, which will lead to instability. The borehole diameter must be small enough that L is long enough to satisfy the following criteria: the width of the reinforced zone must be at least 0.6H, and foundation and overall slope stability must be maintained. The design length is estimated by using the procedure above. The designer must then verify that this length provides adequate factors of safety for sliding, bearing capacity, and slope stability, which require separate analyses.

Suggested soil nail head diameters are given in Hong Kong (2008). The diameter, which varies from 400 to 800 mm, depends on soil strength and slope angle. Figure 10.33 shows a method of calculating the required diameter.

There are other required design checks. Slope (overall) stability is reviewed in FHWA (2003). A simplified, manual analysis is presented there, to precede more sophisticated computer slope stability analysis. Seismic stability is presented in FHWA (2003). This analysis generates a factor of safety against movement due to earthquake forces. Terzaghi et al. (1996) and US Navy (1986) present analyses of base heave, which can occur in soft soils. Here, the soil squeezes out from under the reinforced structure, rather than failing in classical bearing capacity. Other failure modes, explained in (FHWA 2003) are:

1. face strength and deformation,
2. nail head design details,
3. sliding,
4. head stud failure,
5. punching of head stud, and
6. combinations of loadings including seismic, surcharge, ice, and live loads.

FHWA (2003) recommends the factors of safety shown in Table 10.5.

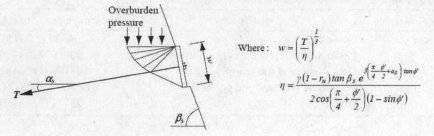

Legend:

w	Size of square soil-nail head (m)
T	Design load of soil nail (kN)
γ	Unit weight of soil (kN/m³)
α_s	Inclination of soil nail (radians)
β_s	Slope angle (radians)
r_u	Pore pressure parameter ($= u / \gamma h$)
u	Pore water pressure (kPa)
h	Depth of overburden directly above point in question (m)
ϕ'	Angle of shearing resistance of soil under effective stress condition (radians)

Note: Method after the UK Department of Transport (DOT, 1994).

Figure 10.33 Soil nail head design method recommended by the UK Department of Transport (DOT 1994).

Table 10.5 Minimum recommended factors of safety for the design of soil nail wall using the ASD method (FHWA 2003)

Failure Mode	Resisting Component	Symbol	Minimum Recommended Factors of Safety		
			Static Loads[1]		Seismic Loads[2] (Temporary and Permanent Structures)
			Temporary Structure	Permanent Structure	
External stability	Global stability (long-term)	FS_G	1.35	1.5[1]	1.1
	Global security (excavation)	FS_G	1.2–1.3[2]		NA
	Sliding	FS_{SL}	1.3	1.5	1.1
	Bearing capacity	FS_H	2.3[3]	3.0[3]	2.3[5]
Internal stability	Pullout resistance	FS_P	2.0		1.5
	Nail bar tensile strength	FS_T	1.8		1.35
Facing strength	Facing flexure	FS_{FF}	1.35	1.5	1.1
	Facing punching shear	FS_{FP}	1.35	1.5	1.1
	H-stud tensile (A307 bolt)	FS_{HT}	1.8	2.0	1.5
	H-stud tensile (A325 bolt)	FS_{HT}	1.5	1.7	1.3

Notes: (1) For noncritical, permanent structures, some agencies may accept a design for static loads and long-term conditions with FS_G = 1.35, when less uncertainty exists due to sufficient geotechnical information and successful local experience on soil nailing.

(2) The second set of safety factors for global stability corresponds to the case of temporary excavation lifts that are unsupported for up to 48 hours before nails are installed. The larger value may be applied to more critical structures or when more uncertainty exists regarding soil conditions.

(3) The safety factors for bearing capacity are applicable when using standard bearing-capacity equations. When using stability analysis programs to evaluate these failure modes, the factors of safety for global stability apply.

Modern soil nail walls are designed using software. Most use limiting equilibrium analysis or finite element methods. It is important that software results be analyzed critically, to confirm coincidence with experience and applicability to the field conditions, particularly the groundwater regime.

FHWA (1994) and NYS DoT (2008) give construction guidance and preliminary design charts to estimate the number and length of soil nails. Various other applications of launched nails are given in FHWA (1994), including horizontal and vertical drains, slope stability, and road widening.

10.5.7 Construction of soil nailed walls

The basic construction sequence for soil nail walls is:

1. excavation to a limited depth,
2. drill the nail hole,
3. insert and grout the nail; install wire mesh and nail heads,
4. apply a temporary shotcrete facing, and
5. repeat steps 1 through 4.

Inspection follows all steps.

A permanent facing may be applied later. Shotcreting is an art and requires skilled craftsmen. North Carolina DoT (2010), for example, has soil nail shotcrete guidance. Figure 10.34 shows shotcreting.

A site investigation must be completed before construction. The investigation must determine soil strata, soil properties, location of the groundwater table, obstructions, and underground utilities.

The process begins with excavation made to a depth less than the stand-up height. Rotary or percussion angle drilling is followed by nail installation and grouting. The hole is angled downward slightly to keep grout from flowing out of the hole during placement. Care must be taken to minimize the amount of soil removed when auger drilling, lest the ground surface settle. The typical hole is 10–30 cm in diameter. Larger diameters are used when

Figure 10.34 Photo of shotcretings a wall face (note protruding nails, and mesh to hold shotcrete in place) (Focus 2011).

greater bond strength is required. The hole may be cleaned to remove drill cuttings before grouting. While cleaning is effective in non-running cohesionless soils, cleaning procedures may loosen cohesive soils, and, hence, are not recommended. Alternatively, self-drilled soil nails may be used. These hollow nails serve as the drill bit and stem. Alternatively, the nails may be driven in. Once in place, grout is pumped through the hollow core. Since these nails are not as robust as conventional drill bits, their usage is limited to softer/looser soils, free of obstructions.

The hole may require casing to prevent collapse before grouting. Hollow stem augers are commonly used to hold the hole open. The soil nail is inserted in the hole, using centralizers, and full length grouted. Nails should not be pushed into the bottom of the borehole. Allow space for grout to fill around the end of the soil nail, providing corrosion protection. Do not delay grouting, as the hole may collapse. Gravity grouting is typical, although pressure grouting is occasionally used in cohesionless soils to increase bond strength.

Drainage must be controlled. A vertical geosynthetic strip drain may be placed on the soil face before wire mesh and shotcreting. The drain is extended during successive excavations. Drain splices must not allow soil penetration. The drain must daylight at the base of the soil nail wall to be effective.

After nail installation, welded wire mesh is placed over the soil nails and attached to the soil. A soil nail head is installed and tightened against the mesh and soil. The mesh opening size, mesh bar diameter and mesh strength are design values. FHWA (2003) provides guidance. The mesh should overlap vertically and horizontally with installed mesh. Additionally, four wire bars, two vertical and two horizontal, may be attached to the soil nail head to increase punching resistance of the soil nail head through the shotcrete.

The entire face is immediately shotcreted for stability. While shotcrete thickness varies, 15 cm is typical. Thicker sections, with additional reinforcement are used around nail heads and for more permanent installations.

Other facings may be attached to the shotcrete facing for aesthetics. These may be precast panels, cast-in-place concrete, or reinforced shotcrete, which may be sculpted and stained (Figure 10.35). All facing systems must incorporate drainage.

Figure 10.35 Photo of sculpted and stained shotcrete wall face (US Department of Transportation 2015).

10.5.8 Nail testing

The soil nail bond strength, estimated from the design procedures, must be field verified by pullout testing of a few nails. The test nail(s) should be chosen at locations where a minimum bond strength is expected. The test nails are not fully grouted. Only about two meters of the bottom part of the soil nail is grouted to reduce the amount of force necessary to do the test. The tested soil nails are wasted. The results are returned to the designer for review and, if needed, redesign. To test, selected soil nails are pulled with hydraulic rams (or other means) until the nail fails, or a specified load is reached, often twice the design load/meter. The hollow loading ram is inserted over the nail and attached with a washer and nut. The ram pushes against the shotcreted face, perhaps with the addition of a bearing plate. Displacement is measured independent of the face to account for face displacement (as opposed to nail displacement). Often, the test is conducted until nail movement is constant, indicating failure. NYS DoT (2008) suggests acceptable deformations. Further details are provided in Hong Kong (2008) and NYS DoT (2008). Incidentally, the test is an indirect measure of the effectiveness of the installation procedures.

Nondestructive testing techniques, including time domain reflectometery (Gong et al. 2006; Lee and OAP 2007) can be used to evaluate grout continuity. North Carolina DoT (2010) provides test design and acceptance criteria.

In addition, proof testing and creep testing are often done on a selected, larger number of nails. For proof testing, the nails are loaded as above, but to a lower load, perhaps 1.5 times the design capacity. For creep testing, movement of the nail, under sustained loading, is measured. Failure criteria are empirical, and established to provide confidence that the structure will not creep. FHWA (2003) suggests criteria.

10.5.9 Corrosion protection

Steel soil nails must be protected from corrosion. Corrosion is exacerbated by soil chlorides and sulfides, extreme pH soils, stray electrical potentials, and high groundwater. Corrosion protection methods include:

1. use of grout around the soil nail,
2. galvanization or epoxy coating of the soil nail,
3. soil nail encapsulation in a corrugated plastic sheath, grouted inside,
4. electrical isolation of the nail from the nail head and from other nails, and
5. galvanic protection.

PTI (1996) provides tieback corrosion protection guidance, which can be used for soil nails in corrosive environments. Hong Kong (2008) gives advice on evaluating corrosion potential based on soil and nail type.

10.5.10 Instrumentation

Instrumentation is used to monitor post-construction displacement performance of the wall and loads in the soil nail. Visual monitoring of the wall face is the minimum performance measurement. Slope inclinometers (horizontal, internal movements) and survey points (vertical and horizontal external movements) are used. Load cells at the soil nail head, or strain gages mounted on the soil nail before insertion, are used to evaluate the distribution of soil nail loads, and the force at the head. A decrease in load readings suggests failure is underway. These readings are compared to design values. Drainage must be effective constantly, as increasing water pressures reduces soil strength and increases loads on the shotcrete facing. Temperature and rainfall measurements add to understanding of wall performance. Measurements during construction, and for at least two post-construction years, are recommended. LiDAR is an elegant way to monitor the movement of the entire wall face.

Remedial measures for walls failing to meet criteria are left to the engineer. These may include:

1. change in drainage,
2. additional soil nails,
3. removal of loads at the crest, and
4. addition of berms at the base.

Example Problem Ex.10.1: Soil Nailing

An 18.3 m vertical cut in a stiff, clayey silt is planned. The site investigation yielded these soil properties: $\phi_U = 40°$, $c_U = 28.7$ kPa, and $\gamma = 18.9$ kN/m^3. The top of the vertical slope will be 0° from horizontal.

The cut is deemed unstable in the long term. Soil nailing is elected. Testing shows that 86.2 kPa of bond strength can be achieved in a 15.3 cm diameter augered hole for the given soil/grout condition. Client constraints require soil nails be spaced on 2.44 m centers, both directions. 345 MPa steel rods will be used for the soil nails. For the anticipated soil/grout conditions, α_N is 0.55 while α_Q is 0.5.

Estimate the required bar size and length needed to produce a factor of safety of 1.5.

Solution:

The solution is in two parts.

First, estimate the required soil nail size; begin by estimating the required internal angle of friction, ϕ_D

$$\phi_D = \tan^{-1}\left((\tan\phi_u)/FS\right) = \tan^{-1}\left((\tan 40)/1.5\right) = \tan^{-1}(0.556) = 29°$$

Calculate C_D

$$C_D = \left(c_u / \left(FS(\gamma)(H)\right)\right) = 28.7 / \left((1.5)(18.9)(18.3)\right) = 0.056$$

Use Figure 10.25 to estimate T_D. $T_D \approx 0.11$

Calculate T_{NN}

$$T_{NN} = \left(T_D S_v S_h \gamma H\right) / \alpha_N = 0.19\,(2.44)\,(2.44)\,(18.9)\,(18.360)/0.55 = 711.4\text{ kN}$$

Calculate the bar size, based on the load it must carry, 711.4 kN.

The bar area, A_B, is the load divided by the yield strength of the bar

$$A_B = T_{NN}/F_y = 711.4\text{ kN}/344747\text{ kPa} = 0.002\text{ m}^2$$

Table 10.1 suggests using a number 18 soil nail, which has an area of 0.00258 m^2.

Specify a no. 18 soil nail.

The second part is estimating the length of the soil nail. All nails will be the same length.

First, calculate Q_U

$$Q_U = (\text{bond strength of soil})(\text{hole circumference})$$

The bond strength is estimated at 86.2 kPa. The hole is 15.3 cm in diameter. Hence

$$Q_U = (86.2 \text{ kPa})(3.14)(0.153 \text{ m}) = 12.4 \text{ kN/m}$$

then

$$Q_D = (\alpha_Q)(Q_U)/(\gamma)(S_\nu)(S_H) = 0.5(12.4)/(18.9)(2.44)(2.44) = 0.055 \text{ m}$$

To get L/H, use Figure 10.29, chart 1B. Enter the chart with

$$T_D/Q_D = 0.11/0.184 = 0.598$$

Using $C_D = 0.056$, and $\tan \phi_D = 0.556$, from the figure,

$$L/H \approx 0.5$$

With H = 18.3 m, L = 0.5H = 0.5(18.3) = **9.1 m.**

10.5.11 Launched soil nails

Soil nails are also installed with compressed air, using giant nail guns. These are called launched soil nails (or ballistic soil nails). The nails may be up to 5.1 cm in diameter and 7.6 m long. The extremely high speed (300+kph) solid or hollow, steel or fiberglass nail penetrates the soil by creating a shock wave that displaces the soil in front and around the nail, until the nail slows down, at which point the soil collapses around the nail, providing anchorage. The nails are shot into the ground and trimmed at the ground surface. The gun is held at the end of an excavator arm, which reaches up (or down) the slope, greatly reducing the size of the working platform, minimizing surficial disturbance. As with conventional nailing, soil stand-up time is required, and the face is shotcreted. The giant nail guns are used for all the same scenarios as drilling and driving soil nails, provided obstructions or bedrock are not in the stabilization zone.

The primary advantages of launched soil nails are the speed of equipment mobilization, speed of installation, elimination of need for grout and a nail head, and minimal ground disturbance. Disadvantages include impossibility of on-nail instrumentation (violent installation technique), not useful in all soil profiles (obstruction and bedrock issues), and depth of penetration (Barrett and Devin 2001).

FHWA (1994) and NYS DoT (2008) give construction guidance and preliminary design charts to estimate the number and length of soil nails. Various other applications of launched nails are given in FHWA (1994), including horizontal and vertical drains, slope stability, and road widening. Figure 10.36 shows a nail launcher on an excavator. Hall (1995) provides an interesting case history of launched soil nailing used in combination with reinforced soil. Steward (1994) provides case histories from the US Forest Service.

10.6 PROBLEMS

10.1 A 3.66 m high retaining wall with a horizontal backslope is planned. Your firm won the bid. You have chosen a reinforced, modular block wall, because of the low unit price, aesthetics, and ground conditions. You will place six different-length geotextile

Figure 10.36 Photo of soil nail launcher on an excavator (Photo courtesy of Geostabilization International).

reinforcement layers every 0.6 m, starting with the subgrade. As part of the design, evaluate the following:

a. length of the highest layer of reinforcement, for a factor of safety in sliding ≥ 1.5. The geosynthetic has an external angle of friction, $\delta = 21°$ with the reinforced sandy soil, which has a $\gamma = 19.5 \text{ kN/m}^3$, and $\phi = 34°$. Sandy soil with $\gamma = 19.2 \text{ kN/m}^3$ and $\phi = 29°$ is the unreinforced soil behind the reinforced section.

b. length of the lowest layer of reinforcement, for a factor of safety in sliding ≥ 1.5, for the properties given in (a). Comment on this length compared to the length obtained in (a).

10.2 A 7.32-m high retaining wall with a horizontal backslope is planned. Your firm won the bid. You have chosen a reinforced, modular block wall, because of the low unit price, aesthetics, and ground conditions. You will place 12 different-length geotextile reinforcement layers every 0.6 m, starting on the subgrade. As part of the design, evaluate the following:

a. length of the highest layer of reinforcement, for a factor of safety in sliding ≥ 1.3. The geosynthetic has an external angle of friction, $\delta = 17°$ with the reinforced sandy soil, which has a $\gamma = 17.3 \text{ kN/m}^3$ and $\phi = 30°$. Sandy soil with $\gamma = 18.9 \text{ kN/m}^3$ and $\phi = 29°$ is the unreinforced soil behind the reinforced section. Will this be the longest or shortest layer?

b. length of the lowest layer of reinforcement, for a factor of safety in sliding ≥ 1.3 and for the properties given in (a) above. Will this be the longest or shortest layer?

10.3 For the wall described in problem 10.1, estimate the required LTDS of the geosynthetic for the layer that is 1.22 m above the subgrade, for a factor of safety in breakage ≈ 1.3.

10.4 For the wall described in problem 10.1, estimate the required LTDS of the geosynthetic for the layer that is 1.83 m above the subgrade, for a factor of safety in breakage ≈ 2.2.

10.5 The wall in problem 10.1 must be analyzed for geosynthetic pullout. Previous testing showed the coefficient of interaction for this geosynthetic and this soil is 0.8. Calculate a $\text{FS}_{\text{resisting}}$ pullout for the 2.44 m long layer of geosynthetic reinforcement, 1.22 m above the subgrade. The failure surface is estimated at 60° from the horizontal, from the heel of the facing. Comment on this design.

10.6   The wall in problem 10.2 must be analyzed for geosynthetic pullout. Previous testing showed the coefficient of interaction for this geosynthetic and this soil is 0.6 (slippery!). Calculate a $FS_{resisting}$ pullout for the 3.05 m long layer of geosynthetic reinforcement, 1.82 m above the subgrade. The failure surface is estimated at 60° from the horizontal, from the heel of the facing. Comment on this design.

10.7   The wall in problem 10.6 has been redesigned with 5.5 m long layers of reinforcement. Calculate a $FS_{resisting}$ pullout. Comment on this design.

10.8   What analyses are required for MSE wall design, beyond pullout and sliding analyses? Explain each, briefly.

10.9   Use the FHWA Geosynthetic Reinforced Soil – Integrated Bridge System procedure to design the required geosynthetic strength for a 3.66 m high wall, with 0.3 m geosynthetic spacings, a retaining soil with $\gamma = 19.3$ kN/m^3, $\phi = 32°$, and a $d_{max} = 5$ cm. No strain limitations are required. The candidate geosynthetic has a 4.07 kN/m failure tensile strength. You wish to use a FS = 2.1 in geosynthetic rupture. What factor of safety does this geosynthetic have in geosynthetic breakage, in this design? What lengths of geosynthetic reinforcement will you specify for the top and bottom layers?

10.10   Use the FHWA Geosynthetic Reinforced Soil – Integrated Bridge System procedure to design the required geosynthetic strength for an 11.6 m high wall, with 0.3 m geosynthetic spacings, a retaining soil with $\gamma = 20.7$ kN/m^3, $\phi = 35°$, and a $d_{max} = 3.1$ cm. No strain limitations are required. The candidate geosynthetic has a 16.27 kN/m failure tensile strength. You wish to use a FS = 2 in geosynthetic rupture. What factor of safety does this geosynthetic have in geosynthetic breakage, in this design? What lengths of geosynthetic reinforcement will you specify for the top and bottom layers? What length(s) of reinforcement will you specify?

10.11   Find a case history of a successful MSE wall. What materials were used? What is the purpose of the wall?

10.12   How important is soil compaction to the success of MSE walls?

10.13   Describe three ways MSE walls can fail. What construction procedures can jeopardize the success of an MSE wall?

10.14   Choose the top three reasons why a client would prefer an MSE wall over a gravity wall. Why did you choose these three? Think like a client.

10.15   Describe geogrids, which can be used for MSE wall reinforcement. What advantage(s) do these have over geotextile reinforcement?

10.16   Find a picture of a wrapped face MSE wall. Name three advantages and disadvantages of this facing.

10.17   MSE walls with large (0.91 m square) facing elements have significant spaces between the elements. What would you do to prevent soil from escaping through these spaces?

10.18   What is a "steepened slope"? Why does it have this name? When would a client be best served with a steepened slope?

10.19   What facing(s) do you recommend for steepened slopes? Why?

10.20   Find a case history of a failed natural slope rebuilt with reinforcement. What type of reinforcement was chosen? Why do you think it was chosen?

10.21   Why would leakage from a pipe inside the slope be cause for concern? Be specific.

10.22   A 13.7 m high, vertical cut in a dry, silty, clayey sand is planned. The site investigation yielded these soil properties: $\phi_U = 32°$, $c_U = 14.4$ kPa, and $\gamma = 16.5$ kN/m^3 The top of the slope will be 10° above horizontal. The cut will be unstable. Soil nailing is elected. Testing shows that 76.6 kPa of bond strength can be achieved in a 15.2 cm diameter augered hole for the given soil/grout condition. Client constraints require soil nails be spaced on 1.83 m centers, both directions. 345 MPa steel rods will be used for soil nails. For the anticipated soil/grout conditions, α_N is 0.55 while α_Q is 0.5. Estimate the required bar size and length needed to produce a factor of safety of 1.5.

10.23 A 9.14 m high, vertical cut in a dry, silty, clayey sand is planned. The site investigation yielded these soil properties: $\phi_U = 28°$, $c_U = 7.18$ kPa, and $\gamma = 17.3$ kN/m³. The top of the slope will be 20° from horizontal. The cut will be unstable. Soil nailing is elected. Testing shows that 47.9 kPa of bond strength can be achieved in a 15.2 cm diameter augered hole for the given soil/grout condition. Client constraints require soil nails be spaced on 3.05 m centers, both directions. 354 MPa steel rods will be used for soil nails. For the anticipated soil/grout conditions, α_N is 0.55 while α_Q is 0.5. Estimate the required bar size and length needed to produce a factor of safety of 1.4.

10.24 The project described in problem 10.1 was let to a contractor who was found to be installing the nails on 2.44 m centers, instead of the specified 1.83 m centers. As the designer, you are contacted to find out if the wall will stand safely. Estimate the factor of safety for this adjusted design. What is your recommendation?

10.25 *(data from problem 10.1, repeated below)* A 13.7 m high, vertical cut in a dry, silty, clayey sand is planned. The site investigation yielded these soil properties: $\phi_U = 32°$, $c_U = 14.4$ kPa, and $\gamma = 16.5$ kN/m³. The top of the slope will be 10° from horizontal. The cut will be unstable. Soil nailing is elected. Testing shows that 76.6 kPa of bond strength can be achieved in a 15.2 cm diameter augered hole for the given soil/grout condition. Client constraints require soil nails be spaced on 1.83 m centers, both directions. 414 MPa steel rods will be used for soil nails. For the anticipated soil/grout conditions, α_N is 0.55 while α_Q is 0.5. Estimate the required bar size and length needed to produce a factor of safety of 1.5.

REFERENCES

ASTM D 5321, 2008. Standard Test Method for Determining the Coefficient of Soil and Geosynthetic or Geosynthetic and Geosynthetic Friction by the Direct Shear Method, American Society of Testing and Materials, West Conshohocken, Pennsylvania, USA

Barrett, C.E., and Devin, S.C., 2011. Shallow Landslide Repair Analysis Using Ballistic Soil Nails: Translating Simple Sliding Wedge Analyses into PC-Based Limit Equilibrium Models. In *Geo-Frontiers 2011: Advances in Geotechnical Engineering* (pp. 1703–1713).

Barrett, R., 2007. Personal Communication, Pigeon Point, Trinidad and Tobago.

Been, D.A., 2011. Personal Communication. Birmingham, AL.

Berg, R.R., Christopher, B.R., Samtani, N.C., and Berg, R.R., 2009. *Design of Mechanically Stabilized Earth Walls and Reinforced Soil Slopes–Volume I*. No. FHWA-NHI-10-024. Federal Highway Administration, Washington, DC, USA

Bernard, S. (ed.), 2010. *Shotcrete: Elements of a System*. CRC Press/Taylor and Francis Group, London.

Bowles, J.E., 1996. *Foundation Analysis and Design*, 5th ed. McGraw-Hill, New York.

Cheung, W.M., and Lo, D.O.K, 2005. Use of Carbon Fibre Reinforced Polymer Reinforcement in Soil Nailing Works. In *Proceedings of the HKIE Geotechnical Division 25th Annual Seminar: Safe and Green Slopes* (pp. 175–184). Hong Kong Institution of Engineers, Hong Kong.

Coduto, D.P., Kitch, W.A., and Yeung, M.C.R., 2001. *Foundation Design: Principles and Practices* (Vol. 2). Prentice Hall, Upper Saddle River, NJ.

Das, B.M., 2015. *Principles of Foundation Engineering*. Cengage Learning, Boston, Massachusetts, USA

Dewey, R.L., 1989. The Bureau of Reclamation Uses Geosynthetics. *Geotechnical News*, 7(2), 39–42.

DOT, (1994). Design Methods for the Reinforcement of Highway Slopes by Reinforced Soil and Soil Nailing Techniques (HA 68/94). Department of Transport, UK.

Duncan, J.M., Buchignani, A.L., and DeWet, M., 1987. *An Engineering Manual for Slope Stability Studies*. Department of Civil Engineering. Geotechnical Engineering, Virginia Polytechnic Institute and State University, Blacksburg, VA.

Elias, V., and Christopher, B.R., 1996. *Mechanically Stabilized Earth Walls and Reinforced Earth Slopes*. Report FHWA-DP. 82–1. Federal Highway Administration, Washington, DC, USA

Elias, V., Christopher, B.R., and Berg, R.R., 2001. *Mechanically Stabilized Earth Walls and Reinforced Soil Slopes: Design and Construction Guidelines (Updated Version)*. No. FHWA-NHI-00-043. Federal Highway Administration, Washington, DC, USA

Elias, V., Fishman, K., Christopher, B.R., and Berg, R.R., 2009. *Corrosion/Degradation of Soil Reinforcements for Mechanically Stabilized Earth Walls and Reinforced Soil Slopes*. No. FHWA-NHI-09-087. National Highway Institute, Washington, DC, USA

Elias, V., and Juran, I., 1991. *Soil Nailing for Stabilization of Highway Slopes and Excavations*. Publication FHWA-RD-89-198. Federal Highway Administration, Washington, DC.

Engemoen, W.O., and Hensley, P.J., 1989. Geogrid Steepened Slopes at Davis Creek Sam. In *Proceedings of Geosynthetics' 89 Conference* (pp. 255–268), Industrial Fabrics Association International, Roseville, Minnesota, USA.

Fang, H.-Y., 1990. *Foundation Engineering Handbook*. van Nostrand Reinhold, New York City, NY, USA

FHWA, 1975. *Retaining Walls Lateral Support Systems and Underpinning, 1,2, and 3*. FHWA-RD-75-128 / 129 / 130.

FHWA, 1994. *Application Guide for Launched Soil Nails, Volume 1*. FHWA-FPL-93-003. US Federal Highway Administration, Washington, DC.

FHWA, 1998. *Manual for Design and Construction of Soil Nail Walls*. FHWA-SA-96-069R. Federal Highway Administration, Washington, DC.

FHWA, 1999. *Demonstration Project 103: Design and Construction Monitoring of Soil Nail Walls, Project Summary Report*. FHWA-IF-99-026. Federal Highway Administration, Washington, DC.

FHWA, 2003. *Soil Nail Walls, Geotechnical Engineering Circular no. 7*. FHWA0-IF-03-017, by Lazarte, C.A., Elias, V., Espinoza, D., and Sabatini, P. US Federal Highway Administration, Washington, DC.

FHWA, 2018. *Design and Construction Guidelines for Geosynthetic Reinforced Soil Abutments and Integrated Bridge Systems*. FHWA-HRT-17-080. US Federal Highway Administration, Washington, DC, 204 pp.

Focus, 2011. *Focus, Accelerating Infrastructure Innovations*. Publication Number: HRT-11-012, September 29, 2014, US Federal Highway Administration, Washington, DC, USA

Gong, J., Jayawickrama, P.W., and Tinkey, Y., 2006. *Nondestructive Evaluation of Installed Soil Nails, Transportation Research Record: Journal of the Transportation Research Board*, No. 1976, pp. 104–113. Transportation Research Board of the National Academies, Washington, DC.

Hall, G.J., 1995. The Joint Use of Ballistic Soil Nailing and Reinforced Soil in Huddersfield. In *The Practice Of Soil Reinforcing in Europe: Proceedings of the Symposium the Practice of Soil Reinforcing in Europe* (pp. 227–240). Organised by the Tenax Group under the auspices of the International Geosynthetics Society, and held at the Institution of Civil Engineers on 18 May 1995. Thomas Telford Publishing, London, UK.

HCL (Halcrow China Limited), 2007. *Study on the Potential Effect of Blockage of Subsurface Drainage by Soil Nailing Works*. GEO Report No. 218. Geotechnica Engineering Office, Civil Engineering and Development Department, Hong Kong, 102 p.

Holtz, R., Kovacs, W., and Sheahan, T., 2011. *An Introduction to Geotechnical Engineering*. Pearson Education Inc., Upper Saddle River, NJ.

Hong Kong, 2008. *Guide to Soil Nail Design and Construction*. Geotechnical Engineering Office, Civil Engineering and Development Department, The Government of the Hong Kong, Special Administrative Region, Homantin, Kowloon, Hong Kong, 97 pp.

Koerner, R.M., 2012a. *Designing with Geosynthetics*, Vol. 2. Xlibris Corporation West Conshohocken, Pennsylvania, USA.

Koerner, R.M., 2012b. Personal Communication.

Kramer, S.L., 1996. *Geotechnical Earthquake Engineering*. Prentice Hall, Upper Saddle River, NJ.

Lang, T.A., 1972. Rock Reinforcement. *Bulletin of the Association of Engineering Geologists*, IX(3), 215–239.

Lazarte, C.A., Elias, V., Espinoza, D., and Sabatini, P., 2003. *Soil Nail Walls, Geotechnical Engineering Circular no. 7*. FHWA0-IF-03-017. US Federal Highway Administration, Washington, DC.

Lazarte, C.A., Robinson, H., Gómez, J.E., Baxter, A., Cadden, A., and Berg, R., 2015. *Geotechnical Engineering Circular no. 7, Soil Nail Walls - Reference Manual*. FHWA-NHI -14–0. US Federal Highway Administration, Washington, DC.

Lee, C.F., and OAP (Ove Arup & Partners Hong Kong Limited), 2007. *Review of Use of Non-Destructive Testing in Quality Control in Soil Nailing Works*. GEO Report No. 219. Geotechnical Engineering Office, Civil Engineering and Development Department, Hong Kong, 109 p.

Miyata, Y., and Bathurst, R.J., 2007. Development of the K-Stiffness Method for Geosynthetic Reinforced Soil Walls Constructed with c-ϕ Soils. *Canadian Geotechnical Journal*, 44(12), 1391–1416.

Murthy, V.N.S., 2003. *Geotechnical Engineering: Principles and Practices of Soil Mechanics and Foundation Engineering (Civil and Environmental Engineering)*. CRC Press, Taylor and Francis, London.

NCHRP, 2011. *Proposed Specifications for LFRD Soil-Nailing Design and Construction*. NCHRP Report 701. National Academies Press, Washington, DC.

NCMA, 1993. *Design Manual for Segmental Retaining Wall*, 1st ed. National Concrete Masonry Association, Herndon, VA.

North Carolina DoT, 2010. *Standard Soil Nail Wall Provision*. North Carolina Department of Transportation, Raleigh, NC, 16pp.

NYS DoT, 2007. *Mechanically Stabilized Earth System Inspection Manual, Geotechnical Engineering Manual*. GEM-16 Revision #2. Geotechnical Engineering Bureau, New York State Department of Transportation, Albany, NY.

NYS DoT, 2008. *Design Procedure for Launched Soil Nail Shallow Slough Treatment, Geotechnical Design Procedure GDP-14*. NY State Department of Transportation, Albany, NY.

PileBuck, Inc. 1990. *Sheetpile Wall Design Manual*. PileBuck, Inc, Jupiter, FL.

PTI, 1996. *Recommendations for Prestressed Rock and Soil Anchors*, 3rd ed. Post-Tensioning Institute, Phoenix, AZ.

Schnabel, H., 2002. *Tiebacks in Foundation Engineering*, 2nd ed. CRC Press/Taylor and Francis Group, London.

Shah, H.J, Lacy, H.S, and van Rensler, M.B., 2008. Mechanically Stabilized Earth for Steep Surcharge Slopes in Proximity of Adjacent Structures to Improve Compressible Soils, Paper 8.02a. In *6th International Conference on Case Histories in Geotechnical Engineering*, Arlington, VA.

Steward, J.E., 1994. Launched soil nails: A new technology for stabilizing failing road shoulders, (Appendix 6.7). In *Slope Stability Reference Guide for National Forests in the United States: Volume 3 (EM-710.13)* (pp. 1064–1091). Forest Service – U.S. Department of Agriculture, Washington, DC.

Terzaghi, K., Peck, R.B., and Mesri, G., 1996. *Soil Mechanics in Engineering Practice*. John Wiley & Sons, Hoboken, New Jersey, USA

US Department of Transportation, 2015. *Soil Nail Walls - Reference Manual*. FHWA-NHI-14-007. Federal Highway Administration FHWA GEC 007, Washington, DC.

US Navy, 1986. *Design Manual—Soil Mechanics, Foundations, and Earth Structures*. NAVFAC DM, 7.01. US Naval Facilities Engineering Command, Alexandria, VA.

USACE, 1993. *Manual EM 1110.2-1901, Seepage Control, Appendix D, Filter Design*. U.S. Army Corps of Engineers, September 30, 1986, revised April 30, 1993.

Vidal, H., 1969. The Principle of Reinforced Earth. *Highway Research Record*, 282, 1–16.

Wisconsin DOT, 2010. *Wisconsin Department of Transportation Bridge Manual*. Wisconsin Department of Transportation, Madison, WI.

Chapter 11

Additional techniques in ground improvement

This chapter addresses the techniques that do not fit squarely in any of the other chapters of this book but deserve attention in discussions of ground improvement. In many cases, these techniques do not fit into other chapters because the technique could actually fit into multiple chapters. In other cases, the methods herein are extensions of methods from other industries, e.g. deep foundations that are applied to solve ground improvement problems. Finally, some of the methods herein are novel approaches to solving ground improvement problems that do not have a wide-enough application to warrant a complete chapter or are so nuanced that the topic deserves a book itself or would be better suited to advanced study. Below is an overview of the topics addressed in this chapter with a short discussion of where these topics could be covered elsewhere in this book and why the topic is included in this chapter.

1. *Jet grouting.* Jet grouting is a method of in situ mixing that is accomplished through erosion of the soil via a high-pressure jet of grout or slurry. This method could be included in the grouting or soil mixing chapters, but since the technique is a combination of both and has unique attributes as well, the topic was moved to this chapter. In a more complete ground improvement book aiming to cover all topics in detail, jet grouting would be deserving of its own chapter and it is the subject of standalone publications.

2. *Ground freezing.* Ground freezing could be addressed as part of a cutoff wall discussion, but since this book is organized by construction methodologies, the cutoff wall discussion is spread amongst multiple chapters: slurry trenching, soil mixing, and grouting. Ground freezing is a very specialized ground improvement technique that is generally reserved for very challenging situations and is therefore the subject of its own publications written by specialty practitioners and designers engaged in its application.

3. *Secant pile walls.* Secant pile walls are an extension of piles used in deep foundations to specific ground improvement problem cases, cutoff walls and excavation support. As with ground freezing, this topic could be addressed in a cutoff wall or excavation support system discussion, but excavation support is not separately addressed in this book. This method is regularly included in deep foundations textbooks and courses as the construction methodology is very similar to the methods used for foundation pile installations, hence the unique name secant pile, with the secant component of the name referring to the overlap of adjacent piles.

4. *Compaction grouting.* Compaction grouting is a method of densifying and compacting soils by displacement with a viscous grout. This topic could be addressed as a subcomponent in the shallow or deep compaction chapters or the grouting chapter since the application applies to both chapters and the construction equipment is very similar to equipment used for other grouting methods.

5. *Explosives in ground improvement.* Explosives can be used for compaction and could be addressed as a subcomponent in the shallow or deep compaction chapters.

There are many other methods, such as bio-stabilization, that could be included in this chapter or as a standalone chapter. Those selected for inclusion in this chapter are widely used and accepted. As has been the case over the history of ground improvement, it is expected that the ground improvement marketplace and technologies will continue to evolve giving rise to additional methods of ground improvement.

11.1 JET GROUTING

11.1.1 Introduction to jet grouting

Jet grouting is an in situ soil mixing process that uses high-pressure jets to simultaneously inject additives and mix them with the in situ soil. As shown in Figure 11.1, jet grouting is generally performed from the bottom up. The process begins with drilling from the surface to the target bottom elevation using a low-pressure and low-flow fluid stream as the drill bit lubricant (left panel of Figure 11.1). Once the target bottom elevation is achieved, the pressure and flow are increased, and the drill rod is slowly withdrawn from the drill hole while rotating the jet. The in situ soil properties, the rate of withdrawal, the rotation rate, the fluid pressure, and flow all influence the diameter of the column created. The center panel of Figure 11.1 shows the injection and mixing as the rod and jets are withdrawn. The right panel shows the process a little further along. This process has many variations but, in principle, consists of injection of additives by mixing with high-pressure jets to form in situ mixed columns of solidified soil.

The jet grouting construction technique combines features found in the soil mixing and grouting chapters, Chapters 6 and 7. The final product and target objectives of jet grouting align more closely with soil mixing than grouting. However, the means and methods of accomplishing the target objectives of jet grouting projects share many similarities with other grouting techniques. In the process of jet grouting, a significant portion of the soil-grout mixture is displaced upward. In some cases, a significant portion of the resultant

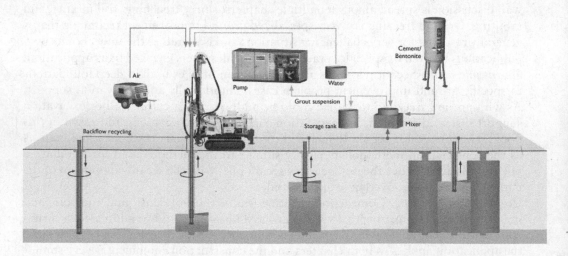

Figure 11.1 Jet grouting schematic (courtesy of Keller; www.Keller-na.com).

mixture in the ground can be made up of grout. For this reason, jet grouting is more of an excavation and replacement methodology rather than a mixing or grouting method.

Jet grouting applications include:

1. retaining walls,
2. cutoff walls,
3. excavation shoring,
4. stabilization of contaminants,
5. shaft and other structural foundations,
6. single pile columns outfitted with steel rebar,
7. horizontal low permeability, in situ liner,
8. under slab reinforcement,
9. structural underpinning, and
10. tunneling.

Jet grouted columns can be installed in numerous arrangements and patterns, including overlapping columns. Advantages of jet grouting include:

1. The technique provides permanent ground improvement.
2. During construction there are few vibrations imparted to the surrounding soil outside the column, which makes it a safe method to use around existing buildings and around construction where it is important not to induce settlement.
3. The technique reduces the permeability and increases the shear strength of the grouted soil.
4. Jet grout columns do not need to be continuous, that is, a specific zone of the subsurface can be improved.
5. Relative to other ground improvement methods, the amount and size of the equipment are small.
6. Jet grout holes can be predrilled if there are existing obstructions in the soil such as boulders or old foundations. The jet grout columns can then be installed above and below these obstructions. This is not feasible with other soil mixing techniques.
7. Pressure buildup from the grout being confined is generally not an issue if the grout return is allowed to exit freely at the surface.

The main disadvantages of jet grouting include:

1. Cost is high compared to alternate technologies.
2. The method produces a large amount of spoil in the form of a waste mixture of soil, water, and grout.

Jet grouting is generally performed using one of three combinations of fluids, termed single, double (dual), or triple fluid (phase) jet grouting. In single-phase jet grouting, the fluid is a combination of the binding reagent(s) and water, i.e. grout. In dual-phase jet grouting, two fluids are used: grout and air. In triple-phase jet grouting, three fluids are used: grout, air, and water. The location of the nozzles delivering each fluid varies by contractor and application and are generally modified throughout the project to maximize jet grouting efficiency, i.e. increase column diameter. Figure 11.2 shows schematics of single-, double-, and triple-phase jet grouting systems.

Double- and triple-phase jet grouting offer advantages over single-phase jet grouting in that the air and/or air/water fluids help to reduce friction and increase the effective penetration distance (increased column diameter) of the grout stream. The air also helps to lift

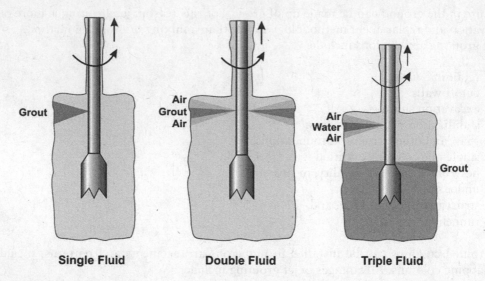

Figure 11.2 Jet grouting types (courtesy of Keller; www.Keller-na.com).

the soil particles out of the column, which can result in a higher percentage of grout in the column and thereby a stronger column. However, dual- and triple-phase grouting are more expensive and relatedly more complex than single-phase grouting.

Jet grouting is most effective in highly erodible soils, less effective in very dense soils, and has very limited effectiveness in cohesive soils. The single- and dual-phase systems, like mechanical soil mixing, are more of an in situ mixing technology. The triple-phase system is more of an excavation and replacement technology where the water-air jets are used to excavate and remove most of the soil and the grout jets are used to backfill the created cavity. Table 11.1 outlines important parameters for the three types of jet grouting.

The column diameter that is achievable via jet grouting is dependent on the jet grouting methodology, the jet grouting parameters, and the soil properties. Figure 11.3, adapted from Modoni et al. (2006), is a representation of equilibrium conditions between the jet fluid and the fluid that escapes the column vertically along the drill rod.

Although empirical equations can be used to estimate starting values for these parameters, most jet grouting contractors rely heavily on experience. Table 11.2 provides guidance in the absence of site specific information.

If possible, a test program should be performed at the project site. In the test program, various parameter configurations should be performed. After the columns have cured sufficiently, the columns should be subjected to in situ sampling, if applicable, and exposed. During the excavation, the relative homogeneity of the column can be evaluated, and the column diameter can be measured. After column excavation, the contractor can more confidently select a set of grouting parameters and column layout. Given the variability of jet grouting, a conservative column layout with redundant coverage is recommended.

11.1.2 Environmental considerations

There are clear advantages to jet grouting which make it a common choice for specific situations. From an environmental impact viewpoint, at least one case study found that jet grouting had no measurable effects on the surrounding environment (Nelson and Reed 2007). Here, the effects of jet grouting on the pH of ground and surface water were

Table 11.1 Variables for single-, dual-, and triple-phase jet grouting

Jet Grouting Variables		
Single Phase	Dual Phase	Triple Phase
-grout pressure	-grout pressure	-grout pressure
-grout flow rate	-grout flow rate	-grout flow rate
-grout makeup	-grout makeup	-grout makeup
-number of grout nozzles	-number of grout nozzles	-number of grout nozzles
-dimensions of grout nozzles	-dimensions of grout nozzles	-dimensions of grout nozzles
-location of grout nozzles	-location of grout nozzles	-location of grout nozzles
-rotation speed	-rotation speed	-rotation speed
-lifting Speed	-lifting Speed	-lifting Speed
	-air pressure	-air pressure
	-air flow rate	-air flow rate
	-number of air nozzles	-number of air nozzles
	-location of air nozzles	-location of air nozzles
		-water pressure
		-water flow rate
		-number of water nozzles
		-dimensions of water nozzles
		-location of water nozzles

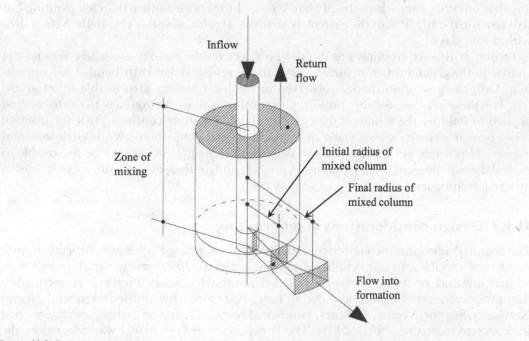

Figure 11.3 Pictorial representation of equilibrium of grout flow during jet grouting.

analyzed at Mormon Island Dam in California. The purpose of these tests was to see if alkaline compounds used in cement would impact the surrounding geochemistry by leaching into the environment or releasing calcium hydroxide, which would increase the pH and the alkalinity of the nearby water and harm ecology. Nelson and Reed (2007) point

Table 11.2 Design parameters for jet grouting

Factor	Jet grouting parameter	Single	Double	Triple
Pressure	Water jet (MPa)	n/a	n/a	30–50
	Grout jet (MPa)	30–50	30–50	<5
	Air jet (MPa)	n/a	~1	~1
Flow	Water jet (L/min)	n/a	n/a	75–100
	Grout jet (L/min)	150–300	150–400	150–500
	Air jet (L/min)	n/a	1–3	1–3
Nozzles	Water jet (mm)	n/a	n/a	2–3
	Grout jet (mm)	2–3	2–3	2–6
	Number of water nozzles	n/a	n/a	1–2
	Number of grout nozzles	1–2	1–2	1
Water/cement mixing	Water/cement ratio	0.8–2.5	0.8–2.5	0.5–2
	Cement use (kg/m)	100–300	100–300	300–900
	Rotating speed (rpm)	10–30	10–30	3–8
	Lifting speed (m/min)	1–2	1–2	1–3
Diameter of column	Course-grained soils (m)	0.8–1.2	1–2	1.5–3
	Fine-grained soils (m)	0.6–0.9	1–1.5	1–2
Compressive strength of column	Course-grained soils (MPa)	2–10	2–10	5–20
	Fine-grained soils (MPa)	1.5–10	1.5–10	1.5–5

out that on other sites where the pH had changed for reasons other than jet grouting but still remained < 10, 50% of the extremely sensitive amphipod species Hyalella Azteca died within four days.

Testing at this jet grouting site showed no effect on the biology and there was no pH change in the ground water or surface water in the tested wetland attributable to jet grouting. Also, there were no changes observed in the local biology attributable to jet grouting. For these reasons, for this project, jet grouting was an environmentally safe method applied to improve the seismic stability of the dam. The study concluded that jet grouting was a benign activity vis-a-vis water quality and aquatic populations. While environmental impacts of possible pH changes are not studied on most projects, it may be reasonable to extend these findings to other jet grouting projects and, indeed, other in situ mixing or even grouting techniques.

11.1.3 Design considerations in jet grouting

The required unconfined compressive strength of a jet grouted column is dependent on a number of variables. In one detailed study (Tinoco et al. 2009), many variables and their relative influence on strength were measured. The variables included soil type, cement addition rate, water-cement ratio, and curing time. Four models were used: Artificial Neural Networks, Support Vector Machines, Functional Networks, and an earlier model developed for concrete structures (NAIS 2005). The Functional Network model was selected by the authors of the study as the preferred model because it was most accurate and the results of the experiment could be easily analyzed by practicing engineers. The other models are difficult to understand on a conceptual level and would therefore be limited for practical use and validation. The results of the Functional Network modeling from this study included various jet grouting parameters and their relative importance in predicting strength. The study found that the amount of sand and cement in the mixture, the time of curing, and the ratio of the mixture porosity and volumetric cement content had the most positive effects on the strength of a jet grouted material.

While contractor's experience and standard industry practice often guide jet grouting projects, quantitative design calculations can also be performed to guide the design. This is not common in practice but working through these types of equations is valuable for understanding the factors that influence the jet grouting process. The equations, below, from Modoni et al. (2006), are useful for design. The following series of equations identify and use key construction, design, and material property parameters to predict key outcomes including the diameter of the jet grouted column.

The following equation estimates the required impact period of the jet at each point to achieve the radius desired:

$$t^* = \frac{ma_1}{2\pi R_1 v_s}$$ (11.1)

where

t^* = impact period on corresponding point of borehole face (i.e. the time during which the fluid stream impacts the soil.)
m = number of nozzles
a_1 = cross sectional area of the jet at the borehole wall
R_1 = borehole radius
V_s = average monitor withdrawal rate ($\Delta s/\Delta t$)
Δs = change in lifting steps

The resistive force of sandy soils against the jet-erosive force can be estimated from:

$$\psi_s = \Omega_2 \left[c' + \sigma' tg(\phi') \right]$$ (11.2)

where

ψ_s = erosion resistance in sand
Ω_2 = calibration factors for gravels, sands, and clays
c' = soil cohesion
σ' = effective stress
t = time
ϕ' = effective soil friction angle

Equation 11.3 can then be used to estimate the minimum jet velocity that would result in soil erosion based on the strength of the soil:

$$v_L = \sqrt{\frac{\Omega_s g}{\gamma_f} \frac{c' + \sigma_z \tan(\phi')}{1 + \Omega_s \left[\frac{\tan(\phi')}{2} \right]}}$$ (11.3)

where

v_L = "limiting velocity" (minimum jet velocity producing soil erosion)
Ω_s = calibration factors for gravels, sands and clays
c' = effective soil cohesion
γ_f = unit weight of injected fluid
ϕ' = effective soil friction angle
σ_z = vertical stresses

Equation 11.4 can be used to estimate the maximum column radius that can be achieved in a sandy soil:

$$R_{max} = \frac{2\Lambda v_{x0} C d_0}{\sqrt{\dfrac{\Omega_s g N}{\gamma_f} \dfrac{c' + \sigma_z \tan(\phi')}{1 + \Omega_s \left[\dfrac{\tan(\phi')}{2}\right]}}} \tag{11.4}$$

where

Λ = calibration factor for jet velocity
R_{max} = maximum jet column radius attainable by erosion
N = relative turbulent kinematic viscosity of injected fluid
d_0 = nozzle diameter. Typical nozzle diameters are 2–4 mm.
c' = effective soil cohesion
γ_f = unit weight of injected fluid
v_{x0} = velocity of injected fluid at nozzle
C = square root of ratio between minimum effective and maximum jet velocity (submerged flow) divided by 2
Ω_s = calibration factors for gravels, sands, and clays
σ_z = overburden stress
ϕ' = effective soil friction angle

Equation 11.5 can be used to estimate the jet erosion in clayey soils based on soil cohesion:

$$\psi_c = \Omega_3 c_u \tag{11.5}$$

where

ψ_c = erosion resistance clays
c_u = undrained soil cohesion
Ω_3 = calibration factors for gravels, sands, and clays

Equation 11.6 can be used to estimate the minimum jet velocity needed to erode clayey soils:

$$v_L = \sqrt{\frac{\Omega_c g}{\gamma_f} c_u} \tag{11.6}$$

where

v_L = minimum value of jet velocity capable of producing erosion
Ω_c = calibration factors for gravels, sands, and clays
c_u = undrained soil cohesion
γ_f = unit weight of injected fluid

Equation 11.7 can be used to estimate the maximum column radius that can be achieved in clayey soil:

$$R_{max} = \frac{2\Lambda C d_0 v_{x0}}{\sqrt{\dfrac{\Omega_c g N c_u}{\gamma_f}}} \tag{11.7}$$

where

R_{max} = maximum jet column radius attainable by erosion

γ_f = unit weight of injected fluid

Ω = calibration factors for clay

d_0 = nozzle diameter. Typical nozzle diameters are 2–4 mm.

c_u = undrained soil cohesion

Λ = calibration factor for jet velocity

N = relative turbulent kinematic viscosity of injected fluid

11.2 GROUND FREEZING

11.2.1 Introduction to ground freezing

A cutoff wall with low hydraulic conductivity and high compressive strength can be formed in situ by freezing the ground. More precisely, it is the frozen water in the voids that produces the properties of low hydraulic conductivity and high strength. This combination of properties means that ground freezing is useful in numerous engineering applications including excavation support, structural underpinning, groundwater control, tunneling, and environmental remediation projects (Andersland and Ladanyi 2003). Ground freezing can be done in a wide variety of saturated soils including clay, sand, peat, gravel, cobbles, and bedrock and can be very effective through man-made obstructions often encountered in urban fill. Ground freezing was originally patented by Friedrich H. Poetsch in Germany in 1883 and first used in South Wales in 1862 to stabilize vertical shafts (Harris 1995). Depths of 900 meters have been reached (Harris 1995). Once the freezing system is removed, the frozen ground returns to the prior in situ state without residual contamination or remnant admixtures.

Frozen ground can be formed in various configurations as needed for different applications including shafts, mass freezing, sensitive site conditions, and tunneling. Examples in the following four paragraphs of projects illustrate the versatility of ground freezing. They are from Schmall and Braun (2007).

The ground improvement needed for excavation of deep shafts is ideally suited for ground freezing. Frozen ground in a cylindrical configuration offers both groundwater control and excavation support without internal bracing as the frozen walls are self-supporting through compression between overlapping freeze elements. For example, ground freezing was used to construct vertical access shafts to depths of up to 40 m of overburden for New York's Water Tunnel No. 3 Project. A photo of a circular shaft surrounded by a frozen wall is shown on the left side of Figure 11.4 and the shaft is shown schematically on the right side of Figure 11.4.

Where the improvement of large volumes of soil is needed, ground freezing can control groundwater flow and provide a strong, stable material for excavation and tunneling. Construction of the Boston Central Artery (a.k.a. the "Big Dig") required jacking of three large tunnels beneath an active rail yard without disrupting rail service. The soil consisted of more than 6 m of fill and man-made obstructions underlain by Boston Blue clay with groundwater about 3 m below the surface. Ground freezing was used to cut off groundwater flow and stabilize the fill and underlying soils permitting the tunnels to be constructed without disruption of rail service.

Excavation support and groundwater control are often needed adjacent to sensitive structures or in sensitive soils. Since ground freezing installation is relatively vibration free and progresses at a moderate rate that can be monitored, the technique is often well suited for

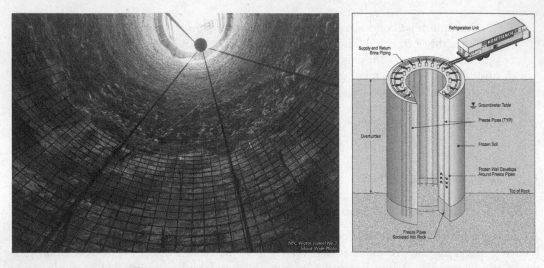

Figure 11.4 Ground freezing for shaft construction (courtesy of Keller; www.Keller-na.com).

Figure 11.5 Ground freezing for earth retention (courtesy of Keller-na.com).

such conditions. At one site, buried fuel tanks needed to be removed from their location immediately adjacent to a hospital wall and close to the operating room. By freezing the ground in the shape of an arch, the over 4 m deep excavation was safely made without causing damaging movements to the sensitive structure. A photo of this project is presented as Figure 11.5.

While there are numerous tunneling techniques, the use of horizontal freezing to construct arches or complete tunnels has been found effective, particularly in difficult soils and fills. On one project in Syracuse, NY, a 37 m long and 3.2 m diameter tunnel for a wastewater outfall was built by freezing the ground in a horizontal cylindrical shape which permitted mining of the unfrozen ground in the frozen tunnel to construct the outfall. The frozen tunnel was 2 m below a railway running 90 trains per day at speeds in excess of 90 kph and did not disrupt service.

Ground freezing has a niche particularly in deep tunneling work because a frozen shaft can be confirmed to be adequately formed (closed) prior to excavation, whereas other ground improvement techniques cannot provide such an assurance. Ground freezing is also advantageous in difficult ground conditions where a substantially thick frozen wall can be created merely by installing a row of small diameter pipes.

11.2.2 Fundamentals of ground freezing

The foregoing examples illustrate the robustness and versatility of ground freezing. Ground freezing uses freeze pipes installed in the ground from which heat energy is extracted by circulating a chilled fluid through double-walled pipes. Depending upon the size and duration of the freeze, the chilled fluid is normally either a brine of calcium chloride ($CaCl_2$) or liquid nitrogen (LN_2). Brine is used for larger projects of longer duration and liquid nitrogen is best for smaller projects of short duration. When using brine, the freeze pipes are connected through a header connected to a refrigeration plant that circulates chilled brine through the pipes in a closed-loop system. The chilled brine solution is pumped down the center of a double-wall freeze pipe to the bottom, where the solution flows upward through the annulus drawing the heat from the soil. Figure 11.6 is a schematic diagram of the freeze pipe system and coolant flow. Liquid nitrogen is not recirculated but fed to the freeze pipes as a liquid and wasted to the atmosphere after extracting the heat from the ground.

The principles of ground freezing are straightforward. Soil is a three-phase system consisting of solid soil particles, porewater, and pore air. In a saturated soil, all of the pores are filled with water and there is no pore air. Along the freeze pipes, heat is extracted from the ground. At $0°$ C the porewater undergoes a phase change to pore ice. Since all of the porewater is not instantaneously converted to pore ice, the ground may exist as a four-phase system that includes both porewater and pore ice. Ground freezing initially occurs immediately adjacent the freeze pipes and, with continued heat extraction, the diameter of the frozen ground grows radially outward. As frozen ground columns grow radially from freeze pipes placed in a row, they eventually intersect to create a frozen wall. With continued heat extraction, the frozen zone continues to grow. Once the desired dimensions are reached, the heat extraction rate is reduced such that heat flowing toward the wall is still removed, but at a reduced rate so the growth of the frozen wall is limited. Upon excavation and exposure of frozen ground to the atmosphere, insulating materials like blankets, concrete, or spray foam, are typically placed to protect the frozen ground to prevent thawing.

The differences between brine and liquid nitrogen freezes are cost and temperature of the coolant. Brine can be chilled and circulated at temperatures as low as $-40°$ C. Once circulated, the brine temperature is increased as a result of the heat extracted from the ground. Brine need only be rechilled before being recirculated. Liquid nitrogen is fed into the freeze pipes at $-196°$ C freezing the ground significantly faster than brine. Thus liquid nitrogen can often freeze the ground within hours and days as compared to days and weeks for brine recirculation. However, liquid nitrogen is not reused but rather undergoes a phase change from liquid to gas as heat is transferred from the ground to the nitrogen. The now-gaseous nitrogen is vented to the atmosphere after use in the freeze pipes. Decreasing the

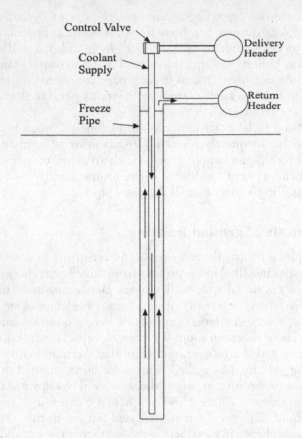

Figure 11.6 Freeze pipe system showing coolant flow.

pipe spacing reduces the time for the freeze to close, regardless of whether the freeze is done with brine or liquid nitrogen.

The heat to be extracted to transform unfrozen ground into frozen ground is of two types. *Sensible heat* is that exchanged between the ground and freeze pipes that simply lowers the ground temperature. *Latent heat* is that associated with the phase change from liquid water to solid water (ice) without a change in temperature. Hence, initial heat extraction requires higher energy consumption as the unfrozen ground is both cooled and converted to frozen ground. The *specific heat capacity* (or just *specific heat*) is the quantity of heat required to change a unit mass of substance by one unit of temperature. In equation form, specific heat is:

$$Q = mc\Delta T \tag{11.8}$$

where
  Q = heat required for temperature change (J)
  m = mass of material (kg)
  c = specific heat capacity (J/kg/K)
  ΔT = change in temperature (K)

To make matters more complicated, the thermal conductivity and specific heat capacities of water and ice are temperature dependent as shown in Figure 11.7. As a result of

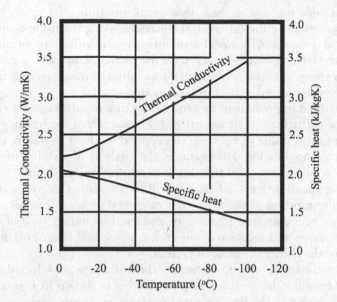

Figure 11.7 Thermal conductivity and specific heat of ice.

these, and geometric complexities, closed-form modeling of ground freezing is not possible. Nomagraphs for design based upon simplifying assumptions have, however, been developed (Sanger and Sayles 1979).

Heat flow to a freeze pipe is analogous to groundwater flow to a well and can be described by the LaPlace equation in three dimensions for unsteady heat flow as:

$$\frac{\partial u}{\partial t} = \frac{\partial^2 u}{\partial x^2} + \frac{\partial^2 u}{\partial y^2} + \frac{\partial^2 u}{\partial z^2} \tag{11.9}$$

where

u = temperature
t = time
$x, y, and z$ = position

The rate of change of temperature with time can be set equal to zero for steady-state conditions. Notice this similarity in form between this equation for heat flow and the equation for groundwater flow (see Chapter 2). For steady heat flow in one-dimension, the heat flow equation can be presented in a form similar to Darcy's law. In the case of one-dimensional heat flow through a unit cross-sectional area, the governing equation can be written in algebraic form as:

$$q = k_T \frac{\Delta T}{\Delta L} \tag{11.10}$$

where

q = heat flow per unit area (W or j/s)
k_T = thermal conductivity (K/m)
ΔT = change in temperature (K)
ΔL = distance over which temperature changes (m)

Software is available for solving heat flow problems (Pimentel et al. 2011 and Krahn 2004). Some programs solve the differential equations using the finite element method and use computer-aided design (CAD) tools for inputting geotechnical conditions. Programs can also account for the latent heat associated with the phase change of water into ice and vice versa. Figure 11.8 shows the results of thermal modelling of mass freezing of saturated sand with liquid nitrogen conducted as part of liquefaction studies.

The design of ground improvement by ground freezing requires consideration of two distinct aspects of the performance: structural and thermal. First, the frozen ground and associated supports must be stable in resisting the applied loads. The nature of these analyses will depend upon the application. For example, the analysis of stability of a frozen circular vertical shaft is very different than that for a horizontally bored tunnel. One method of analysis for one application (vertical shafts) will be presented as an example. All analyses require an understanding of the properties of frozen ground. The second aspect of the design is the design of the ground freezing system itself including type of system (brine or liquid nitrogen), spacing and location of the freeze pipes, and determination of the needed cooling capacity of the chiller or nitrogen system.

As a first approximation, and to illustrate the influence of selected parameters, the required freezing time for different freeze pipe spacing is shown in Figure 11.9. Notice as the temperature goes down, the time required for freezing at any given spacing goes down, as expected. The figure also reveals clay freezes more slowly than sand. Finally, the figure shows that the higher the water content, the more time it takes to freeze the ground. Note that some form of Figure 11.9 appears in several publications including Hausmann (1990), and Jessberger (1987 and 2012) and reportedly originated with Stoss (1976).

The refrigerating plant is an important component of the design and should be designed to be flexible and be able to provide adequate cooling. The pipes are installed using various drilling methods depending on the ground conditions. The monitoring system should take measurements of the temperature of the coolant and the ground. These readings can be had from embedded probes or RTDs (resistance temperature detectors). The dimensions of the frozen wall are normally determined using RTDs. Once the ground is frozen and excavation of unfrozen ground begins, ground deformations may need to be monitored at ground survey points and with inclinometers. It is important to monitor the project during all stages, including the initial installation of pipes, refrigeration (freezing), excavation, and thawing.

The design of a ground freezing system needs to recognize the distinct phases of the process as the frozen ground columns grow independently around freeze pipes, and then the frozen ground columns grow together to form a continuous wall which continues to thicken with time. If a double row of freeze pipes has been deployed, the two separate walls grow together to form a single wall.

11.2.3 Properties of frozen ground

The thermal, hydraulic, and mechanical properties of frozen ground are complex and influenced by numerous factors such as soil type, moisture content, salinity, freezing rate, and freezing direction. It is the frozen water (ice) that increases the strength of frozen ground. In granular soils, the porewater freezes rapidly in the absence of strong groundwater gradients and flow. As a result, for granular soil, strength increases rapidly as the porewater is frozen and the temperature drops to just below 0°C. With further declines in temperature, there is some strength increase. Silts and clays, on the other hand, are more dependent on temperature and continue to strengthen as the temperature drops further below the freezing point. This is because the polar water molecules are in a more complex relationship with the soil particles. Researchers measured the unfrozen water content in silt as a function of

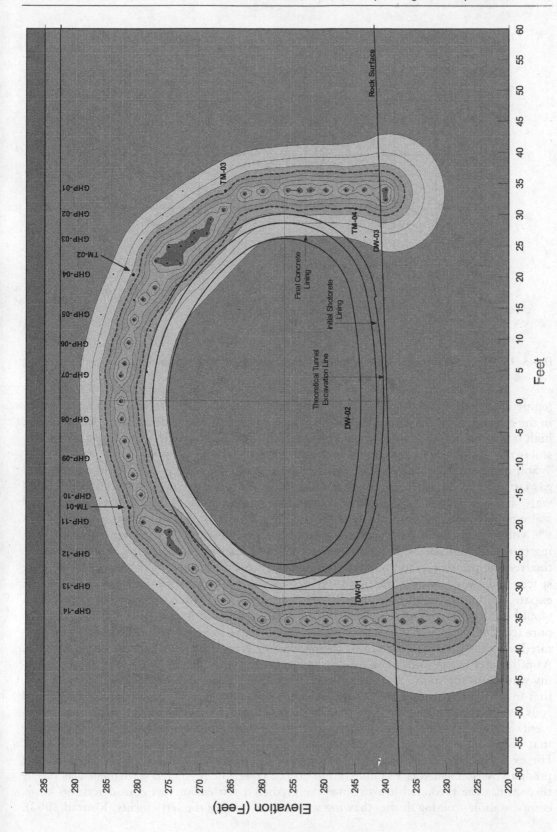

Figure 11.8 Modeling ground freezing for tunnel arch support (courtesy of Keller, www.Keller-na.com).

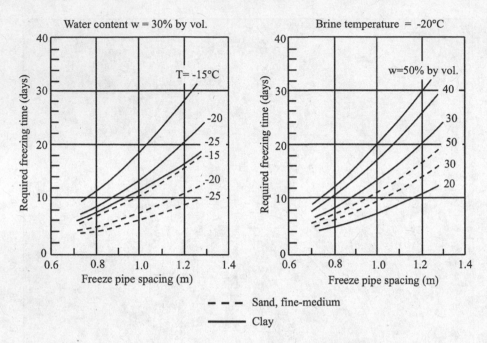

Figure 11.9 Effect of freeze pipe spacing, coolant temperature, soil type, and soil water content on freezing time.

temperature and found that even at temperatures as low as −20°C unfrozen water remained in the pore spaces (Christ and Kim 2009). Thus, a very low temperature is needed to obtain high strengths in fine-grained soils. Powers et al. (2007) indicate that the moisture in clay soil is not completely frozen until it is cooled down to −40°C.

Soil in a natural, unfrozen state consists of three phases: solids (soil), liquid (water), and gas (air). When soil is frozen, another solid phase, frozen water (ice), is introduced. Liquid water has a density of 1.00 g/cm³ at 0°C and ice has a density of 0.92 at 0°C. A saturated soil would expand substantially with freezing in an undrained condition as a result of this 9% volume expansion. However, the volume increase also results in water expulsion during freezing (McRoberts and Morgenstern 1975). Under most geotechnical conditions, as the freezing progresses and porewater expands, excess porewater is pushed out of the pore spaces and away from the freezing front. As a result, in a granular soil there is no volume expansion due to ground freezing. In a fine-grained soil, volume expansion is less than would result from a simple consideration of the expansion of the porewater as it changes to pore ice. The surface manifestation of this volume expansion is known as heave. In unsaturated soils, the pore fluid movement during freezing is reversed from that for saturated soils. As porewater in the voids of an unsaturated soil freezes, a suction gradient is induced pulling water into the voids. This water flowing into the voids increases the water content over that initially present prior to freezing. A similar phenomenon occurs in some fine-grained soils, which can result in the formation of ice lenses which may translate into ground movements and heave. When heave occurs with freezing there is a corresponding volume decrease that occurs upon thawing. The surface manifestation of this volume decrease is settlement. For example, on a tunneling project in Japan, the predicted heave was 4.8 cm while the predicted settlement was 13 cm (Konrad 2002). Here, the predicted settlement was nearly three times the predicted heave. Since the predicted settlement was considered too large, compensation grouting during thawing was used to control the settlements (Konrad 2002).

A large horizontal frozen arch was utilized for both ground support and groundwater cutoff for a tunnel in New York City requiring heave control measures and thaw settlement control measures to protect an overlying active subway tunnel. Heave and settlement control were required as the ground was frozen through an appreciable thickness of varved silt, a soil susceptible to ice lensing. A process referred to as soil extraction was developed to remove soil between the frozen ground and the overlying structure in order to mitigate heave. Compensation grouting was also utilized to balance out the thaw-induced settlement (Schmall 2015). Figure 11.10 is a photo of the excavated tunnel supported by the frozen ground arch.

It is important to understand the fundamental behavior of the soil-water-air system subjected to artificial ground freezing to provide context for the examination of the material properties of frozen ground. Frozen ground is subject to creep and frozen soil parameters chosen for design must take this into consideration. Frozen soil testing is warranted for many ground freezing projects to determine design ground temperatures and corresponding design strengths with acceptable deformation. In one study (Bragg and Andersland 1981), the tensile and compression properties of frozen sand were investigated to determine the effects of strain rate, temperature, and sample size. This study showed that the peak compressive strength and initial tangent modulus all increased with decreasing temperatures and increasing strain rates while tensile strengths were independent of deformation rate. Importantly, the failure mode was observed to change from plastic to brittle as strain rates increased. The authors of that study hypothesized that, at lower strain rates, melting of ice crystals under pressure was followed by water migration and refreezing, thus producing larger failure strains.

11.2.4 Containment of contaminated soils

Ground freezing to provide groundwater control and excavation stability has been used at contaminated sites where the remediation strategy is excavation and removal for treatment

Figure 11.10 Northern Avenue Boulevard Tunnel (courtesy of Keller; www.Keller-na.com).

or landfill disposal. As ground freezing will provide a positive groundwater cutoff, it eliminates the need for continuous groundwater pumping, an important consideration for sites with contaminated groundwater.

The advantages of using a frozen barrier for confining hazardous waste are that it can be used in all soil types, provided water is present, barrier thicknesses can be controlled by temperature, barrier gaps can be filled easily by the addition of new pipes, and the barrier can be frozen into any configuration and at any depth (Sayles and Iskandar 1998). Disadvantages are that sufficient groundwater must be available, refrigeration systems require continuous inspection and maintenance, and frozen ground is not an effective barrier for all contaminants. However, in most cases, this approach can be quite effective and environmentally friendly as the ground is minimally disturbed and the process can be stopped at any time.

11.2.5 Limitations of ground freezing

The most difficult ground condition for ground freezing is moving groundwater. The seepage velocity limit for ground freezing using brine is taken to be approximately 2 m/day (Schmall et al. 2007). Gradients producing these groundwater seepage velocities are rare under natural conditions but as the freeze progresses toward closure, the cross-sectional area for groundwater flow decreases and seepage velocities increase. With increasing seepage velocity between freeze pipes, the cooling effect is distributed down the gradient resulting in elliptically shaped columns of frozen ground. In extreme conditions, the cooling capacity of the freeze pipes may be insufficient for closure of the freeze wall. For example, in one project where ground freezing was used to form a frozen arch to aid in tunneling, the heat associated with flowing water was greater than the capacity of the freeze pipes to remove that heat and closure of the freeze was prevented (Schmall and Dawson 2017). Typical remedial measures for such conditions are the reduction in hydraulic conductivity by grouting or additional freezing through added freeze pipes and/or using a different freezing medium such as liquid nitrogen.

The detection of the presence or absence of unfrozen zones requires careful monitoring of water pressures and temperatures. Plots of temperature versus depth along freeze pipes will reveal zones of higher temperatures. Piezometer readings within a shaft freeze will rise upon closure. As noted for the case study just discussed, when an unfrozen zone is detected, the most common approaches to address this are to grout or to use secondary or tertiary freeze pipes in the unfrozen zone.

While not a limitation per se, it is important to control ground movements during freezing and thawing. The allowable ground movement can vary significantly with the nature of the adjacent structures and infrastructure. For example, the Big Dig in Boston required a mass ground freeze that provided a complete groundwater cutoff and a stable unsupported excavation height of 11.5 m while maintaining the overlying active rail lines in service (Donohoe et al. 2001) which necessitated very tight movement tolerances. Other projects may be more tolerant of ground movements. Ground movements can also be controlled by the sequencing of freezing. Ground movements during thawing can be controlled with compensation grouting.

11.2.6 Conclusions regarding ground freezing

Ground freezing is an established ground improvement method for controlling groundwater flow and solidifying soil in situ for tunneling or excavation stability. As space for construction operations decreases, urban environmental concerns become more prominent, and geotechnical projects get increasingly complex, ground freezing offers an approach that has the potential to reduce the risks and provide effective results. Ground freezing improves the properties of soil by increasing shear strength, increasing tensile strength, increasing the stiffness, and decreasing the hydraulic conductivity. Ground freezing is minimally intrusive and can be removed (i.e. with thawing) with minimal long-term effects. Alternative means for ground freezing mean the technique can be deployed rapidly for short-term ground improvement projects using liquid nitrogen or for longer and larger projects using brine chillers. Finally, ground freezing is relatively quiet compared to other ground improvement and foundation engineering techniques adding value to its use in urban environments. For additional reading, see Chapter 24 of Powers et al. (2007).

11.3 SECANT PILE WALLS

Secant pile walls are cutoff and/or excavation support features that are constructed using alternating "piles" installed in a primary/secondary overlapping pattern to create a continous wall structure. These systems are a natural extension of pile type deep foundation construction approaches reapplied to solve ground improvement objectives including excavation support and hydraulic containment. This method is incorporated in this book because these secant pile systems are often designed and constructed as part of the overall soil system and because the technique, in the applications discussed herein, is not explicitly a foundation construction approach.

Secant piles are typically installed using continous flight auger (CFA) or large diameter auger (LDA) excavation techniques with the excavated cylindrical hole filled with concrete or flowable fill, with or without reinforcement. Drilling is often completed with a casing to keep the drilled hole open between excavation and backfilling. Figure 11.11 shows secant piles used to control groundwater flow and maintain excavation stability at the excavations for the World Trade Center in 2008. Notice the two rows of walers and tieback anchors (external bracing). Also notice the equipment operating adjacent to the top of the wall, indicating the substantial capacity of these secant piles.

Secant piles are installed in a primary and secondary pattern of piles. Two sets of secant piles are shown in plan view in Figure 11.12. The upper set of piles shows reinforcement with an I-beam in the secondary piles and the lower with rebar reinforcement. In each set of five secant piles, three primary and two secondary piles are shown. Sometimes the primary columns are backfilled with a weaker material, commonly cement-bentonite (CB) slurry, to facilitate easier installation of the secondary piles. In this approach, the column spacing may be so close that the secondary elements are essentially tangential and therefore the primary elements are almost completely drilled out in the process of installing the secondary elements. In this case, the secondary elements likely have a larger diameter than the primary elements. Alternatively, the primary and secondary elements can be backfilled with the same material, e.g. concrete, with little overlap between elements. In this case, the amount of element overlap is determined based on the verticality limitations of the drilling equipment and the cutoff wall objectives, i.e. the overlap is selected to ensure that if one or more piles deviate from vertical, there is still adequate connection between piles. In both of these installation scenarios, rebar cages, or other forms of tensile reinforcement, can be installed in the drilled hole prior to backfilling.

Advances in drilling techniques have improved the speed of installation of these systems, and the maximum diameter of piles, thereby reducing the cost and increasing the applicability of this approach. However, secant pile walls, in most construction markets, remain an excavation support and cutoff wall technique that is most commonly applied to sites where other techniques will not work. These installations can be cost-effective in limited access situations, e.g. urban environments, especially where there is a need for both excavation

Figure 11.11 Secant pile wall for earth retention.

support and hydraulic containment. This method is also particularly well suited for installations through obstructions, as the drilling (excavation) techniques can be switched to accommodate a variety of obstructive materials, e.g. steel, reinforced concrete, boulders (rock), debris, and wood, that would result in issues with other methods. Relative to other methods of installing cutoff walls and/or excavation support systems, the secant pile method offers advantages of low noise, minimal vibration, and can accommodate logistical complications. Pile diameters range from 0.3 m to 1.2 m. Depths of up to 30 m are common with depths of 60 m possible in the right conditions.

Once the secant pile wall is installed, the excavation can be performed using a top-down approach wherein internal or external bracing can be installed as the excavation proceeds and as the project conditions dictate. Circular configurations of secant pile walls can also be designed as self-supported systems without the need for internal or external bracing. Finally, secant pile walls containing reinforcement can be designed as cantilever walls with the secant piles generally extending two times the depth of the excavation below the bottom of the excavation.

11.4 COMPACTION GROUTING

11.4.1 Introduction and history

Compaction grouting is a technique that forces a thick viscous semisolid into the ground displacing the in situ material and thus compacting the soil in the vicinity of the compaction

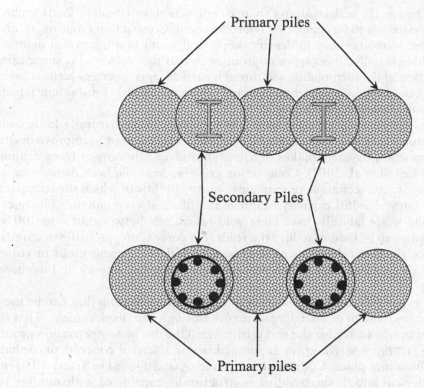

Primary piles

Secondary Piles

Primary piles

Figure 11.12 Plan view of secant pile wall for earth retention.

grout injection point. The technique is particularly useful in counteracting potential settlement as it both densifies the soil in the vicinity of the grout injection and creates a composite between the in situ soil and the stronger injected grout material.

According to *Compaction grouting – the first 30 years* by James Warner (1982), the earliest applications of compaction grouting were for accomplishing remedial objectives on the West Coast of the United States. In fact, this technique was used almost exclusively for remedial activities for the first 30 years (through the late 1970s or early 1980s). In the late 1970s, compaction grouting started gaining acceptance away from the West Coast, e.g. for settlement and loosening control for urban tunneling in soft ground, as described in Baker et al. (1983). Compaction grouting is now recognized as a valuable tool for subway construction and has been used in major cities all over the United States.

In the 1980s, compaction grouting was first used for site improvement in conjunction with deep dynamic compaction at a coal-fired power plant in Florida and shortly thereafter at a nuclear submarine servicing facility in Georgia. In these early site development applications, compaction grouting was used to densify liquefiable granular soils that could not be cost-effectively managed with other techniques.

11.4.2 Uses

Ground improvement associated with compaction is discussed in more detail in Chapter 4. Shallow compaction results from the application of energy (static or vibratory) at the surface of the soil being compacted. Deep compaction is accomplished through the application of a very large impact load at the surface (deep dynamic compaction) or vibratory loads applied deeper within the materials to be compacted (vibroflotation). Compaction grouting, or displacement grouting, is another way to densify deeper soil strata in a controlled manner. Compaction grouting is unlike most types of grouting in that the main goal is specifically strength improvement of the surrounding soil through particle rearrangement (densification or consolidation). For example, compaction grouting is now used widely for sinkhole repair, site improvement, and sinkhole prevention.

The main use of compaction grouting is to compact loose fill or naturally loose soils to underpin structures that have experienced differential settlement or to improve in situ soils against liquefaction by earthquakes or vibration induced movement from drilling or excavation (El-Kelesh et al. 2001). Compaction grouting has even been demonstrated as an effective compaction technique in municipal waste landfills in which the technique can be used to improve landfill capacity, stabilize the fill, and even enhance biological activity. Since solid waste landfills have large void ratios, e.g. between 30% to 100%, compaction grouting can be used to reduce the voids. For compaction grouting in municipal landfill wastes, the grout does not need to be a "standard" cement grout mixture; it can be made from sewage sludge which will improve biological activity and methane production.

Compensation grouting is a sub-application of compaction grouting that can be used to arrest or correct movement of a settling structure through soil displacement. That is, compaction grout *compensates* for the soil settlement. One use of compensation grouting is called *slab jacking* wherein grout is pumped below a settled concrete or asphalt slab to "jack" it back into place. Compensation grouting is also used to arrest differential building settlement and, if the building is structurally capable of withstanding it, even correcting issues associated with differential settlement. If carefully performed, compensation grouting can even be used to arrest or correct excessive total settlement. A specific example of compensation grouting is Soil Fracture Grouting (Soilfrac) which creates subsurface fractures by injecting a low-viscosity fluid grout at pressures in excess of the hydro-fracture pressure, causing heave of the overlying soil (Essler et al. 2000).

One of the principal advantages of compaction grouting is the relatively small size of the drilling equipment and associated kit to perform compaction grouting in the field. Shown on the left side of Figure 11.13 is a drilling rig performing compaction grouting beneath the foundation of a relatively small building. On the right side of Figure 11.13 is the grout mixing truck pulling the grout pump.

Figure 11.13 Compaction grouting drill rig (left) and supporting equipment (right) (courtesy of Keller-na.com).

11.4.3 Design

Soil properties before and after compaction grouting can be obtained through standard penetration tests (SPTs) and cone penetration tests (CPTs), as well as sampling undisturbed soil at depth for laboratory testing. These samples are tested for moisture content, grain size distribution, fines contents, and Atterberg limits (Miller and Roycroft 2004). Prior to grouting, the results of these tests allow engineers to create a subsurface profile. This profile is then used to determine if grouting is a feasible alternative. Once this is known, the spacing of grout holes must be calculated, as well as the expected volume of grout take and pumping pressure. SPTs and CPTs are run before and after compaction grouting to see if the necessary bearing capacity and factor of safety have been attained. Generally, compaction grouting is best used for loose sands where void ratios are high. Compaction grouting is not effective for low void ratio sands or high cohesion clays.

Many geotechnical engineers have proposed methods to quantify the effects of grouting on soil. In order to optimize the way grouting is performed, theoretical models have been created and manipulated to identify ways to effectively reproduce grouting success through calculations rather than by experience alone. Using only experience often results in factors of safety that are much higher than necessary. Also, empirical methods would be expected to assist with quantifying the amount of grout necessary to attain needed densities. Wilder et al. (2005) discusses the difficulties facing engineers using compaction grouting. These include limited success in evaluation of grouting using SPTs, CPTs, and dilatometer testing, as well as difficulties interpreting in situ test results. In order to determine whether a compaction grouting project is successful, a post-construction observation of the site is generally necessary.

Much work has been done to create models that will replicate what has been observed in the field. El-Kelesh et al. (2001) presents a theoretical model that compares well with their data. This model is based on the assumptions that the growing grout bulb is always spherical and ground surface upheaval is associated with shearing above the bulb. The model begins with soil data and injection depth to find the limiting pressure, p_{lim}, and the limiting bulb radius, R_{lim}. The limiting pressure is then used to find the average volumetric strain, Δ. The bulb radius and the volumetric strain are then used to find the radius of the plastic zone, R_p. Using the volumetric strain, one can then determine the improvement in terms of relative compaction, %R. Using the plastic zone radius, one can then determine the spacing needed for grout holes, S, and using the limiting bulb radius one can find the grout quantity needed, Q.

The limiting radius is based on plastic equilibrium and the Mohr-Coulomb criterion. In order to find the volumetric strain, the volume change of the cavity is set equal to the sum of the plastic zone and elastic zone volume changes. The radius of the plastic zone is obtained using the rigidity index which represents the ratio of shear modulus to initial shear strength. In order to find the limiting pressure, the cohesion, angle of internal friction, and the plastic radius are manipulated. The limiting pressure can either be the upheaval pressure or the excessive plastic deformation pressure found using the yield stress. The grout hole spacing is just the diameter of the plastic zone radius. Compaction is assessed by finding the actual relative compaction achieved using the initial and final dry unit weights (El-Kelesh et al. 2001).

Example Problem Ex.11.1: Compaction grouting

Calculate the vertical total and effective stress at a depth of 10 m in a sandy soil profile with the soil having a unit weight of 20 kN/m³ and with the water table located at the ground surface. Would ground heave occur if the pressure used in compaction grouting was 3500 kPa at this depth? Why or why not?

Solution: The total stress at (10 m)(20 kN/m³) = 200 kPa and the effective stress is 200 kPa – (9.8 kN/m³)(10 m) = 102 kPa. To address the second part of the problem, first consider that the pressure during compaction grouting is isotropic. The pressure is isotropic in that the grout is behaving as a liquid without shear stresses to produce anisotropy. To lift the overlying material, the grouting pressure would have to exceed the effective stress. Computations show this to be the case. However, the overlying material also has shear strength that would need to be overcome for ground heave to occur. Modeling of grouting pressures needed to overcome both effective vertical effective stress and shear strength is complicated and imprecise. As a result, ground heave is monitored. So, the answer is that ground heave would not occur solely because the grout pressure is greater than the effective stress but for the given information, it is not possible to definitely say if ground heave would or would not occur. Ground heave usually occurs only after large volumes of grout are injected in a zone. For this reason, compaction grouting of a 0.6 m to 1 m meter zone of a grout hole is limited based on the first of the following three criteria: maximum injection pressure (usually around 3500 kPa) is reached, a maximum target grout volume is injected, or any sign of ground heave as monitored with laser levels.

11.4.4 Construction

Compaction grouting is performed using a high viscosity (low mobility) grout suspension that is pumped into a drilled borehole under enough pressure to physically displace the soil particles bordering the drilled hole. This movement causes the soil particles to compact and densify, in turn increasing the overall strength of the soil. Compaction grouting is based on the displacement of soil, and, thus, the viscosity of the grouting material must be high enough to prevent significant permeation of the grout into the surrounding soil.

Most compaction grouting is performed with similar procedures by monitoring grout consistency, the rate and pressure at which the grout is pumped, and varying grout hole spacing, pipe size, and diameter. The injected grout typically has a slump of less than 25 mm (one inch) and consists of soil and approximately 12% cement in water (El-Kelesh et al. 2001), e.g. similar to weak concrete made with only fine aggregate. The soil contains fine materials, both silts and clays, to provide plasticity, and sand to develop internal friction (El-Kelesh et al. 2001). The left side of Figure 11.14 shows compaction grout as it is delivered by the grout pump. A photo on the right side of Figure 11.14 aids in the visualization of the 25 mm slump.

Compaction grouting can be performed using an *ascending* (up-hole) or *descending* (down-hole) approach. Grout holes are usually grouted for the entire depth of the weak or loose soil

Figure 11.14 Compaction grout as pumped (left) and field slump test for compaction grout (right) (courtesy of Keller; www.Keller-na.com).

stratum. In ascending compaction grouting, a cased borehole (75 mm to 150 mm in diameter) is first drilled to the lowest elevation deemed necessary to compact. A variation on ascending compaction grouting is to first grout the near-surface materials, then go to the bottom of the hole and ascend. As the grouting process nears the ground surface, the near-surface grouted material (now hardened) provides something stiff for the ascending grout to push against, allowing greater grouting pressure. In all methods, the thick, viscous grout materials push the formation material outward, downward, and upwards, causing compaction of the surrounding soils. Pumping is performed until the design measurements are achieved, with design typically controlled by three measurements. First, the quantity of grout is measured by tracking the number of strokes of a piston pump with a calibrated volume per stroke. Second, injection pressure is measured. Finally, ground surface elevations are measured to assure that the grout is staying within the formation being compacted and therefore grout flow is not a result of ground heave. Once the design parameters have been achieved at a specific depth, the casing is withdrawn in 0.3 m to 1.0 m stages with the grouting process repeated at each stage. This allows multiple grout bulbs to be created per location, forming a column (Perkins and Harris 2003). Grouting locations follow a grid pattern in which initial primary center-to-center spacing is typically 3 m to 15 m between each grout hole. Typical final grout hole spacing can be as low as 1 m to 3 m center to center using a square, rectangular, or triangular grid. Holes are grouted using the split spacing method where grout pressures and grout takes are compared between the grouting of the primary, secondary, and, if necessary, tertiary grout hole locations (see Chapter 7 for discussion of primary, secondary and tertiary grout holes). Based on the initial relative density of the soil to be compacted, total grout volumes in the range of 5% to 15% of the volume of soil being grouted are possible.

11.5 EXPLOSIVES IN GROUND IMPROVEMENT

11.5.1 Introduction

Explosives can be used to modify ground conditions through forced particle rearrangement and densification. Explosives work by applying a large amount of shock and gas energy. The explosives, set off underground, cause vibrations that alter the ground conditions, or, on a nearer surface scale, can be used to create a crater that allows for placement of more suitable soils. Explosive compaction and excavation are relatively inexpensive, rapid, easy to use in remote areas, and can affect a large volume of soil in a short time.

Explosives have been used, albeit sparingly compared to other ground modification techniques, for at least eighty years (since the ~1940s) to densify loose sands (Ivanov 1967; Lyman 1941; Fordham et al. 1991; Wild 1961). More recently, explosives have also been applied to other ground improvement problems.

11.5.2 Applications of explosives

The three main applications of explosives in ground improvement are soil compaction, mass excavation, and in situ pile creation. The suitability of explosives for each of these applications depends upon the soil conditions. The deployment methodology of the explosives varies with application. Soil compaction is the most common application of explosives in ground improvement and will be the focus of a latter part of this chapter.

Compaction of deep, cohesionless soil deposits to increase shear strength is the primary application of explosives in geotechnical engineering. Explosives are applicable, and likely to be economical, for ground modification in the following scenarios:

1. when deep liquefaction abatement of loose sands by densification is needed. Explosive compaction has been used in loose deposits over 45 m deep (Shakeran et al. 2016).
2. when compaction of deep, loose deposits is not economical because the cost of conventional heavy equipment access is prohibitive. Projects in remote areas, on steep terrain, or on very soft soils are good candidates for explosive compaction.
3. when the project opportunity cost is high. Explosive compaction is very quick compared to any other method. Equipment is minimal, can be deployed rapidly, and can be executed swiftly. Stabilization of recent debris flows threatening life or safety, for example, are candidates for explosive compaction.
4. when the site is distant from the built environment. Vibrations and, to a lesser extent, noise, associated with explosive compaction may lead to litigation over damage to structures and psychological damage to humans.

Due to the empirical nature of explosive compaction, the availability of an experienced explosive compaction contractor is paramount.

Explosives are also used for rapid excavation; typically, the removal of very soft soils. Zhu et al. (2003) describe a method of creating a gravel road base by explosively removing soft soils and replacing with gravel. Their one-step procedure consists of setting off the charges (here, very large charges) that displace the soil upwards and away from a large pile of gravel adjacent to the crater created by the explosion. The gravel pile then slides down into the crater, perhaps even with enough impact to embed in the soft clay at the base of the crater loosened by the explosion. Yan and Chu (2005) also describe a similar process. More traditionally, though, the crater is backfilled with suitable soil by bulldozers. Jin and Shi (1999) describe an underwater project of excavation using explosives.

In situ pile creation, reported by Gohl et al. (2001), consists of drilling a shallow hole, perhaps three meters deep, and loading with a continuous, low-energy explosive. After explosion, a cylindrical hole is created which can be backfilled with concrete, or other materials, to form a pile for supporting surface loads.

11.5.3 Ground conditions favorable to explosives for compaction

Loose, cohesionless materials compact when vibrated. These materials are the most suitable for explosive compaction since the explosion creates shear and compression waves that break existing soil particle bonds and rearrange the particles under the overburden pressure. The process is similar to what would occur in an earthquake. The effects of earthquakes on loose, saturated cohesionless soils are well documented: liquefaction resulting in large settlements and displacements. Liquefaction typically occurs in saturated soils, although it can occur in loose, unsaturated soils. The effect is the same, only the pore air pressure (as opposed to the porewater pressure in saturated soils) is increased until the soil becomes a fluviated bed. As in liquefaction, the effective stress (see Chapter 2) approaches zero which means the shear strength is negligible, which results in settlement and densification. Clayey soils are far less suitable for modification by explosives.

11.5.4 Construction practice for compaction by explosives

Explosive compaction practice generally consists of drilling a hole and setting explosive charges at different elevations therein. Many holes are drilled, typically in a grid pattern. The charges are activated sequentially: from bottom to top in each hole, and from row to row across the grid, creating a traveling wave across the site. Several applications of charge may be needed (Gohl et al. 2001). Even with repeated treatments, explosive compaction may be faster for treatment of deep cohesionless soils than alternatives such as dynamic compaction or vibroflotation.

When done properly, the surface manifestations of explosive compaction are minimal - a small heave, perhaps a small cloud of dust, followed by settlement that can occur over a period of days. Sand boils may be observed in sites where liquefaction has occurred. Typical vertical strains due to liquefaction caused by blasting or earthquakes are in the range of 5% to 7% of the thickness of the liquefying layer. Ten percent strains have been observed (Tokimatsu and Seed 1987). Gohl et al. (2001) propose a method for predicting the vertical strain.

11.5.5 Post explosion evaluations

Evaluation of explosive compaction is done using soil penetrometers with the SPT (ASTM 2011) and CPT (ASTM 2000) being common choices. The relative density of cohesionless soils can be increased by up to 70% using explosive compaction (ASTM 2006; Murray et al. 2006). Often, penetration tests are done before and after to assess the need for another round of treatment. Practitioners occasionally report settlement but decreased penetration resistance. Finno et al. (2016) present an explanation, based on the saturation of the groundwater with N_2 released by the explosive. N_2 is less soluble in water than CO_2 (also released during blasting), thus, reaching saturation quickly. The remaining N_2, unable to go into solution with the groundwater, remains as independent bubbles in the soil. Since bubbles have very low shear strength, penetration resistance is decreased until the gas eventually dissipates into the adjacent soil or surfaces. This may take months, depending on site conditions, so it is important to understand the length of time it will take to reach the final strength and settlement properties.

11.5.6 Collateral concerns with the use of explosives

Setting off an explosion in the built environment may have undesirable collateral effects. For instance, an explosive shock near fisheries may cause fish kills. Similarly, large noises (which may result when explosive compaction is done improperly) have been known to lead to the early demise of small birds and animals, e.g. chickens and rabbits.

Surface and subsurface structures may be damaged by explosive compaction; for example, buildings, tunnels, pipelines, and canals. Particle velocity and acceleration monitoring are highly recommended to mitigate the effects of litigation.

Explosive compaction and excavation have also caused landslides. This hazard must be assessed. Using explosives near recent volcanic activity should be given particular attention to prevent triggering landslides in the volcanic debris, which is often cohesionless and very loose. Hachey et al. (1994) describe densification of the Mt. St. Helens debris flows.

Highly sensitive clays may liquefy with very small movements, including the small movements caused by explosive compaction. One such example is a marine deposited, freshwater leached clay that will become a viscous fluid when disturbed. Lacasse (2013) describes several quick clay landslides, including the Kattmarka slide, inadvertently triggered by explosive excavation. For this reason, areas with highly sensitive clays are not candidates for explosive compaction.

Soil above the zone of explosive compaction may arch (Elliot et al. 2009) resulting in significant, rapid vertical displacement at significant times after explosive compaction. The soil above the zone of explosive compaction is pushed upwards and may not come back down completely immediately after the explosion. The lack of immediate, significant settlement suggests this may have occurred. In order to assess and remediate this, further compaction is required with dynamic compaction being the recommended approach.

11.5.7 Case studies

Vega-Posada et al. (2014) describe the use of explosives to densify loose sands for liquefaction mitigation. The site, in South Carolina, was a municipal solid waste landfill underlain by liquefiable loose sand to depths of up to 12 m below the surface. Grid patterns were used for boreholes. Successive blasts were undertaken with approximately seven days between blasts. Ground surface settlements were measured as large as 0.5 m. These results were used to compute post-blast relative density. Initial relative density was estimated to be between 12% and 24% (from CPT). Final relative density values were computed to be between 65% and 91%. Post-blast relative densities mean the sand is no longer considered liquefiable. Despite the unquestionable densification as determined by the settlement data, a decrease in penetration resistance was found post-blast. As mentioned earlier, this is not uncommon.

Finno et al. (2016), Shakeran et al. (2016), and Van Court and Mitchell (1995) identify other case studies of explosives use in geotechnical engineering.

11.6 PROBLEMS

11.1 Using definitions and phase diagram principles from Chapter 2, compute the expected surface settlement for a 5 m thick loose sand layer ($e_0 = 1.0$) to be densified by blasting.

11.2 Why is it difficult to freeze flowing water?

11.3 The results of the treatability study for an environmental cleanup site indicate that it will take at least 10% cement by dry weight of soil to achieve the cleanup objectives. Due to underground and overhead obstructions, jet grouting has been selected as the mixing/delivery method. The total density of the soil is 1,800 kg/m³ and the moisture content is 20%. What is the minimum grout flow rate needed to ensure the appropriate amount of cement is added to a 1 m diameter column if each liter of grout holds 750 g of cement? The test program indicates that it will take a flow rate of at least 150 liter/min at a lift rate of 0.3 m/min to create a 1 m column. What lift rate and pump rate should you use to install a 1 m column meeting the cement dosage criteria?

11.4 A jet grouting test program has been performed and the table below shows the summary information collected. What flow rate and lift rate would you recommend to achieve 100% coverage of an area if the unit cost of materials in the grout is $5/liter and the unit cost of consumables for the jet grouting is $1000/minute?

Flow Rate	Lift Rate = 0.3 m/min	Lift Rate = 0.6 m/min
150 l/min	1	0.75
200 l/min	1.5	1
300 l/min	2	1.5

11.5 Ground freezing may be done with chilled brine or liquid nitrogen. What site conditions would lead to choosing one over the other?

11.6 A secant pile wall is needed to serve as excavation support and a cutoff wall for a deep excavation. In order to achieve the structural and hydraulic containment objectives, the secant piles must overlap enough to achieve a minimum 1 m minimum width. The primary and secondary piles are planned to be the same diameter, 3 m. If a center-to-center spacing of 2.9 m achieves a minimum width of 0.77 m, 2.8 m achieves a minimum width of 1.08 m, 2.7 m achieves a minimum width of 1.31 m, and 2.6 m achieves a minimum width of 1.5 m, what center-to-center spacing would you recommend if the drill's maximum deviation is 1% from vertical and the pile height is 10 m?

11.7 Compaction grouting is planned for a sandy soil site. The radius of influence for each compaction grout column is planned to be 1 m. If the initial porosity of the sand is 50%, what is the final porosity within the radius of influence if 1 m^3 of compaction grout is used for every lineal meter?

11.8 What soil types are most susceptible to explosive compaction? What are the least? Suggest why this is.

11.9 What gasses are released when dynamite explodes? What is the significance of this for the geotechnical engineer involved with explosive compaction?

11.10 Describe the scenario when you would want to trigger a landslide using explosives. Tell what precautions you would require.

11.11 What is explosive compaction called when done improperly?

11.12 How much heat is needed per liter of water at 15°C to form ice at 0°C?

11.13 Using Darcy's law, show why the velocity of the water in the ground increases as the cross-sectional area of flow decreases during ground freezing.

REFERENCES

Andersland, O.B. and Ladanyi, B., 2003. *Frozen ground engineering.* John Wiley & Sons, New York.

ASTM, 2000. *ASTM D5778 (04.08). Test method for performing electronic friction cone and piezocone penetration testing of soils.* American Society for Testing and Materials, West Conshohocken, PA.

ASTM Committee D-18 on Soil and Rock, 2006. *Standard test methods for minimum index density and unit weight of soils and calculation of relative density.* Conshohocken, PA: ASTM International.

ASTM D1586-11, 2011. *Standard test method for standard penetration test (SPT) and split-barrel sampling of soils.* Conshohocken, PA: ASTM International.

Baker, W.H., Cording, E.J. and MacPherson, H.H., 1983. Compaction grouting to control ground movements during tunneling. In *Underground space* (pp. 205–212), Vol. 7. Elmsford, NY: Permagon Press.

Bragg, R.A. and Andersland, O.B., 1981. Strain rate, temperature, and sample size effects on compression and tensile properties of frozen sand. *Engineering Geology, 18*(1–4), 35–46.

Christ, M. and Kim, Y.C., 2009. Experimental study on the physical-mechanical properties of frozen silt. *KSCE Journal of Civil Engineering, 13*(5), (pp.317–324).

Donohoe, J.F., Corwin, A.B., Schmall, P.C. and Maishman, D., 2001. Ground freezing for Boston Central Artery contract section C 09 A 4, jacking of tunnel boxes. In *2001 rapid excavation and tunneling conference* (pp. 337–344). San Diego, CA: Society for Mining, Metallurgy, and Exploration.

El-Kelesh, A.M., Mossaad, M.E. and Basha, I.M., 2001. Model of compaction grouting. *Journal of Geotechnical and Geoenvironmental Engineering, 127*(11), 955–964.

Elliott, R.J., Clarke, L., Gohl, B., Fulop, E., Singh, N.K., Berger, K.C. and Huber, F., 2009, February. Explosive compaction of foundation soils for the seismic upgrade of the Seymour Falls Dam. In *Proceedings of the 35th annual conference on explosives and blasting technique*, Denver, CO, January.

Essler, R.D., Drooff, E.R. and Falk, E., 2000. Compensation grouting: concept, theory and practice. In *Advances in grouting and ground modification* (pp. 1–15), Denver, CO, August 2000. Reston, VA: ASCE.

Finnickey, J.C. and Pensive, N.A., 2017. Blasting induced migraine headaches. *Journal of Irreproducible Results, 88*(9), 1122–1129.

Finno, R.J., Gallant, A.P. and Sabatini, P.J., 2016. Evaluating ground improvement after blast densi-fication: performance at the Oakridge landfill. *Journal of Geotechnical and Geoenvironmental Engineering, 142*(1), 10–1061.

Fordham, C.J., McRoberts, E.C., Purcell, B.C. and McLaughlin, P.D., 1991. Practical and theoretical problems associated with blast densification of loose sands. In *Proceedings of the 44th Canadian geotechnical conference* (Vol. 2, pp. 92–1). Calgary, Canada, September.

Gohl, W.B., Howie, J.A. and Rea, C.E., 2001. Use of controlled detonation of explosives for liquefaction testing. In *Proceedings of fourth international conference on recent advances in geotechnical earthquake engineering and soil dynamics*, Missouri University of Science and Technology, paper 9–13 (pp. 1–9). San Diego, CA, March 2001.

Hachey, J.E., Plum, R.L., Byrne, R.J., Kilian, A.P. and Jenkins, D.V., 1994. Blast densification of a thick, loose debris flow at Mt. St. Helen's, Washington. In *Vertical and horizontal deformations of foundations and embankments* (pp. 502–512). ASCE, College Station, TX, June 1994. Reston, VA: ASCE.

Harris, J.S., 1995. *Ground freezing in practice.* Thomas Telford, London.

Hausmann, M., 1990. *Engineering principles of ground modification.* New York, NY: McGraw-Hill Publications.

Ivanov, P.L., 1967. *Compaction of noncohesive soils by explosions* (translated from Russian). National Technical Information Service Report No. TT70-57221. US Department of Commerce, Springfield, VA, 211.

Jessberger, H.L., 1987. Artificial freezing of ground for construction purposes. Chapter 31 in Bell, F.G. and Bell, F.G. (Eds.), *Ground engineer's reference book* (Vol. 59). Butterworths, London.

Jessberger, H.L. ed., 2012. *Ground freezing* (Vol. 26). Amsterdam: Elsevier.

Jin, L. and Shi, F., 1999. Blasting techniques for underwater soft clay improvement. *China Harbour Engineering, 2*, 10–17.

Konrad, J.M., 2002. Prediction of freezing-induced movements for an underground construction project in Japan. *Canadian Geotechnical Journal, 39*(6), 1231–1242.

Konrad, J.M., 2008. Freezing-induced water migration in compacted base-course materials. *Canadian Geotechnical Journal, 45*(7), 895–909.

Krahn, J., 2004. *Thermal modeling with TEMP/W: an engineering methodology.* Calgary, Canada: GEO-SLOPE.

Lacasse, S., 2013. 8th Terzaghi Oration Protecting Society from landslides–the role of the geotechnical engineer. In *Proceedings of the 18th international conference on soil mechanics and geotechnical engineering*, Paris (pp. 15–34).

Lyman, A.K.B., 1941. Compaction of cohesionless foundation soils by explosives. In *Proceedings of the American society of civil engineers*, Vol. 67, No. 5 (pp. 769–780). New York, NY: ASCE.

McRoberts, E.C. and Morgenstern, N.R., 1975. Pore water expulsion during freezing. *Canadian Geotechnical Journal, 12*(1), 130–141.

Miller, E.A. and Roycroft, G.A., 2004. Compaction grouting test program for liquefaction control. *Journal of Geotechnical and Geoenvironmental Engineering, 130*(4), 355–361.

Modoni, G., Croce, P. and Mongiovi, L., 2006. Theoretical modelling of jet grouting. *Géotechnique, 56*(5), 335–347.

Murray, L., Singh, N.K., Huber, F., Siu, D. and District, G.V.R., 2006, July. Explosive compaction for the Seymour Falls dam seismic upgrade. In *Proceedings on the 59th Canadian geotechnical conference*, Vancouver, Canada, October.

NAIS, 2005. *National Standards Authority of Ireland. Eurocode 2: Design of concrete structures – Parts 1–1: General rules and rules for buildings*, Dublin, Ireland: National Standards Authority of Ireland.

Nelson, S.M. and Reed, G., 2007. *Effects of jet grouting on wetland invertebrates at Mormon Island Auxiliary Dam, Folsom, California.* US Department of the Interior, Bureau of Reclamation, Technical Service Center, Washington, DC.

Perkins, S.W. and Harris, J., 2003. Using the grouting intensity number (GIN) to Assess Compaction Grouting Performance. In *Third international conference on grouting and ground treatment* (pp. 991–1009), New Orleans, LA, February 10–12, 2003.

Pimentel, E., Papakonstantinou, S. and Anagnostou, G., 2011. Case studies of artificial ground freezing simulations for urban tunnels. In *Proceedings of ITA-AITES World Tunnel Congress 2011 "Underground spaces in the service of a sustainable society society & 37th ITA-AITES General Assembly*. Helsinki, Finland, May 05–10, 2011.

Powers, J.P., Corwin, A.B., Schmall, P.C. and Kaeck, W.E., 2007. *Construction dewatering and groundwater control: new methods and applications*. Hoboken, NJ: John Wiley & Sons.

Sayles, F.N. and Iskandar, I.K., 1998. *Ground freezing for containment of hazardous waste*. No. DOE/OR/22141-T3; CONF-9508244. Corps of Engineers, Cold Regions Research and Engineering Lab., Hanover, NH.

Sanger, F.J. and Sayles, F.H., 1979. Thermal and rheological computations for artificially frozen ground construction. *Engineering Geology, 13*(1–4), 311–337.

Schmall, P.C. and Braun, B., 2007. Ground freezing—a viable and versatile construction technique. In *Proceedings of 13th international conference on cold regions engineering* (pp. 29–40), Orono, ME.

Schmall, P.C., Corwin, A.B., & Spiteri, L.P., 2007. Ground freezing under the most adverse conditions: moving groundwater. In Traylor, M.T. & Townsend, J.W., Littleton CO (Eds.), *Proceedings of rapid excavation & tunneling conference* (pp. 360–368), June, 2007. Toronto, ON: Society for Mining, Metallurgy & Exploration.

Schmall, P., Curry, A., Perrone, F., and Rice, J., 2015. Compensation grouting for the east side access Northern Boulevard crossing. In *Proceedings from the 2015 rapid excavation and tunneling conference*, New Orleans, LA, June.

Schmall, P. and Dawson, A., 2017. Ground-freezing experience on the east side access Northern Boulevard crossing, New York. *Proceedings of the Institution of Civil Engineers: Ground Improvement, 170*(3), 159–172.

Shakeran, M., Eslami, A. and Ahmadpour, M., 2016. Geotechnical aspects of explosive compaction. *Shock and Vibration*, 2016, Article ID 6719271, 14 pages. https://doi.org/10.1155/2016/6719271.

Stoss, K., 1976. *Die Anwendbarkeit der Bodenvereisung zur Sicherung und Abdichtung von Baugruben*. Gesellschaft für Technik und Wirtschaft, Dortmund.

Tinoco, J., Correia, A.G. and Cortez, P., 2009, December. A data mining approach for jet grouting uniaxial compressive strength prediction. In *2009 world congress on nature & biologically inspired computing (NaBIC)* (pp. 553–558), Coimbatore, India, December. IEEE.

Tokimatsu, K. and Seed, H.B., 1987. Evaluation of settlements in sands due to earthquake shaking. *Journal of Geotechnical Engineering, 113*(8), 861–878.

Van Court, W.A.N. and Mitchell, J.K., 1995. New insights into explosive compaction of loose, saturated, cohesionless soils. *Geotechnical Special Publication (GSP) 49* (pp.51–65). Sessions in conjunction with the ASCE Conference. San Diego, CA, October 22–26, 1995.

Vega-Posada, C.A., Zapata-Medina, D.G. and García Aristizabal, E.F., 2014. Ground surface settlement of loose sands densified with explosives. *Revista Facultad de Ingeniería – Universidad de Antioquia, 70*, 9–17.

Warner, J., 1982, February. Compaction grouting-the first thirty years. In *Grouting in geotechnical engineering* (pp. 694–707). ASCE.

Wild, P.A., 1961. Tower foundations compacted with explosives. *Electrical World, 66*, 36–38.

Wilder, D., Smith, G.C. and Gómez, J., 2005. Issues in design and evaluation of compaction grouting for foundation repair. In Schaefer, Vernon R., Bruce, Donald A., and Byle, Michael J. (Eds.), *Innovations in grouting and soil improvement* (pp. 1–12), Geo-Frontiers Congress 2005, Austin, TX, January 24–26, 2005.

Yan, S.W., and Chu, J., 2005. Use of explosion in soil improvement projects. In Indraratna, B., and Chu, J. (Eds.), Volume 3, Chapter 38 of *Ground improvement and case histories* (pp. 1085–1096). Amsterdam: Elsevier Geo-Engineering Book Series.

Zhu, B., Chen, R.-P., and Chen, Y.-M., 2003. Transient response of piles-bridge under horizontal excitation. *Journal of Zhejiang University. SCIENCE. Part A, 4*(1), 28–34.

Chapter 12

The future of ground improvement engineering

12.1 INTRODUCTION

The field of ground improvement engineering has grown enormously since the terms ground improvement, ground modification, and similar terms entered our lexicon in the later third of the 20th century. The first conference on the subject was "Placement and Improvement of Soil to Support Structures", held in Cambridge, Massachusetts, in 1968. This first conference was sponsored by the Division of Soil Mechanics and Foundation Engineering of the American Society of Civil Engineers (ASCE 1968). Hausmann (1990) published the first comprehensive textbook on the subject. While ground improvement engineering is a relatively new field within geotechnical engineering, many ground improvement technologies have matured and, at the same time, new developments are occurring at a rapid pace.

The future of ground improvement will be marked by continued invention of new equipment, and development of existing equipment, particularly through computer controls, and sensors, to improve precision and construction productivity. Artificial intelligence will aid the field engineer in decision making. Developments in instrumentation-related performance monitoring and quality control are inevitable. Development of new materials and material combinations will undoubtedly result in superior performance of the improved ground as measured by strength, compressibility, and permeability as well as bearing capacity, reduced settlement, and improved liquefaction resistance. Ground improvement methods will also encompass new approaches that will include the use of biological methods. One such biological method showing potential is microbial induced calcite precipitation to increase the strength and decrease the compressibility and permeability of sandy soils.

Historically, geotechnical engineers were primarily concerned with providing an adequate factor of safety against failure, controlling settlements and movements of the ground, and cost. More recently, environmental and sustainability considerations have become an important part of the *decision* process. Construction noise, historically and/or architecturally important structures, archeological finds, and inconvenience to the public have all become essential environmental considerations when employing ground improvement. It seems clear that future projects (beyond 2021) will need to explicitly consider sustainability and legacy effects in the decision-making process.

In the development of technology, it takes time for ideas to be formulated, research to be conducted and disseminated, for early adopters to utilize the new technology, and finally for use of the technology to become widespread. Further, in reality, those processes occur along a continuum with inevitable overlap. Hence, for the purposes of this chapter, ground improvement methods that are not in widespread use are discussed with recognition that there has been limited deployment for some technologies or only research and development for others.

This chapter on the future methods in ground improvement is organized in a principled way to include developments in biogeotechnical methods, in materials, and in construction monitoring. A separate section on sustainability follows.

12.2 BIOGEOTECHNICAL METHODS FOR GROUND IMPROVEMENT

Biogeotechnical ground improvement, also known as microbial geotechnology, is an emerging area that synergistically combines biological processes and geochemistry to produce improvements in soil properties. Mineral precipitation, biopolymer generation, mineral transformation, and gas production are all biomediated processes that have potential in ground improvement. Mineral precipitation can be used to improve bearing capacity, reduce settlement, and improve liquefaction resistance. Biopolymer generation (biofilm growth) can be used to reduce permeability and thus for subsurface hydraulic barriers, underground seepage control, and corrosion protection. Mineral transformation can be used to reduce soil expansion and improve slope stability. Gas production from microbes can be used to reduce liquefaction potential (O'Donnell et al. 2017a, 2017b).

Figure 12.1 is a schematic of some of the possible uses of biogeotechnical ground improvement including soil stabilization for tunneling, controlling groundwater infiltration from runoff, stabilizing a slope against erosion, adding strength to a slope to improve slope stability, and stabilizing a subgrade.

It is estimated that there are billions of bacteria per gram of soil in the near-surface underground environment. These living creatures are actively engaged in life's processes including participation in biogeochemical reactions and in reproduction. Engineers can learn from natural biogeotechnical processes that alter the properties of soil in a way that is beneficial to their engineering properties. For example, these natural processes include the carbonate cementation of sand, the formation of a desert crust, and the mineral transformation of smectitic clays to illites. Similarly, engineers can learn from natural biogeotechnical processes that are detrimental to the constructed environment including mineral scaling of piping and bioclogging of drainage systems in landfills and dams. While these processes may be detrimental when they occur uncontrolled and in the wrong place, a properly engineered biogeotechnical system can harness these processes for beneficial use.

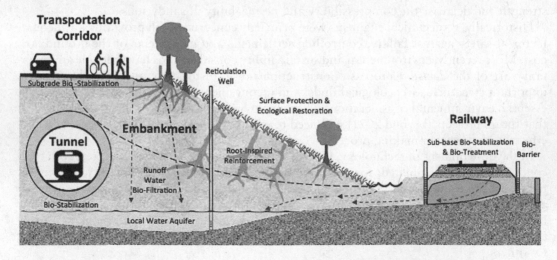

Figure 12.1 Biogeotechnical applications for ground improvement (from DeJong et al. 2011).

As with any new technology, there are challenges to overcome. Social challenges include educating owners, engineers, regulators, and the general public to the benefits of biogeotechnical ground improvements. Technical challenges include further development of biogeotechnical means and methods, scaling laboratory studies to the field, reaction products, development of theoretical analytical models incorporating both biological and geochemical processes, and possible reversibility of the improvement. That said, tremendous advances are being made in this area. Shown in Figure 12.2 is a biocemented sand column made in the Arizona State University test pit as part of the research undertaken by the Center for Bio-mediated and Bio-inspired Geotechnics. The column was created using enzyme induced carbonate precipitation in just three working days using three injection cycles at 24-hour intervals. The cemented column was then exposed by using a vacuum extractor to remove the surrounding uncemented sand.

12.2.1 Biocementation

Biocementation is defined as the biological production of materials that bind soil grains together resulting in increased shear strength and reduced compressibility. Microorganisms living in the subsurface can induce calcium carbonate (calcite) precipitation, also termed

Figure 12.2 Bio-cemented column made in Arizona State University test pit (courtesy of Kimberly Martin, Center for Bio-mediated and Bio-inspired Geotechnics).

biomineralization. In so doing, the calcite binds (cements) soil particles together resulting in increased strength, stiffness, and liquefaction resistance, as well as reduced permeability. *Microbial Induced Calcite Precipitation* (MICP) bacteria, such as Sporosarcina Pastueurii, break down nutrients such as urea by hydrolysis resulting in an increased pH and the precipitation of calcium carbonate. The chemical reaction for bacterial ureolysis is as follows:

$$CO(NH_2)_2 + 2H_2O \rightarrow 2NH_4 + CO_3^{2-} \tag{12.1}$$

$$Ca + CO_3^2 \rightarrow CaCO_3 \tag{12.2}$$

Precipitates form at the particle-to-particle contact points and within the voids to bond the particles resulting in an increased density and reduced permeability. This method has also been termed biogrout. With these property changes, it is possible to envision applications where improved bearing capacity and reduced settlement are sought, where greater liquefaction resistance is needed, or where reductions in hydraulic conductivity are required such as in geoenvironmental applications. Obvious advantages of this technique over other ground improvement methods discussed earlier in the book are that it is minimally intrusive and minimally disruptive. In the future, biogeotechnical ground improvement methods are projected to be both cost effective and sustainable. Current research is aiming to overcome the challenges including the management and control of complex biological, hydrological, and geochemical processes while avoiding unintended consequences and side effects. Laboratory testing demonstrates that MICP improves mechanical properties of sand including increased strength and stiffness (O'Donnell and Kavazanjian 2017). It has also been found that surfactants may aid in achieving a uniform distribution of bacteria and thus of precipitated calcite when using MICP (Dawoud et al. 2014).

MICP can be produced through multiple and different processes. Shown in Table 12.1 are end products of four microbial processes. Undesirable attributes of the end products are shown as well. While the end product attributes shown in the third column of Table 12.1 are generally considered undesirable, the formation of gas has been shown to reduce pore pressure build up during earthquake loading and thus improves liquefaction resistance.

So how much calcium carbonate precipitation is needed to measurably improve the soil? Whiffin et al. (2007) experimented with a 5 m sand column treated with bacteria and reagents that were injected without clogging of the sand. Calcium carbonate was found over the entire column length. Peak compressive strength was improved for samples with calcium carbonate concentrations greater than 60 kg/m³. The threshold calcium carbonate content of 60 kg/m³ corresponds to 3.6% cement by weight based on the dry unit weight for the sand used in these experiments (1.65 g/cm³). While somewhat higher, this compares favorably with reported increases in peak strength with as low as 0.5% cement by weight for portland cement mixed with sand ex situ.

Table 12.1 MICP Processes, End Products, and Attributes

Microbial process	End products	Undesirable side-effect
Bacterial ureolysis	NH_3 (ammonia)	Toxic gas
	NH_4^+ (ammonium)	Forms toxic salts
Sulfate reduction	H_2S (hydrogen sulfide)	Toxic gas
Fermentation (fatty acids)	CH_4 (methane)	Explosive gas
Denitrification	N_2 (nitrogen)	None

While calcium carbonate concentration clearly has an impact on the resulting strength gain, it is not quite that simple. Al Qabany and Soga (2014) found that lower solution concentrations produced larger strength gains. They attributed this to the more homogeneous distribution of calcite precipitation as shown in the scanning electron microscopy images in Figure 12.3. Notice the small but well-distributed crystals on the left side of Figure 12.3 compared to the larger but more randomly distributed crystals on the right side of Figure 12.3. Hence it is not only the calcium carbonate content but the nature and distribution of the crystals that influences the strength gain.

MICP can also be used to reduce the potential for internal erosion. In one study (Jiang et al. 2014) MICP treatment of a sand-kaolin mixture was found to increase the critical hydraulic gradient and the critical shear stress needed to cause internal erosion. The total internal erosion mass was reduced for samples modified through MICP as compared to control samples that were unmodified. Studies such as this demonstrate the potential for MICP in ground improvement.

At the time of this writing, MICP research continues to inform the efficacy of the process for ground improvement. In addition to strength gain, MICP will also reduce the permeability of sand. This may or may not be a desirable characteristic, particularly when the delivery of nutrients to the process is inhibited.

12.2.2 Bioclogging to reduce hydraulic conductivity

Bioclogging is defined as the biological production of materials that reduce the effective porosity and hydraulic conductivity. Bonal and Reddi (1998) noted slime, biomass, and biogenic gas bubbles can accumulate in soil pores causing a reduction in the soil's ability to transmit water. Ivanov and Chu (2008) summarized the microbial processes that can lead to bioclogging. Bioclogging is the result of different organisms and mechanisms including algae and cynobacteria, nitrifying and ammonifying bacteria as well as bacteria that produce slime. Each bacteria type requires certain conditions such as light, nutrients, oxygen, salts, and/or urea to produce bioclogging. Of course, under certain field situations, such as well filters, aquifer recharge, and sand filtration; bioclogging is an undesirable outcome. In the future of ground improvement, the reduction in hydraulic conductivity by controlled use of bioclogging will require additional research in laboratory and scale up to the field. The question of permanence may be the most challenging to address.

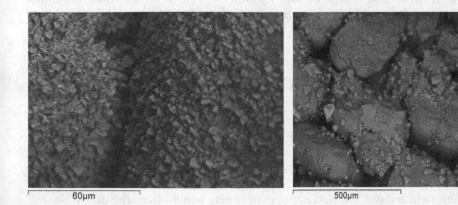

60μm   500μm

Figure 12.3 Scanning electron microscopy images of samples with different chemical treatments (left side) 0.25 M at 80 kg/m3 and (right side) 1 M at 65 kg/m3 (from Al Qabany et al. 2013 with permission from ASCE).

The foregoing discussion of bioclogging is based upon the microbes producing by-products that lower the hydraulic conductivity. Adding biopolymers directly to the subsurface is a different approach. One of the earliest applications was the use of guar gum slurry in a biopolymer slurry trenching process for the construction of a drainage trench (Hanford and Day 1988). Since that time, biopolymer slurry trenches have been used for installation of a chimney filter for a dam (Jairaj and Wesley 1995) and in permeable reactive barrier construction (Wilkin et al. 2014). While these uses are well established for temporary reduction in hydraulic conductivity, research is needed to examine the potential for long-term use.

Numerous laboratory studies have demonstrated the efficacy of biopolymers to reduce the hydraulic conductivity of sands. One study (Bouazza et al. 2009) compared the effectiveness of guar gum, sodium alginate, and xanthan gum in reducing the hydraulic conductivity of silty sand and found xantham gum the most effective. The hydraulic conductivity was reduced four to five orders of magnitude and the suggested use is in temporary seepage barriers. Importantly, the authors note "a large amount of additional research work is needed before practical application can be considered."

12.2.3 Bio-methods for liquefaction mitigation

Liquefaction potential can be reduced by adding a cementitious component, such as Portland cement, to loose sand. It has been demonstrated that this same benefit can be achieved using MICP (Montoya et al. 2012). Bio-mediated ground improvement via MICP can both increase the resistance to liquefaction triggering as well as reduce the consequence of liquefaction should it occur. In centrifuge modeling tests, Montoya et al. (2014) showed that the resistance to liquefaction increased and there were substantial decreases both in the excess porewater pressure development and in the settlement of a model structure.

It has long been known that induced partial saturation (desaturation) can reduce liquefaction potential (Yegian et al. 2007). In more recent studies of microbially induced desaturation and precipitation (MIDP), partial saturation was bio-mediated through the employment of denitrifying bacteria (O'Donnell et al. 2017a). Stimulated denitrifying bacteria produce nitrogen and carbon dioxide as a by-product of their bio-stimulation resulting in desaturation thus reducing liquefaction potential. This desaturation is short term, but, during the process, the pH and carbonate alkalinity are altered in a way that causes MIDP. Thus, this second stage of MIDP results in cementation, void filling, and increase in particle roughness which, in turn, reduces liquefaction potential (O'Donnell et al. 2017b).

12.3 NEW MATERIALS FOR GROUND IMPROVEMENT

Many ground improvement methods employ reagents in their process. For example, as discussed in Chapter 8, vertical barriers of soil-bentonite (SB) use sodium bentonite to prepare the slurry and to mix with the backfill to aid in constructability and reduce hydraulic conductivity. Recent research has shown SB walls can be constructed of modified calcium bentonites (Yang et al. 2018). No doubt other material changes can be developed to improve performance and/or reduce costs. Ground improvement may benefit from the use of "smart" materials, that is, materials whose properties can change in response to external stimuli. This section of the text discusses a few areas of research that offer promise for future material developments in ground improvement engineering.

12.3.1 MgO cement

A number of ground improvement methods incorporate ordinary portland cement (OPC) into the process to increase strength, reduce compressibility, and reduce hydraulic conductivity. Examples include vertical barriers, deep soil mixing, jet grouting, stabilization/solidification, and vibratory concrete columns. In all of these cases, there is no benefit to the shrinkage that is inevitable with cementitious mixtures. In some cases, such as vertical barriers, shrinkage can reduce the stresses and result in cracking (Cermak et al. 2012). However, magnesia (MgO) has been found to work as a *shrinkage-compensating* additive without detrimentally affecting other properties.

In one study the effect of MgO addition to mixtures used for the in situ construction of soil mixed barrier walls was investigated (Al-Tabbaa et al. 2014). Typical cement-bentonite mixtures containing portland cement, bentonite, and water experienced hydration and thermal shrinkage and axial strains of approximately –0.5 mm/m (shrinkage). Given the longitudinal nature of barrier walls and the friction along both sides of the wall which provides restraint against movements, cracking can result from such shrinkage (Evans and Jefferis 2014). In contrast, samples containing MgO as an additive were found to experience expansion both initially at a high rate and also throughout the curing period at a lower rate. The resulting linear strains were greater than 1.0 mm/m (expansion). Interestingly, when granulated ground blast furnace slag was added to the mixture, samples experienced some initial shrinkage, but this was negated over time due to the presence of small amounts of MgO present in the slag. Effects of MgO upon other properties such as hydraulic conductivity and strength have yet to be studied.

12.3.2 Polymers

Drilling support fluids are needed for a wide range of ground improvement and subsurface construction techniques including cutoff wall construction, tunneling, directional drilling, site investigations, well installations, and permeable reactive barriers to name a few. The fluid of choice for many years has been bentonite-water slurry (drilling mud). There is a long history of the use of bentonite-water slurry. An example of the use of bentonite-water slurry and the necessary properties is described in Chapter 8 in the discussion of SB slurry trench cutoff walls.

In the 1990s, water-soluble polymers saw increasing use. Initially, most polymers were biodegradable natural polymers such as guar gum. Uses of the natural polymers were described for slurry trenching (Tallard 1992) and for subsurface drain installation (Day and Ryan 1992). More recently, the use of synthetic polymers has been increasing. A comprehensive examination of polymer support fluids for civil engineering is given by Lam and Jefferis (2018). As developments in polymer science continue, and as the availability of quality bentonite decreases, is it expected that polymer support fluids will find increasing usage in ground improvement applications.

12.3.3 Smart and self-healing materials

Smart materials have properties that adapt in response to changes in their environment. Hence, a smart structure is a system with components that can sense, control, and respond to environmental changes (Cao et al. 1999). For example, piezoceramic actuators can be used for vibration suppression in structures (Song et al. 2006). There is a wide range of smart materials including piezoelectric materials that respond to an applied voltage, shape-memory alloys that respond to temperature or stress changes, and photomechanical materials that change shape in response to light. Some smart materials relevant to ground improvement engineering include temperature-responsive materials such as those proposed to aid closure in ground freezing walls, and self-healing materials which might repair cracking that may occur in barrier walls. The idea of self-healing is taken from living systems and is sometimes termed biomimetics.

The occurrence of self-healing of concrete has been recognized for many years by observing cracks in concrete structures that have been found filled with white crystals of calcium carbonate (Li and Yang 2007). Self-healing has been demonstrated through the embedment of microbes that convert nutrients into what is effectively a limestone (Jonkers 2007). Acid-producing bacteria are used for self-healing of cracks. These types of bacteria can remain viable for over 200 years under dry conditions. In the crack healing process, these bacteria act as a catalyst. While the concept of self-healing for concrete is well studied, the use of self-healing for ground improvement is in early days and represents a likely development for the future.

12.4 TECHNOLOGY DEVELOPMENTS IN GROUND IMPROVEMENT: DRONES, SENSORS, AND ARTIFICIAL INTELLIGENCE

Verifying the quality of ground improvement processes often requires a significant level of effort, destructive testing, post-construction investigations, and reliance on process monitoring rather than actual verification of the in situ product. Drones are already used to some extent in construction monitoring as well as in air and water pollution monitoring. In construction, drones provide a fast and accurate visual report of construction progress, if such progress is observable from above. Coupled with GPS systems, drones can provide quantitative data as well in terms of topographic mapping and survey results. Other uses include safety and security monitoring. Drones may be used to access remote sites, and, with adequate technology, gather data telemetrically from sensors buried in modified ground. Water content, strain, and temperature sensors are currently available. The future development requires adaptation of drones to monitor the quality of ground improvement processes.

Structural health monitoring, listed in order of increasing difficulty, includes detecting damage, locating damage, identifying the type of damage, and assessing the severity of the damage. In this process, sensors are first installed, monitoring data is then transmitted (sometimes wirelessly), after which data is acquired via a data acquisition system, and, finally, signal processing and data analysis occurs. Just as building health is monitored, monitoring of geotechnical structures and ground conditions is viable. For example, tunnels have been monitored and monitoring with prototype wireless sensor networks has been done (Stajano et al. 2010). In another project, the stresses and pore pressures were monitored with sensors hard wired to a data acquisition system and wirelessly transmitted to a dashboard for analysis (Marchiori et al. 2019). This project used legacy instrumentation with a modern *Internet of Things* (IoT) data collection and visualization pipeline. In another example, fiber optic instrumentation has been used to monitor the bending and circumferential hoop strains for a deep shaft for a construction project in London (Schwamb et al. 2014). Smart infrastructure and construction is clearly an emerging field with great potential to transform construction (Mair et al. 2016) and ground improvement engineering more specifically. Future buried sensors could include chemical analyses and a sonar-based system for monitoring internal erosion in, for example, embankment dams or slurry walls.

Looking beyond extensions of current technology, ground modification sensors that measure ground modification set time for cement-treated soils, in situ shear strength, strain during loading of modified soils, and groundwater flow rates are envisioned.

Future ground modification equipment may contain technology for better imaging the subsurface during ground modification. Sensors may measure soil densities below, say, a mixing head, allowing the operator (which may be an artificial intelligence computer) to alter the energy inputs to the mixing head. Sensors similar to the CT scans used in current medical practice may be used. The sensor would allow the operator to see obstructions, cracks developing, the presence of voids, and perhaps, even, predict the loss of drilling fluids or ground modification slurry.

As databases of operational records become more public, artificial intelligence may be used to select the type of ground modification needed, the type of slurry (if any) and its characteristics, and provide improved initial time and cost estimates, as well as cost estimates on the fly. The use of "big data", which has current use in the medical field, will likely find application in ground improvement.

12.5 EQUIPMENT DEVELOPMENTS

The developments in sensors, data acquisition, and data analysis are embraced by the construction equipment industry. For example, construction of pneumatic caissons puts workers at risk due to their lengthy exposure to hyperbaric working conditions. Peng et al. (2019) describe the use of a *remote-controlled* construction process that eliminated this risk to workers. The process included a remote-controlled excavator, a vertical screw conveyor for soil removal and an integrated, real-time, monitoring system. The skill of operators of remote-controlled excavators can be evaluated using a remodeled remote-controlled excavator and virtual reality technology (Sekizuka et al. 2020). Remote-controlled excavators are available and offer added safety for operation in conditions where it is desirable to remove the operator from the machine (see Figure 12.4).

If the work at Wolf Creek dam is any indication of future projects, the use of multiple methods for the same project are likely to find increasing usage. A total of nine techniques were used to meet the project goals while maintaining the safety of the dam including sonic

Figure 12.4 A remote-controlled D7R-II dozer removes rubble to create an open lane for vehicles to travel through during experimental testing at the Caterpillar Edwards Demonstration and Learning Center, October 25, 2018, Edwards, Ill. (US Army Photo by Sgt. 1st Class Jason Proseus/416th TEC).

Figure 12.5 Clamshell excavation at Wolf Creek Dam.

drilling, high mobility grouting, low mobility grouting, clamshell excavation, hydromill excavation, directional drilling, auger/bucket drilling, reverse circulation drilling, and verification coring (Bruce 2016). A photo of just one of the techniques, clamshell excavation, is presented in Figure 12.5. One way to use multiple techniques is to build a composite cutoff wall where grouting is combined with a cutoff wall (Bruce et al. 2012).

12.6 SUSTAINABILITY IN GROUND IMPROVEMENT

12.6.1 Introduction to sustainable ground improvement

Current practice in ground improvement engineering requires an assessment of the subsurface conditions, development of alternatives, and selection of the appropriate alternative as judged by performance, cost, and schedule criteria. In the future, *sustainability* will be commonly considered in the evaluation process.

Sustainability in geotechnical engineering can manifest itself in many different ways. For example, use of urban underground space for transportation can free up space on the surface for green areas, reduce surface noise, and improve surface air quality. In a recent study, a model for the geological suitability for underground space development was developed which included topography and landforms, engineering geology, and hydrology (Wang and Peng 2014). Consider the selection of the appropriate vertical barrier for environmental containment. Current practice would consider site geology, cost, project requirements, and regulatory requirements but not the global environmental impact. In one study, four vertical barrier alternatives were assessed using the triple bottom line approach considering economic, environmental, and social sustainability (Evans et al. 2020). The four alternatives were soil-bentonite, cement-bentonite, soil-mixed, and sheet pile walls. A sustainability index was computed using the Spanish integrated value model for sustainability assessment as described by Reddy et al. (2018). For the case study project, the soil-bentonite method was the most sustainable. The outcome, however, is site specific. Importantly, during the phase of the project during which alternatives are evaluated, sustainability will likely be one of the criteria used in the future.

12.6.2 Sustainable materials

Ground improvement using conventional compaction techniques and materials is widely used and well understood (see Chapter 4). However, there are alternatives to building exclusively with virgin construction materials (a non-renewable resource). These include mixing waste materials with the fill soil or constructing the fill of materials other than soil such as foamed glass aggregate, expanded polystyrene, tire shreds, or foamed concrete. There are a number of reasons to include materials other than soil in a compacted fill or to build a compacted fill out of alternative materials to soil. First, sustainable construction practice dictates that recycled/reused materials are preferred over virgin construction materials from non-renewable resources. Second, soils are rather dense materials and lighter weight fill materials can reduce settlement of the foundation material. Lastly, many recycled/reused materials have very high permeabilities compared to soil.

The inclusion of *shredded scrap tires* offers a means to produce a lightweight fill and to recycle/reuse waste materials in lieu of non-renewable construction materials (Bosscher et al. 1997). The engineering properties of tire chip and sand mixtures (Masad et al. 1996) or clay mixtures (Cetin et al. 2006) are promising for the properties they possess as a lightweight highway fill material. Concerns regarding inclusions usually relate to the environmental impact. There are concerns about the properties of the compacted fill, as compared to a fill without inclusions. A comprehensive review of related studies found it reasonable to recommend the use of recycled scrap tires in civil engineering applications (Liu et al. 2000). In one study both laboratory and field investigations were undertaken to assess the feasibility of a shredded tire fill for highway embankments. It was found that pure tire chips were highly compressive whereas sand-chip mixtures could have moduli comparable to pure sand fills and that existing flexible and rigid pavement design methods could be used with the modulus of tire chips and tire chip-soil mixtures (Bosscher et al. 1997). In another study of sand mixtures, the optimum tire shred content leading to the maximum shear strength was 35% (Zornberg et al. 2004).

There are a number of compacted fill applications that might benefit from the inclusion of shredded scrap tires including:

1. subgrade fill and embankments over weak and compressible soils,
2. backfill of retaining walls and bridge abutments,
3. subgrade insulation against freezing and frost heave, and
4. landfill applications including gas venting systems and leachate collection systems.

Whenever a waste material or by-product is used in the environment, it is essential to determine if the material will have a detrimental effect on the environment. In the case of ground improvement projects, the concern is normally with groundwater impacts. In one study of the groundwater impacts of shredded tires (Humphry et al. 1997), no impact was found for any of the primary drinking water standards. It was found that some secondary standards, such as iron and aluminum, may be elevated under certain pH conditions. Tests for organics found the results all below the detection limits. Two more recent studies (Yoon et al. 2006; Hennebert et al. 2014) also found that embankments with shredded tire fill had negligible impacts on the environment. It should be noted that, in the case of fire, both the fire and the residue may have unacceptable environmental impacts.

Figure 12.6 Foamed Glass Aggregates (courtesy of AeroAggregates of North America LLC).

Lightweight fill is an excellent alternative to conventional fill in cases where it is beneficial to minimize subgrade settlement. A product meeting this need is *foamed glass aggregate* (FGA) made from recycled glass having a unit weight of 0.13 to 0.4 tonne/m^3 (Loux 2018). FGA lightweight fill is substantially lighter than conventional soil fill that typically has a unit weight greater than 1.6 tonne/m^3. A photo of FGA is shown in Figure 12.6. Environmental assessment of FGA shows leachate of the material is far lower than the drinking water standards and the material has a much lower energy consumption than conventional aggregate-cement material typically used in construction (Arulrajah et al. 2015). FGA can also be used as an aggregate in lightweight concrete (Khatib et al. 2012). An added benefit is an increase in the ductility of the concrete, for instance in reinforced concrete beams.

The generation of electricity by burning coal has generated, and currently generates, significant quantities of coal combustion residuals (CCR). The bulk of the CCR is flyash which is a fine, predominantly silt-sized material composed mostly of silica. In terms of sustainability, the reuse of CCR reduces the need for landfill construction and waste disposal and reduces the need for other virgin resources. In terms of engineering, the reuse of CCR can improve the strength and durability of engineered materials. Examples of encapsulated uses of CCR include concrete, brick, and roofing materials. Unencapsulated uses include embankments and structural fills. To be clear, CCRs, and in particular flyash, have been studied and used intermittently for a long time (Raymon 1961; Mateos and Davidson 1962; Digioia and Nuzzo 1972; Gray and Lin 1972). While CCR continues to be widely used in concrete, geotechnical or geoenvironmental usage of CCR is not widespread. Future material evaluations should include a sustainability component and it is expected CCR will be more widely used in ground improvement projects.

12.7 CROSSOVER INFORMATION IN GROUND IMPROVEMENT

As the world grows smaller, as knowledge grows cheaper, and as information becomes more readily available, crossover knowledge becomes more common. It is expected that ground improvement will be no exception. New mix designs, components, and procedures, based on materials already in use in other fields of human endeavor, will come to light. Medical and biological technology will be a fertile source of information for the ground improvement industry.

12.8 SUMMARY OF FUTURE DEVELOPMENTS IN GROUND IMPROVEMENT

Ground improvement is one of the two substantial parad-shifting technologies in civil engineering in the last forty years (the other is geosynthetics). Ground improvement will see more interesting, useful, cost-effective, paradigm-shifting developments in the future thanks to big data, better communication, more experience, and inputs from other fields.

12.9 PROBLEMS

12.1 Increasing the shear strength of sand using MICP shows promise. How might one deliver microbes and nutrients in situ to ensure homogeneous calcite precipitation and strength gain?

12.2 What are the advantages/disadvantages of stimulating indigenous microbes compared with microbe injection?

12.3 MICP is being considered for ground improvement for a site on a flood plain adjacent to a river where the silty, clayey sand subsurface material is weak and compressible. Considering the depositional history of flood plain deposits (the geology that is), identify the major issues associated with successful MICP ground improvement at this site.

12.4 What fields of human endeavor, besides medicine, will contribute to improved ground improvement? How?

12.5 If you had a magic wand, what improvements in ground improvement would you wish for?

12.6 What have you seen, or thought of, that could be applied to ground improvement?

12.7 Suggest ways in which a smartphone can be used for ground improvement.

12.8 Choose a topic representing the future of ground improvement and prepare a brief paper and presentation on the topic.

REFERENCES

Al Qabany, A. and Soga, K., 2014. Effect of chemical treatment used in MICP on engineering properties of cemented soils. In *Bio-and Chemo-Mechanical Processes in Geotechnical Engineering: Géotechnique Symposium in Print 2013* (pp. 107–115). ICE Publishing, London, UK.

Al-Tabbaa, A., O'Connor, D. and Abunada, Z., 2014. Field trials for deep mixing in land remediation: Execution, monitoring, QC and lessons learnt. In *International Conference on Piling & Deep Foundations*, Stockholm, Sweden.

American Society of Civil Engineers, 1968. *Specialty Conference on Placement and Improvement of Soil to Support Structures*. ASCE Soil Mechanics and Foundations Division, Cambridge, MA, August 26–28, 440 pp.

Arulrajah, A., Disfani, M.M., Maghoolpilehrood, F., Horpibulsuk, S., Udonchai, A., Imteaz, M. and Du, Y.J., 2015. Engineering and environmental properties of foamed recycled glass as a lightweight engineering material. *Journal of Cleaner Production*, 94, 369–375.

Berechman, J., 2003. Transportation—economic aspects of Roman highway development: The case of Via Appia. *Transportation Research Part A: Policy and Practice*, 37(5), 453–478.

Bonala, M.V.S., and Reddi, L.N., 1998. Physicochemical and biological mechanisms of soil clogging: An overview. *ASCE Geotech Spec Publ 78*, 43–68.

Bosscher, P.J., Edil, T.B. and Kuraoka, S., 1997. Design of highway embankments using tire chips. *Journal of Geotechnical and Geoenvironmental Engineering*, 123(4), 295–304.

Bouazza, A., Gates, W.P. and Ranjith, P.G., 2009. Hydraulic conductivity of biopolymer-treated silty sand. *Géotechnique*, 59(1), 71–72.

Bruce, D.A., 2016. Remedial cutoff walls for dams: Great Leaps and Wolf Creek. In *Proceedings of the DFI International Conference on Deep Foundations, Seepage Control and Remediation*, New York, Deep Fooundations Institute, Inc., Hawthorne, NJ.

Bruce, D.A., Dreese, T.L., Harris, M.C. and Heenan, D.M., 2012. Composite cut-off walls for existing dams: theory and practice. In *Grouting and Deep Mixing 2012* (pp. 1248–1264). Deep Fooundations Institute, Inc., Hawthorne, NJ.

Cao, W., Cudney, H.H. and Waser, R., 1999. Smart materials and structures. *Proceedings of the National Academy of Sciences of the United States of America*, 96(15), 8330–8331.

Cermak, J., Evans, J. and Tamaro, G.J., 2012. Evaluation of soil-cement-bentonite wall performance-effects of backfill shrinkage. In *Grouting and Deep Mixing 2012* (pp. 502–511). New Orleans, LA, ASCE, Reston, VA.

Cetin, H., Fener, M. and Gunaydin, O., 2006. Geotechnical properties of tire-cohesive clayey soil mixtures as a fill material. *Engineering Geology*, 88(1–2), 110–120.

Dawoud, O., Chen, C.Y. and Soga, K., 2014. Microbial-induced calcite precipitation (MICP) using surfactants. In *Geo-Congress 2014: Geo-Characterization and Modeling for Sustainability* (pp. 1635–1643). Atlanta, GA: ASCE, Reston, VA.

Day, S.R. and Ryan, C.R., 1992. State of the art in bio-polymer drain construction. In *Slurry Walls: Design, Construction, and Quality Control*. (pp. 333–343). Philadelphia, PA: ASTM International.

DeJong, J.T., Fritzges, M.B. and Nüslein, K., 2006. Microbially induced cementation to control sand response to undrained shear. *Journal of Geotechnical and Geoenvironmental Engineering*, 132(11), 1381–1392.

DeJong, J.T., Mortensen, B.M., Martinez, B.C. and Nelson, D.C., 2010. Bio-mediated soil improvement. *Ecological Engineering*, 36(2), 197–210.

DeJong, J.T., Mortensen, B., Soga, K., Banwart, S.A., Whalley, W.R., Martinez, B. and Kavazanjian Jr, E., 2011. Harnessing Bio-Geotechnical Systems for Sustainable Ground Modification. *Geo-Strata Magazine, ASCE*.

DeJong, J.T., Soga, K., Kavazanjian, E., Burns, S., Van Paassen, L.A., Al Qabany, A., Aydilek, A., Bang, S.S., Burbank, M., Caslake, L.F. and Chen, C.Y., 2014. Biogeochemical processes and geotechnical applications: Progress, opportunities and challenges. In *Bio-and Chemo-Mechanical Processes in Geotechnical Engineering: Géotechnique Symposium in Print 2013* (pp. 143–157). London, UK: Ice Publishing.

DiGioia, A.M. and Nuzzo, W.L., 1972. Fly ash as structural fill. *Journal of the Power Division*, 98(1), 77–92.

Evans, J.C. and Jefferis, S.A., 2014. Volume change characteristics of cutoff wall materials. In *Proceedings of the 7th International Congress on Environmental Geotechnics* (pp. 10–14). London, UK: ICE.

Evans, J. C., Ruffing, D. G., Reddy, K. R., Kumar, G., and Chetri, J. K. (2020). Sustainability of Vertical Barriers for Environmental Containment. In *Sustainable Environmental Geotechnics* (pp. 271–283). Springer, Cham.

Gray, D.H. and Lin, Y.K., 1972. Engineering properties of compacted fly ash. *Journal of the Soil Mechanics & Foundations Division*, 98(sm4), 361–380.

Hanford, R.W. and Day, S.R., 1988. Installation of a deep drainage trench by the bio-polymer slurry drain technique. In *Proceedings of the Second National Outdoor Action Conference on Aquifer Restoration, Ground Water Monitoring and Geophysical Methods*. (Vol. 3). Las Vegas, NV: National Water Well Association, Dublin, OH.

Hausmann, M.R., 1990. *Engineering principles of ground modification*. New York: McGraw-Hill Publishing Company.

Hennebert, Pierre, Stephane Lambert, Fabien Fouillen, and Benoit Charrasse. 2014. "Assessing the environmental impact of shredded tires as embankment fill material." *Canadian Geotechnical Journal* 51(5), 469–478.

Humphrey, D.N., Katz, L.E. and Blumenthal, M., 1997. Water quality effects of tire chip fills placed above the groundwater table. In *Testing Soil Mixed with Waste or Recycled Materials* (pp. 299–313). New Orleans, LA: ASTM International.

Ivanov, V. and Chu, J., 2008. Applications of microorganisms to geotechnical engineering for bioclogging and biocementation of soil in situ. *Reviews in Environmental Science and Bio/Technology*, 7(2), 139–153.

Jairaj, V. and Wesley, L.D., 1995. Construction of a chimney drain using bio-polymer slurry at Hays Creek dam. In *IPENZ Annual Conference 1995, Proceedings of: Innovative Technology; Volume 1*; Papers presented in the technical programme of the IPENZ Annual Conference held in Palmerston North, February 10–14, 1995 (p. 229). Palmerston North, NZ: Institution of Professional Engineers New Zealand.

Jiang, N.J., Soga, K. and Dawoud, O., 2014. Experimental study of the mitigation of soil internal erosion by microbially induced calcite precipitation. In *Geo-Congress 2014: Geo-Characterization and Modeling for Sustainability* (pp. 1586–1595). Reston, VA: ASCE.

Jonkers, H.M., 2007. Self healing concrete: A biological approach. In S. van der Zwaag (Ed.) *Self Healing Materials* (pp. 195–204). Springer, Dordrecht.

Khatib, J.M., Shariff, S. and Negim, E.M., 2012. Effect of incorporating foamed glass on the flexural behaviour of reinforced concrete beams. *World Applied Sciences Journal*, 19(1), 47–51.

Lam, C. and Jefferis, S.A., 2018. *Polymer Support Fluids in Civil Engineering*. London: Ice Publishing.

Li, V.C. and Yang, E.H., 2007. Self healing in concrete materials. In S. van der Zwaag (Ed.) *Self Healing Materials* (pp. 161–193). Springer, Dordrecht.

Liu, H.S., Mead, J.L. and Stacer, R.G., 2000. Environmental effects of recycled rubber in light-fill applications. *Rubber Chemistry and Technology*, 73(3), 551–564.

Loux, T.A. The new lightweight contender: ultra-lightweight foamed glass aggregate finds the US market. *Geo-Strata—Geo Institute of ASCE*, 22(5), 30–34, 36, 38-39.

Mair, R.J., Soga, K., Jin, Y., Parlikad, A.K. and Schooling, J.M., 2017, February. Transforming the future of infrastructure through smarter information. In *Proceedings of the International Conference on Smart Infrastructure and Construction*, 27–29 June 2016. In *Proceedings of the Institution of Civil Engineers - Civil Engineering* (Vol. 170, No. 1, pp. 39–47). London, UK: ICE.

Marchiori, A., Li, Y. and Evans, J., 2019. Design and evaluation of IoT-enabled instrumentation for a soil-bentonite slurry trench cutoff wall. *Infrastructures*, 4(1), 5.

Martinez, B.C., DeJong, J.T., Ginn, T.R., Montoya, B.M., Barkouki, T.H., Hunt, C., Tanyu, B. and Major, D., 2013. Experimental optimization of microbial-induced carbonate precipitation for soil improvement. *Journal of Geotechnical and Geoenvironmental Engineering*, 139(4), 587–598.

Masad, E., Taha, R., Ho, C. and Papagiannakis, T., 1996. Engineering properties of tire/soil mixtures as a lightweight fill material. *Geotechnical Testing Journal*, 19(3), 297–304.

Mateos, M. and Davidson, D.T., 1962. Lime and fly ash proportions in soil, lime and fly ash mixtures, and some aspects of soil lime stabilization. *Highway Research Board Bulletin*, 335, 40–64.

Mitchell, J.K. and Santamarina, J.C., 2005. Biological considerations in geotechnical engineering. *Journal of Geotechnical and Geoenvironmental Engineering*, 131(10), 1222–1233.

Montoya, B.M., DeJong, J.T. and Boulanger, R.W., 2014. Dynamic response of liquefiable sand improved by microbial-induced calcite precipitation. In *Bio-and Chemo-Mechanical Processes in Geotechnical Engineering: Géotechnique Symposium in Print 2013* (pp. 125–135). London, UK: ICE Publishing.

Montoya, B.M., DeJong, J.T., Boulanger, R.W., Wilson, D.W., Gerhard, R., Ganchenko, A. and Chou, J.C., 2012. Liquefaction mitigation using microbial induced calcite precipitation. In *GeoCongress 2012: State of the Art and Practice in Geotechnical Engineering* (pp. 1918–1927). ASCE, Reston, VA.

O'Donnell, S.T. and Kavazanjian Jr, E., 2015. Stiffness and dilatancy improvements in uncemented sands treated through MICP. *Journal of Geotechnical and Geoenvironmental Engineering*, 141(11), 02815004.

O'Donnell, S.T., Kavazanjian Jr, E. and Rittmann, B.E., 2017b. MIDP: liquefaction mitigation via microbial denitrification as a two-stage process. II: MICP. *Journal of Geotechnical and Geoenvironmental Engineering*, 143(12), 04017095.

O'Donnell, S.T., Rittmann, B.E. and Kavazanjian Jr, E., 2017a. MIDP: Liquefaction mitigation via microbial denitrification as a two-stage process. I: desaturation. *Journal of Geotechnical and Geoenvironmental Engineering*, 143(12), 04017094.

Peng, F.L., Dong, Y.H., Wang, H.L., Jia, J.W. and Li, Y.L., 2019. Remote-control technology performance for excavation with pneumatic caisson in soft ground. *Automation in Construction*, 105, 102834.

Raymon, S., 1961. Pulverized fuel ash as embankment material. *Proceedings of the Institution of Civil Engineers*, 19(4), 515–536.

Reddy, K.R., Chetri, J.K. and Kiser, K., 2018. Quantitative sustainability assessment of various remediation alternatives for contaminated lake sediments: Case study. *Sustainability: The Journal of Record*, 11(6), 307–321.

Schwamb, T., Soga, K., Mair, R.J., Elshafie, M.Z., Sutherden, R., Boquet, C. and Greenwood, J., 2014. Fibre optic monitoring of a deep circular excavation. *Proceedings of the Institution of Civil Engineers: Geotechnical Engineering*, 167(2), 144–154.

Sekizuka, R., Ito, M., Saiki, S., Yamazaki, Y. and Kurita, Y., 2020. System to evaluate the skill of operating hydraulic excavators using a remote controlled excavator and virtual reality. *Frontiers in Robotics and AI*, 6, 142.

Song, G., Sethi, V. and Li, H.N., 2006. Vibration control of civil structures using piezoceramic smart materials: a review. *Engineering Structures*, 28(11), 1513–1524.

Stajano, F., Hoult, N., Wassell, I., Bennett, P., Middleton, C. and Soga, K., 2010. Smart bridges, smart tunnels: Transforming wireless sensor networks from research prototypes into robust engineering infrastructure. *Ad Hoc Networks*, 8(8), 872–888.

Tallard, G.R., 1992. New trenching method using synthetic bio-polymers. In *Slurry Walls: Design, Construction, and Quality Control*. (pp. 86–102). Philadelphia, PA: ASTM International. ASTM, Conshohocken, PA.

van Paassen, L.A., Ghose, R., van der Linden, T.J., van der Star, W.R. and van Loosdrecht, M.C., 2010. Quantifying biomediated ground improvement by ureolysis: Large-scale biogrout experiment. *Journal of Geotechnical and Geoenvironmental Engineering*, 136(12), 1721–1728.

Wang, Y., and Peng, F. L. (2014). Evaluation of urban underground space based on the geological conditions: A feasibility study. In *New Frontiers in Geotechnical Engineering* (pp. 187–197). Reston, VA: ASCE.

Whiffin, V.S., Van Paassen, L.A. and Harkes, M.P., 2007. Microbial carbonate precipitation as a soil improvement technique. *Geomicrobiology Journal*, 24(5), 417–423.

Wilkin, R.T., Acree, S.D., Ross, R.R., Puls, R.W., Lee, T.R. and Woods, L.L., 2014. Fifteen-year assessment of a permeable reactive barrier for treatment of chromate and trichloroethylene in groundwater. *Science of the Total Environment*, 468, 186–194.

Yang, Y.L., Reddy, K.R., Du, Y.J. and Fan, R.D., 2018. Sodium hexametaphosphate (SHMP)-amended calcium bentonite for slurry trench cutoff walls: Workability and microstructure characteristics. *Canadian Geotechnical Journal*, 55(4), 528–537.

Yang, Z., Cheng, X. and Li, M., 2011. Engineering properties of MICP-bonded sandstones used for historical masonry building restoration. In *Geo-Frontiers 2011: Advances in Geotechnical Engineering* (pp. 4031–4040). Reston, VA: ASCE.

Yegian, M.K., Eseller-Bayat, E., Alshawabkeh, A. and Ali, S., 2007. Induced-partial saturation for liquefaction mitigation: Experimental investigation. *Journal of Geotechnical and Geoenvironmental Engineering*, 133(4), 372–380.

Yoon, Y. W., Heo, S. B. and Kim, K. S. 2008. Geotechnical performance of waste tires for soil reinforcement from chamber tests. *Geotextiles and Geomembranes*, 26(1), 100–107.

Zornberg, J.G., Cabral, A.R. and Viratjandr, C., 2004. Behaviour of tire shred sand mixtures. *Canadian Geotechnical Journal*, 41(2), 227–241.

Index

Printed in the United States
by Baker & Taylor Publisher Services